# Industrial Hygiene Simplified

## A Guide to Anticipation, Recognition, Evaluation, and Control of Workplace Hazards

FRANK R. SPELLMAN

Government Institutes
An imprint of
The Scarecrow Press, Inc.
Lanham, Maryland • Toronto • Oxford
2006

Published in the United States of America
by Government Institutes, an imprint of The Scarecrow Press, Inc.
A wholly owned subsidiary of
The Rowman & Littlefield Publishing Group, Inc.
4501 Forbes Boulevard, Suite 200
Lanham, Maryland 20706
http://govinst.scarecrowpress.com

PO Box 317
Oxford
OX2 9RU, UK

Copyright © 2006 by Government Institutes

*All rights reserved.* No part of this publication may be reproduced, stored in a retrieval system, or transmitted in any form or by any means, electronic, mechanical, photocopying, recording, or otherwise, without the prior permission of the publisher.

The reader should not rely on this publication to address specific questions that apply to a particular set of facts. The author and the publisher make no representation or warranty, express or implied, as to the completeness, correctness, or utility of the information in this publication. In addition, the author and the publisher assume no liability of any kind whatsoever resulting from the use of or reliance upon the contents of this book.

British Library Cataloguing in Publication Information Available

**Library of Congress Cataloging-in-Publication Data**

Spellman, Frank R.
  Industrial hygiene simplified : a guide to anticipation, recognition, evaluation, and control of workplace hazards / Frank R. Spellman.
     p. cm.
  Includes bibliographical references and index.
  ISBN-13: 978-0-86587-019-2 (pbk. : alk. paper)
  ISBN-10: 0-86587-019-5 (pbk. : alk. paper)
  1. Industrial hygiene.   2. Industrial safety.
  [DNLM: 1. Occupational Health.   2. Environmental Health.   3. Safety Management.   4. Workplace—organization & administration.   WA 400 S743i 2006]   I. Government Institutes.   II. Title.
  RC967.S64   2006
  616.9'803—dc22
                                    2005029817

∞ ™ The paper used in this publication meets the minimum requirements of American National Standard for Information Sciences—Permanence of Paper for Printed Library Materials, ANSI/NISO Z39.48-1992.
Manufactured in the United States of America.

# Contents

*Preface*   *v*

| | | |
|---|---|---|
| 1 | What Is Industrial Hygiene? | 1 |
| 2 | Industrial Hygiene/Safety Terminology | 25 |
| 3 | Hazard Communication, Occupational Environmental Limits, and Air Monitoring and Sampling | 49 |
| 4 | Indoor Air Quality and Mold Control | 93 |
| 5 | Noise and Vibration | 125 |
| 6 | Radiation | 159 |
| 7 | Thermal Stress | 177 |
| 8 | Ventilation | 197 |
| 9 | Personal Protective Equipment | 223 |
| 10 | Toxicology: Biological and Chemical Hazards | 265 |
| 11 | Ergonomics | 305 |
| 12 | Engineering Design and Controls | 339 |

*Index*   *349*

*About the Author*   *357*

# Preface

The field of industrial hygiene has undergone significant change over the past three decades. There are many reasons for this. Some of the more prominent include the following: technological changes that have introduced new hazards in the workplace; proliferation of health and safety legislation and corresponding regulations; increased pressure from assertive regulatory agencies; realization by industrial executives that a safe and healthy workplace is typically a more productive and litigious-free workplace; skyrocketing health care and workers' compensation costs; increased pressure from environmental groups and the public; a growing interest in ethics and corporate responsibility; and professionalization of industrial hygiene practitioners.

All of these factors, when taken together, have made the job of the modern industrial hygienist more challenging and more important than it has ever been. These factors have also created a need for an up-to-date, condensed and concise, plain English, user-friendly book in industrial hygiene that contains the latest information needed by people who will practice this profession in the age of high technology and escalating on-the-job injuries with accompanying increased health care costs.

This book was written in response to the need for a hands-on, practical resource that focuses on the needs of modern industrial hygiene practice. It is intended for in-field use and for corporate training settings. *Industrial Hygiene Simplified* is valuable and accessible for use by those involved in such disciplines as industrial technology, manufacturing technology, industrial engineering, engineering technology, occupation safety, management, and supervision. This book is ideal for those needing a refresher on industrial hygiene concepts and practices they may not use regularly, as well as those practitioners preparing for the Certified Industrial Hygiene (CIH) exam. The direct, straightforward presentation of material focuses on making the theories and principles of industrial hygiene practical and useful in a real-world setting.

Frank R. Spellman
Norfolk, Virginia

# 1
# What Is Industrial Hygiene?

According to the Occupational Safety and Health Administration (OSHA), industrial hygiene is the science of anticipating, recognizing, evaluating, and controlling workplace conditions that may cause worker injury or illness. Industrial hygienists use environmental monitoring and analytical methods to detect the extent of worker exposure. They also evaluate employee engineering, administrative controls, and other methods, such as personal protective equipment (PPE), designed to control or guard against potential health hazards in the workplace (OSHA, 1998).

Do you remember 9/11? How about the post office anthrax mess? Dumb questions, right? The proper question should be: How can we ever forget? It started with the word being passed around about airplanes crashing into the Twin Towers. News coverage was everywhere. Remember those TV shots with the planes crashing; the towers falling? Over and over again those shots were replayed, etched into our human memory chips. We watched mesmerized; like fire watchers or falling water gazers, hypnotized. We literally could not believe what our eyes were seeing.

Later, during the frantic hunt for survivors, TV coverage continued. We saw the brave police-, fire-, and emergency-responders doing what they do best—rescuing survivors. We saw construction workers and unidentified others climbing over and crawling through the tangled, smoking mess, helping where they could. We saw others too. For instance, do you recall seeing folks walking around in what looked like space suits, instruments in hand? The average TV viewer, watching these space-suited people moving cautiously and deliberately through the smoking mass of death and destruction, had no idea who those dedicated professionals were. Professionals doing what they do best; monitoring and testing the area to make sure it was safe for the responders and everyone else. For example, ensuring it was safe for a president, arm around a hero, who stood there on the rubble in the smoldering mess and spoke those resolute words we all needed to hear; words the terrorists needed to hear; words they are still hearing.

Who were those space-suited individuals who not only appeared on our TV screens during the aftermath of 9/11 but were also prominent figures in footage of post offices trashed by anthrax?

They were the industrial hygienists.

Terrorism and bioterrorism might be new buzzwords in the American vernacular, but responding to hazard sites is nothing new to those space-suited folks whom most people, in regards to profession, can't even identify.

Times have changed, but the need for fully-trained professional industrial hygienists has not.

Is the industrial hygienist also a safety professional?

It depends. The safety profession and industrial hygiene have commonly been thought of as separate entities (this is especially the view taken by many safety professionals and industrial hygienists). In fact, over the years, a considerable amount of debate and argument has risen between those in the safety and industrial hygiene professions on many areas concerning safety and health issues in the workplace—and on exactly who is best qualified to administer a workplace safety program.

Historically, the safety professional had the upper hand in this argument—that is, prior to the enactment of the Occupational Safety and Health Act (OSH Act), which mandated formation of an administrative entity, OSHA. Until OSHA went into effect, industrial hygiene was not a topic that many professionals thought about, cared about, or had any understanding of. Safety was safety—and job safety included health protection—and that was that.

After the OSH Act, however, things changed, and so did perceptions. In particular, people began to look at work injuries and work-related illnesses differently. In the past, they were regarded as separate problems.

Why?

The primary reason for this view was obvious—and not so obvious. Obvious was work-related injury. Work injures occurred suddenly and their agent (i.e., the electrical source, chemical, machine, tool, work or walking surface, or whatever unsafe element caused the injury) was readily obvious.

Not so obvious were the workplace agents (occupational hazardous substances; e.g., lead, asbestos, formaldehyde, etc.) that caused work-related illnesses. Again, why not so obvious? Because most occupational illnesses/diseases develop rather slowly, over time. In asbestos exposure, for example, workers who abate (remove) asbestos-containing materials without the proper training (awareness) and PPE are subject to exposure. Typically, asbestos exposure may be either a one time exposure event (the silver bullet syndrome) or the exposure may go on for several years. No matter the length of exposure, one thing is certain; with asbestos contamination, pathological change occurs slowly—some time will pass before the worker notices a difference in his or her pulmonary function. Disease from asbestos exposure has a latency period that may be as long as 20 to 30 years before the effects are realized (or diagnosed, in some cases). The point? Any exposure to asbestos, short term or long term, may eventually lead to a chronic disease (in this case, restrictive lung disease) that is irreversible (e.g., asbestosis). Of course, many other types of workplace toxic exposures can affect workers' health. The prevention, evaluation, and control of such occurrences is the role of the industrial hygienist.

Thus, because of the OSH Act, and also because of increasing public awareness and involvement by unions in industrial health matters, the role of the industrial hygienist has continued to grow over the years. Certain colleges and universities have incorporated industrial hygiene majors into environmental health programs.

Another result of the OSH Act has been, in effect (though many practitioners in the field disagree with this view), an ongoing tendency toward uniting safety and industrial hygiene into one entity.

This trend presents a problem with definition. When we combine safety and industrial hygiene, do we combine them into one specific title or profession? Debate on this question continues. In attempting to find a solution, we need to consider actual experience gained from practice in the real world(s) of safety and industrial hygiene. Knowledge about safety is widespread (often rooted in common sense, which unfortunately is not always so common), however, the industrial hygiene profession is not as well known. We will take a brief look at its history to gain understanding.

## HISTORY OF INDUSTRIAL HYGIENE (OSHA, 1998)

There has been an awareness of industrial hygiene since antiquity. The environment and its relation to worker health was recognized as early as the fourth century BC when Hippocrates noted lead toxicity in the mining industry. In the first century AD, Pliny the-elder, a Roman scholar, perceived health risks to those working with zinc and sulfur. He devised a face mask made from an animal bladder to protect workers from exposure to dust and lead fumes. In the second century AD, the Greek physician, Galen, accurately described the pathology of lead poisoning and also recognized the hazardous exposures of copper miners to acid mists.

In the Middle Ages, guilds worked at assisting sick workers and their families. In 1556 the German scholar, Agricola, advanced the science of industrial hygiene even further when, in his book *De Re Metallica* (On the Nature of Metals), he described the disease of miners and prescribed preventive measures. The book included suggestions for mine ventilation and worker protection, discussed mining accidents, and described diseases associated with mining occupations such as silicosis.

Industrial hygiene gained further respectability in 1700 when Bernardo Ramazzini, known as the "father of industrial medicine," published in Italy the first comprehensive book on industrial medicine, *De Morbis Artificum Diatriba* (The Diseases of Workmen). The book contained accurate descriptions of the occupational diseases of most of the workers of his time. Ramazzini greatly affected the future of industrial hygiene because he asserted that occupational diseases should be studied in the work environment rather than in hospital wards.

Industrial hygiene received another major boost in 1743 when Ulrich Ellenborg published a pamphlet on occupational diseases and injuries among gold miners. Ellenborg also wrote about the toxicity of carbon monoxide, mercury, lead, and nitric acid.

In England in the eighteenth century, Percival Pott, as a result of his findings on the insidious effects of soot on chimney sweepers, was a major force in getting the British parliament to pass the Chimney-Sweepers Act of 1788. The passage of the English Factory Acts beginning in 1833 marked the first effective legislative acts in the field of industrial safety. The Acts, however, were intended to provide compensation for accidents rather than to control their causes. Later, various other European nations developed workers' compensation acts, which stimulated the adoption of increased factory safety precautions and the establishment of medical services within industrial plants.

In the early twentieth century in the United States, Dr. Alice Hamilton led efforts to improve industrial hygiene. She observed industrial conditions firsthand and startled mine owners, factory managers, and state officials with evidence of a correlation between worker illness and their exposure to toxins. She also presented definitive proposals for eliminating unhealthful working conditions.

At about the same time, U.S. federal and state agencies began investigating health conditions in industry. In 1908, the public's awareness of occupationally related diseases stimulated the passage of compensation acts for certain civil employees. States passed the first workers' compensation laws in 1911. And in 1913, the New York Department of Labor and the Ohio Department of Health established the first state industrial hygiene programs. All states enacted such legislation by 1948. In most states, there is some compensation coverage for workers contracting occupational disease.

The U.S. Congress has passed three landmark pieces of legislation for safeguarding workers' health: (1) the Metal and Nonmetallic Mines Safety Act of 1966, (2) the Federal Coal Mine Safety and Health Act of 1969, and (3) the Occupational Safety and Health Act of 1970. Today, nearly every employer is required to implement the elements of an industrial hygiene and safety, occupational health, or hazard communication program and to be responsive to the Occupational Safety and Health Administration (OSHA) and the Act and its regulations.

## A MATTER OF PERCEPTION AND FUNCTION

After the preceding brief history lesson on the industrial hygiene profession, let's look at the factors that allow us to answer the earlier questions. That is, when we combine safety and industrial hygiene, do we combine them into one specific title or profession? If so, which one?

The safety professional must be a generalist—a jack-of-all-trades. How about the industrial hygienist? What exactly are industrial hygienists required to do and to know in order to be effective in the workplace? Let's take a look at what a typical industrial hygienist does, and from this description, you should be able to determine the level of knowledge industrial hygienists should have.

The primary mission of the industrial hygienist is to examine the workplace environment and its environs by studying work operations and processes. From these studies, he or she is able to obtain details related to the nature of the work, materials and equipment used, products and by-products, number of employees, hours of work, and so on. At the same time, appropriate measurements are made to determine the magnitude of exposure or nuisance (if any) to workers and the public (e.g., the presence and source of an airborne contaminant). The hygienist's next step is to interpret the results of the examination and measurements (metrics) of the workplace environment and environs, in terms of ability to impair worker health (i.e., Is there a health hazard in the workplace that must be mitigated?). With examination results in hand, the industrial hygienist then presents specific conclusions and recommendations to the appropriate managerial authority.

Is the process described above completed? Yes and no. In many organizations, the industrial hygienist's involvement stops here. But remember, discovering a problem is only half the battle. Knowing a problem exists, but not taking steps to mitigate it leaves the job half done. In light of this, the industrial hygienist normally will make specific recommendations for control measures—an important part of industrial hygiene's anticipate-recognize-evaluate-control paradigm.

Any further involvement in working toward permanently correcting the hazard depends on the industrial hygienist's role in a particular organization. Organization size is one of two main factors that determine the industrial hygienist's role. Obviously, an organization that consists of several hundred (or more) workers will increase the work requirements for the industrial hygienist.

A second factor has to do with what the organization actually produces. For example, does it produce computerized accounting records? Does it perform a sales or telemarketing function? Does it provide office supplies to those businesses requiring such service? If the organization accomplishes any of these functions (or many similar functions), the organization probably does not, because of the nature of the work performed, require the services of a full-time industrial hygienist. On the other hand, if the organization handles, stores, or produces hazardous materials or hazardous wastes, if the organization is a petroleum refinery, if the organization is a large environmental laboratory, or if the organization produces or possesses any other major workplace hazard, there may be a real need for the services of a full-time industrial hygienist.

Let's look at some of these elements in combination. Assume, for example, that an organization employs 5,000 full-time workers in the production of chemical products. In this situation, not only are the services of an industrial hygienist required, but the company probably needs more than just one full-time industrial hygienist—perhaps several. In this case, each industrial hygienist might be assigned duties that are important but narrow in scope, with limited responsibility, such as performing air monitoring of a chemical process or processes. In short, each industrial hygienist performs specialized work.

What this all means, of course, is that the size of the organization and the type of work performed may point to the need for an industrial hygienist to monitor safe operations. Each organization's needs set the extent of its staffing requirements.

Another important area of responsibility for the industrial hygienist (one often overlooked by many in the real work world) is providing or conducting worker training. If the industrial hygienist examines a workplace and ancillary environs for occupational hazards, and discovers one or more, he or she will normally perform the anticipate-recognize-evaluate-control actions. However, an important part of the "control" element of the industrial hygiene paradigm is information—training those exposed to the hazards so they can avoid the hazards. Based on experience, in the field of industrial hygiene in general, proper preparation in this vital area and emphasis on its importance in the workplace has been somewhat lacking or just simply overlooked.

What does all this mean? Good question—one answered best by pointing out that safety professionals are (or should be) generalists, while industrial hygienists typically are specialists. This simple straightforward statement sums up the main difference between the two professions—one is a generalist and the other a specialist.

Which one is best? Another good question. Actually, which one is "best" is not the issue. Instead, the question should be which one is better suited to perform the functions required of the safety and health profession—to protect workers and the organization from harm. Based on personal experience, I profess the need for generalization (this was stated earlier and continues to be throughout this text). The safety professional (the industrial hygienist included) needs to be well-versed in all aspects of safety and health, laws and regulations, and several of the scientific disciplines (e.g., chemistry, engineering, biology, toxicology, occupational health, management aspects, etc.).

Presently, the problem of differentiating between the safety and industrial hygiene professions lies in personal perception and function. The safety professional views his role as all-encompassing (as he should), and the industrial hygienist views herself as a step above the safety professional, occupying that lofty position known as "specialist."

Which of these views is correct? Neither. Again, it is a matter of perception. On the function side of the issue, the safety professional is usually employed to manage an organization's safety program (as safety manager, safety engineer, safety professional, safety director, safety coordinator, or another similar title), while the industrial hygienist is usually hired to fill the position of organizational industrial hygienist only.

Can you see the difference? Believe it when I say that practicing safety professionals and industrial hygienists not only see the difference, they feel the difference—and learn to work with the difference—sometimes as colleagues and sometimes as competitors.

So, what is the solution to this dilemma? (Yes, it is a dilemma—especially for those practitioners actively working in the real world of occupational health and safety.) Based on experience I feel the solution lies in merging the two disciplines into one—as safety professionals. As such, the safety professional must be well-grounded in industrial hygiene, but must also have a wide range of knowledge in many other fields. Again, the safety professional must be a generalist—a jack-of-many-trades.

Does this scenario work? In many cases, it already has. Of course, traditionalists still argue against a joining of forces. However, in the real world, the argument really comes down to only one thing (as it does in many cases): economics (in this case, personal economics). Based on experience, safety professionals have had the opportunity to earn a better living than many industrial hygienists. Why? Because safety professionals, as generalists with a wide range of knowledge in several specialties, seem to be better positioned for promotion to management. In the work world, management is where the money is. Why not the industrial hygienist as manager? Though there are industrial hygienists who have climbed the ladder to management positions, this is not generally the case. Simply put, an industrial hygienist specialist often times has little exposure to managing, while the safety professional is often involved both in and with management by the very nature of the job's generalist activities.

The bottom line: if you own an environmental organization that employs generalists and specialists, the organization's safety coordinator (and not the organization's industrial hygienist) is more likely to be promoted to the safety director's position. (Notwithstanding the tendency to promote the "best employee"—the one who has demonstrated ability and top performance on-the-job.)

What do I base this assessment on? First I refer you to a statement by Olishifski and Plog: "Most safety professionals are already involved in some aspects of industrial hygiene. . . . After the industrial hygienist surveys the plant, makes recommendations, and suggests certain control measures, it may become the safety professional's responsibility to see that proper control measures are being applied and followed." Second, for management positions, a well-rounded generalist accustomed to accepting authority and to dealing with employees (a people-person) is more useful in upper management than is the narrowly focused specialist. How do I know this is the way it commonly is? I have been there. However, one thing is certain; the safety professional that does not have a thorough knowledge of and background in industrial hygiene is limited in his or her ability. That is the purpose of this text. To make sure that the generalist has a good background in the fundamental aspects of industrial hygiene and that the specialist in industrial hygiene has a good review of those fundamentals.

## OSHA/NIOSH AND INDUSTRIAL HYGIENE

The principal piece of federal legislation relating to industrial hygiene is the Occupational Safety and Health Act of 1970 (OSH Act) as amended. Under the Act, the Occupational Safety and Health Administration (OSHA) develops and sets mandatory occupational safety and health requirements applicable to the more than six million workplaces in the United States. OSHA relies on, among others, industrial hygienists to evaluate jobs for potential health hazards. Developing and setting mandatory occupational safety and health standards involves determining the extent of employee exposure to hazards and deciding what is needed to control these hazards, thereby protecting the workers. As mentioned, industrial hygienists are trained to anticipate, recognize, evaluate, and recommend controls for environmental and physical hazards that can affect the health and well-being of workers. More than 40 percent of the OSHA compliance officers who inspect America's workplaces are industrial hygienists. Industrial hygienists also play a major role in developing and issuing OSHA standards to protect workers from health hazards associated with toxic chemicals, biological hazards, and harmful physical agents. They also provide technical assistance and support to the agency's national and regional offices. OSHA also employs industrial hygienists who assist in setting up field enforcement procedures, and who issue technical interpretations of OSHA regulations and standards. Industrial hygienists analyze, identify, and measure workplace hazards or stressors that can cause sickness, impaired health, or significant discomfort in workers through chemical, physical, ergonomic, or biological exposures. Two roles of the OSHA industrial hygienist are to spot those conditions and help eliminate or control them through appropriate measures (OSHA, 1998).

The OSH Act sets forth the following requirements relating to industrial hygiene:

- Use of warning labels and other means to make employees aware of potential hazards, symptoms of exposure, precautions, and emergency treatment.
- Prescription of appropriate personal protective equipment and other technological preventive measures (29 CFR Subpart I, 1910.133 and 1910.134).
- Provision of medical tests to determine the effect on employees of exposure to environmental stressors.
- Maintenance of accurate records of employee exposures to environmental stressors that are required to be measured or monitored.
- Making monitoring tests and measurement activities open to the observation of employees.
- Making records of monitoring tests and measurement activities available to employees on request.
- Notification of employees who have been exposed to environmental stressors at a level beyond the recommended threshold and corrective action being taken (Goetsch, 1996).

Effective management of worker safety and health protection is a decisive factor in reducing the extent and severity of work-related injuries and illnesses and their related costs. To assist employers and employees in developing effective safety and health programs, OSHA published recommended Safety and Health Program Management Guidelines (Federal Register 54 (18):3908–3916, January 26, 1989). These voluntary guidelines apply to all places of employment covered by OSHA. The guidelines identify four general elements that are critical to the development of successful safety and health program management:

- management commitment and employee involvement
- worksite analysis

- hazard prevention and control
- safety and health training

The National Institute for Occupational Safety and Health (NIOSH) is part of the Department of Health and Human Services (DHHS). This agency is important to industrial hygiene professionals. The main focus of the agency's research is on toxicity levels and human tolerance levels of hazardous substances. NIOSH prepares recommendations for OSHA standards dealing with hazardous substances, and NIOSH studies are made available to employers.

## WORKPLACE STRESSORS

The industrial hygienist focuses on evaluating the healthfulness of the workplace environment, either for short periods or for a work-life of exposure. When required, the industrial hygienist recommends corrective procedures to protect health, based on solid quantitative data, experience, and knowledge. Recommended control measures may include isolation of a work process, substitution of a less harmful chemical or material, and/or other measures designed solely to increase the healthfulness of the work environment.

To ensure a healthy workplace environment and associated environs, the industrial hygienist focuses on the recognition, evaluation, and control of chemical, physical, or biological and ergonomic stressors that can cause sickness, impaired health, or significant discomfort to workers.

The key word just mentioned was *stressors*, or simply, *stress*—the stress caused by the workplace external environment demands placed upon a worker. Increases in external stressors beyond a worker's tolerance level affect his or her job performance and overall health.

The industrial hygienist must not only understand that workplace stressors exist, but also that they are sometimes cumulative (additive). For example, studies have shown that some assembly-line processes are little affected by either low illumination or vibration; however, when these two stressors are combined, assembly-line performance deteriorates.

Other cases have shown just the opposite effect. For example, the worker who has had little sleep and then is exposed to a work area where noise levels are high actually benefits (to a degree, depending on the intensity of the noise level and the worker's exhaustion level) from the increased arousal level; a lack of sleep combined with a high noise level is compensatory.

In order to recognize environmental stressors and other factors that influence worker health, the industrial hygienist's new employee orientation process should include an overview of all pertinent company work operations and processes. Obviously, the industrial

# WHAT IS INDUSTRIAL HYGIENE?

hygienist who has not been fully indoctrinated on company work operations and processes not only is unprepared to study the environmental affects of such processes, but also suffers from another disability—lack of credibility with supervisors and workers. This point cannot be emphasized strongly enough—know your organization and what it is all about.

*Note:* Woe be it to the rookie industrial hygienist or safety professional who has the audacity (and downright stupidity) to walk up to any supervisor (or any worker with experience at his or her task) and announce that he or she, the company's brand new industrial hygienist, is going to find out everything that is unhealthy (and thus injurious to workers) about their work process, even though having no idea of how the process operates, what it does, or what it is all about! As an industrial hygienist, you must understand the work process to the point that you could almost operate the system efficiently and safely yourself before you dictate how anyone else is to correctly and safely do the same. To be effective, the industrial hygienist must be credible. To be credible, the industrial hygienist must fully understand the type of work the workers are doing.

What are the workplace stressors the industrial hygienist should be concerned with? The industrial hygienist should be concerned with those workplace stressors that are likely to accelerate the aging process, cause significant discomfort and inefficiency, or pose immediate danger to life and health (Spellman, 1998). Several stressors fall into these categories, including:

- Chemical stressors: gases, dusts, fumes, mists, liquids, or vapors.
- Physical stressors: noise, vibration, extremes of pressure and temperature, and electromagnetic and ionizing radiation.
- Biological stressors: bacteria, fungi, molds, yeasts, insects, mites, and viruses.
- Ergonomic stressors: repetitive motion, work pressure, fatigue, body position in relation to work activity, monotony/boredom, and worry.

OSHA lists these health stressors as major job risks. Each is explained further in the following (OSHA, 1998).

*Chemical stressors* are harmful chemical compounds in the form of solids, liquids, gases, mists, dusts, fumes, and vapors that exert toxic effects by inhalation (breathing), absorption (though direct contact with the skin), or ingestion (eating or drinking). Airborne chemical hazards exist as concentrations of mists, vapors, gases, fumes, or solids. Some are toxic through inhalation and some of them irritate the skin on contact; some can be toxic by absorption through the skin or through ingestion, and some are corrosive to living tissue.

The degree of worker risk from exposure to any given substance depends on the nature and potency of the toxic effects and the magnitude and duration of exposure.

Information on the risk to workers from chemical hazards can be obtained from the Material Safety Data Sheet (MSDS) that OSHA's Hazard Communication Standard requires be supplied by the manufacturer or importer to the purchaser of all hazardous materials. The MSDS is a summary of the important health, safety, and toxicological information on the chemical or the mixture's ingredients. Other provisions of the Hazard Communication Standard require that all containers of hazardous substances in the workplace have appropriate warning and identification labels.

*Physical stressors* include excessive levels of ionizing and nonionizing electromagnetic radiation, noise, vibration, illumination, and temperature.

In occupations where there is exposure to ionizing radiation, time, distance, and shielding are important tools in ensuring worker safety. Danger from radiation increases with the amount of time one is exposed to it; hence, the shorter the time of exposure the smaller the radiation danger.

Distance also is a valuable tool in controlling exposure to both ionizing and nonionizing radiation. Radiation levels from some sources can be estimated by comparing the squares of the distances between the work and the source. For example, at a reference point of 10 feet from a source, the radiation is 1/100 of the intensity at 1 foot from the source.

Shielding also is a way to protect against radiation. The greater the protective mass between a radioactive source and the worker, the lower the radiation exposure.

Nonionizing radiation also is dealt with by shielding workers from the source. Sometimes limiting exposure times to nonionizing radiation or increasing the distance is not effective. Laser radiation, for example, cannot be controlled effectively by imposing time limits. An exposure can be hazardous that is faster than the blink of an eye. Increasing the distance from a laser source may require miles before the energy level reaches a point where the exposure would not be harmful.

Noise, another significant physical hazard, can be controlled by various measures. Noise can be reduced by installing equipment and systems that have been engineered to operate quietly; by enclosing or shielding noisy equipment; by making certain that equipment is in good repair and properly maintained with all worn or unbalanced parts replaced; by mounting noisy equipment on special mounts to reduce vibration; and by installing silencers, mufflers, or baffles.

Substituting quiet work methods for noisy ones is another significant way to reduce noise, for example, welding parts rather than riveting them. Treating floors, ceilings, and walls with acoustical material can reduce reflected or reverberant noise. Erecting sound barriers at adjacent work stations around noisy operations will reduce worker exposure to noise generated at adjacent work stations.

It is also possible to reduce noise exposure by increasing the distance between the source and the receiver, by isolating workers in acoustical booths, limiting workers' exposure time to noise, and by providing hearing protection. OSHA requires that workers in noisy surroundings be periodically tested as a precaution against hearing loss.

Another physical hazard, radiant heat exposure in factories such as steel mills, can be controlled by installing reflective shields and by providing protective clothing.

*Biological stressors* include bacteria, viruses, fungi, and other living organisms that can cause acute and chronic infections by entering the body either directly or through breaks in the skin. Occupations that deal with plants or animals or their products or with food and food processing may expose workers to biological hazards. Laboratory and medical personnel also can be exposed to biological hazards. Any occupations that result in contact with bodily fluids pose a risk of biological hazards to workers.

In occupations where animals are involved, biological hazards are dealt with by preventing and controlling diseases in the animal population as well as proper care and handling of infected animals. Also, effective personal hygiene, particularly proper attention to minor cuts and scratches, especially those on the hands and forearms, helps keep worker risks to a minimum.

In occupations where there is potential exposure to biological hazards, workers should practice proper personal hygiene, particularly hand washing. Hospitals should provide proper ventilation, proper personal protective equipment such as gloves and respirators, adequate infectious waste disposal systems, and appropriate controls including isolation in instances of particularly contagious diseases such as tuberculosis.

*Ergonomic stressors* include a full range of tasks such as lifting, holding, pushing, walking, and reaching. Many ergonomic problems result from technological changes such as increased assembly line speeds, adding specialized tasks, and increased repetition; some problems arise from poorly designed job tasks. Any of those conditions can cause ergonomic hazards such as excessive vibration and noise, eye strain, repetitive motion, and heavy lifting problems. Improperly designed tools or work areas also can be ergonomic hazards. Repetitive motions or repeated shocks over prolonged periods of time as in jobs involving sorting, assembling, and data entry can often cause irritation and inflammation of the tendon sheath of the hands and arms, a condition known as carpal tunnel syndrome.

Ergonomic hazards are avoided primarily by the effective design of a job or jobsite and better designed tools or equipment that meet workers' needs in terms of job task and physical environment. Through thorough worksite analyses, employers can set up procedures to correct or control ergonomic hazards by using the appropriate engineering controls (e.g., designing or redesigning work stations, lighting, tools, and equipment); teaching correct work practices (e.g., proper lifting methods); employing proper administrative controls (e.g., shifting workers among several different tasks, reducing production demand, and increasing rest breaks); and, if necessary, providing and mandating personal protective equipment. Evaluating working conditions from an ergonomics standpoint involves looking at the total physiological and psychological demands of the job on the worker.

Overall, industrial hygienists point out that the benefits of a well-designed, ergonomic work environment can include increased efficiency, fewer accidents, lower operating costs, and more effective use of personnel.

In the workplace, the industrial hygienist should review the following to anticipate potential health stressors:

- Raw materials
- Support materials
- Chemical reactions
- Chemical interactions
- Products
- By-products
- Waste products
- Equipment
- Operating procedures

## AREAS OF CONCERN

From the list of health stressors above, it can be seen that the industrial hygienist has many areas of concern related to protecting the health of workers on the job. In this section, the focus is on the major areas that the industrial hygienist typically is concerned with in the workplace. Important areas of industrial toxicology and industrial health hazards are also discussed. Later industrial noise, vibration, and environmental control are covered. All of these areas are important to the industrial hygienist (and to the worker, of course), but they are not all inclusive; the industrial hygienist also is concerned with many other areas—ionizing and nonionizing radiation, for example.

### Industrial Toxicology

The practice of industrial toxicology (especially in the area of industrial poisons) owes its genesis in the United States to the work of Alice Hamilton (1869–1970). In 1919 Hamilton became the first woman to be appointed to the staff at Harvard Medical School. She also did studies on industrial pollution for the federal government and the United Nations. She wrote several books including *Industrial Poisons in the United States* (1925), *Industrial Toxicology* (1934), and *Exploring the Dangerous Trades* (1943).

Currently, we have witnessed unprecedented industrialization, explosive population growth, and a massive introduction of new chemical agents into the workplace. Unfortunately, we lag far behind in our understanding of the impact that many of these new chemicals, particularly mixtures, have on the health of workers and other members of our ecosystem.

Though we devote a full chapter to toxicology (see chapter 10) in this text, it is important to provide an overview of this vital topic early in the presentation. Why? The simple answer: the industrial hygienist must be well-versed and knowledgeable in toxicology. The compound answer: the industrial hygienist must be a practitioner who constantly studies the nature and effects of poison and their treatment in the workplace. The need and importance of the industrial hygienist possessing a full understanding of industrial toxicology can't be overstated.

Consider this: normally, we give little thought to the materials (chemical substances, for example) that we are exposed to on a daily (almost constant) basis, unless they interfere with our lifestyle, irritate us, or noticeably physically affect us. Keep in mind, however, that all chemical substances have the potential for being injurious at some sufficiently high concentration and level of exposure. Again, to prevent the lethal effects of overexposure for workers, the industrial hygienist must have an adequate understanding and knowledge of general toxicology.

## What Is Toxicology?

Toxicology is a very broad, interdisciplinary science that studies the adverse effects of chemicals on living organisms, using knowledge and research methods drawn from virtually all areas of the biomedical sciences. It deals with chemicals used in industry, drugs, food, and cosmetics, as well as those occurring naturally in the environment. Toxicology is the science that deals with the poisonous or toxic properties of substances. The primary objective of industrial toxicology is the prevention of adverse health effects in workers exposed to chemicals in the workplace. The industrial hygienist's responsibility is to consider all types of exposure and the subsequent effects on workers. The industrial hygienist uses toxicity information to prescribe safety measures for protecting workers. Following the prescribed precautionary measures and limitations placed on exposure to certain chemical substances is the worker's responsibility.

To gain better appreciation for what industrial toxicology is all about, you must understand some basic terms and factors—many of which contribute to determining the degree of hazard particular chemicals present. You must also differentiate between toxicity and hazard. *Toxicity* is the intrinsic ability of a substance to produce an unwanted effect on humans and other living organisms when the chemical has reached a sufficient concentration at a certain site in the body. *Hazard* is the probability that a substance will produce harm under specific conditions. The industrial hygienist and other safety professionals employ the opposite of hazard—safety—that is, the probability that harm will not occur under specific conditions. A toxic chemical—used under safety conditions—may not be hazardous.

Basically, all toxicological considerations are based on the *dose-response relationship*, another toxicological concept important to the industrial hygienist. In its simplest terms, the dose of a chemical to the body resulting from exposure is directly related to the degree of harm. This relationship means that the toxicologist is able to determine a *threshold level* of exposure for a given chemical—the highest amount of a chemical substance to which one can be exposed with no resulting adverse health effect. Stated differently, chemicals present a threshold of effect, or a no-effect level.

Threshold levels are critically important parameters. For instance, under the OSHA Act, threshold limits have been established for the air contaminants most frequently found in the workplace. The contaminants are listed in three tables in 29 CFR 1910 subpart Z—Toxic and Hazardous Substances. The threshold limit values listed in these tables are drawn from values published by the American Conference of Governmental Industrial Hygienists (ACGIH) and from the "Standards of Acceptable Concentrations of Toxic Dusts and Gases," issued by the American National Standards Institute (ANSI).

An important and necessary consideration when determining levels of safety for exposure to contaminants is their effect over a period of time. For example, during an 8-hour work shift, a worker may be exposed to a concentration of Substance A (with a 10 ppm [parts per million—analogous to a full shot glass of water taken from a standard-size in-ground swimming pool] TWA [time-weighted average], 25 ppm ceiling and 50 ppm peak) above 25 ppm (but never above 50 ppm) only for a maximum period of 10 minutes. Such exposure must be compensated by exposures to concentrations less than 10 ppm, so that the cumulative exposure for the entire 8-hour work shift does not exceed a weighted average of 10 ppm. Formulas are provided in the regulations for computing the cumulative effects of exposures in such instances. Note that the computed cumulative exposure to a contaminant may not exceed the limit value specified for it.

## Air Contaminants

One of the primary categories of industrial health hazards that the industrial hygienist must deal with is airborne contaminants. Air contaminants are commonly classified as either particulate or gas and vapor contaminants. The most common particulate contaminants include dusts, fumes, mists, aerosols, and fibers. Note that these particulate contaminants are briefly discussed below, but are discussed in greater detail later in the text.

*Dusts* are solid particles that are formed or generated from solid organic or inorganic materials by reducing their size through mechanical processes such as crushing, grinding, drilling, abrading, or blasting.

Industrial atmospheric contaminants exist in virtually every workplace. Sometimes they are readily apparent to workers, because of their odor, or because they can actually be seen. Industrial hygienists, however, can't rely on odor or vision to detect or measure airborne contaminants. They must rely on measurements taken by monitoring, sampling, or detection devices.

*Fumes* are formed when material from a volatilized solid condenses in cool air. In most cases, the solid particles resulting from the condensation react with air to form an oxide.

The term *mist* is applied to a finely divided liquid suspended in the atmosphere. Mists are generated by liquids condensing from a vapor back to a liquid or by breaking up a liquid into a dispersed state such as by splashing, foaming, or atomizing. Aerosols are a form of a mist characterized by highly respirable, minute liquid particles.

*Fibers* are solid particles whose length is several times greater than their diameter.

*Gases* are formless fluids that expand to occupy the space or enclosure in which they are confined. Examples are welding gases such as acetylene, nitrogen, helium, and argon; and carbon monoxide generated from the operation of internal combustion engines or by its use as a reducing gas in a heat treating operation. Another example is hydrogen sulfide, which is formed wherever there is decomposition of materials containing sulfur under reducing conditions.

*Liquids* change into vapors and mix with the surrounding atmosphere through evaporation. Vapors are the volatile gaseous form of substances that are normally in a solid or liquid state at room temperature and pressure. They are formed by evaporation from a liquid or solid and can be found where parts cleaning and painting takes place and where solvents are used.

Although air contaminant values are useful as a guide for determining conditions that may be hazardous and may demand improved control measures, the industrial hygienist must recognize that the susceptibility of workers varies.

Even though it is essential not to permit exposures to exceed the stated values for substances, note that even careful adherence to the suggested values for any substance will not assure an absolutely harmless exposure. Thus, the air contaminant concentration values should only serve as a tool for indicating harmful exposures, rather than the absolute reference on which to base control measures.

## Routes of Entry

For a chemical substance to cause or produce a harmful effect, it must reach the appropriate site in the body (usually via the bloodstream) at a concentration (and for a length of time) sufficient to produce an adverse effect. Toxic injury can occur at the first point of contact between the toxicant and the body, or in later, systemic injuries to various organs deep in the body. Common routes of entry are ingestion, injection, skin absorption, and inhalation. However, entry into the body can occur by more than one route (e.g., inhalation of a substance also absorbed through the skin).

*Ingestion* of toxic substances is not a common problem in industry—most workers do not deliberately swallow substances they handle in the workplace. However, ingestion does sometimes occur either directly or indirectly. Industrial exposure to harmful substances through ingestion may occur when workers eat lunch, drink coffee, chew tobacco, apply cosmetics, or smoke in a contaminated work area. The substances may exert their toxic effect on the intestinal tract or at specific organ sites.

*Injection* of toxic substances may occur just about anywhere in the body where a needle can be inserted, but is a rare event in the industrial workplace.

*Skin absorption* or contact is an important route of entry in terms of occupational exposure. While the skin (the largest organ in the human body) may act as a barrier to some harmful agents, other materials may irritate or sensitize the skin and eyes, or travel through the skin into the bloodstream, thereby impacting specific organs.

*Inhalation* is the most common route of entry for harmful substances in industrial exposures. Nearly all substances that are airborne can be inhaled. Dusts, fumes, mists, gases, vapors, and other airborne substances may enter the body via the lungs and may produce local effects on the lungs, or may be transported by the blood to specific organs in the body.

Upon finding a route of entry into the body, chemicals and other substances may exert their harmful effects on specific organs of the body, such as the lungs, liver, kidneys, central nervous system, and skin. These specific organs are termed target organs and will vary with the chemical of concern (see Table 1.1).

The toxic action of a substance can be divided into short-term (acute) and long-term (chronic) effects. Short-term adverse effects are usually related to an accident where expo-

Table 1.1. Selected Toxic Contaminants and the Target Organs They Endanger

| Target organs | Toxic contaminants |
| --- | --- |
| Blood | Benzene<br>Carbon monoxide<br>Arsenic<br>Aniline<br>Toluene |
| Kidneys | Mercury<br>Chloroform |
| Heart | Aniline |
| Brain | Lead<br>Mercury<br>Benzene<br>Manganese<br>Acetaldehyde |
| Eyes | Cresol<br>Acrolein<br>Benzyl chloride<br>Butyl alcohol |
| Skin | Nickel<br>Phenol<br>Trichloroethylene |
| Lungs | Asbestos<br>Chromium<br>Hydrogen sulfide<br>Mica<br>Nitrogen dioxide |
| Liver | Chloroform<br>Carbon tetrachloride<br>Toluene |

Source: Data from Spellman, 1998.

sure symptoms (effects) may occur within a short time period following either a single exposure or multiple exposures to a chemical. Long-term adverse effects usually occur slowly after a long period of time, following exposures to small quantities of a substance (as lung disease may follow cigarette smoking). Chronic effects may sometimes occur following short-term exposures to certain substances.

## INDUSTRIAL HEALTH HAZARDS

NIOSH and OSHA's Occupational Health Guidelines for Chemical Hazards, DHHS (NIOSH) Publication No. 81–123 (Washington, D.C.: Superintendent of Documents, U.S. Government Printing Office, current edition) summarizes information on permissible exposure limits, chemical and physical properties, and health hazards. It provides recommendations for medical surveillance, respiratory protection, and personal protection and sanitation practices for specific chemicals that have federal occupational safety and health regulations. These recommendations reflect good industrial hygiene practices, and their implementation will assist development and maintenance of an effective occupational health program. The practicing industrial hygienist should maintain a current copy of this important and useful document within easy reach. It is also available in digital format.

Generally, determining if a substance is hazardous or not is simple, if the following is known: (1) what the agent is and what form it is in; (2) the concentration; and (3) the duration and form of exposure.

However, because of the dynamic (ever changing) characteristics of the chemical and product industries, the practicing industrial hygienist comes face to face with trying to determine the uncertain toxicity of new chemical products that are frequently introduced into the workplace each year. A related problem occurs when manufacturers develop chemical products with unfamiliar trade names, and do not properly label them to indicate the chemical constituents of the compounds. (Of course, under OSHA's Hazard Communication Program, 29 CFR 1910.1200, this practice is illegal.) Many commercially available instruments permit the detection and concentration evaluation of different contaminants. Some of these instruments are so simple that nearly any worker can learn to properly operate them. A note of caution, however; the untrained worker may receive an instrument reading that seems to indicate a higher degree of safety than may actually exist. Thus, the qualitative and quantitative measurement of atmospheric contaminants generally is the job of the industrial hygienist. Any samples taken should also be representative—that is, samples should be taken of the actual air the workers breathe, at the point they inhale them, in their breathing zone (roughly, between the top of the head and the shoulders).

## ENVIRONMENTAL CONTROLS

Industrial hygienists recognize that engineering controls, work practice controls, administrative controls, and personal protective equipment are the primary means of reducing employee exposure to occupational hazards. Workplace exposure to toxic materials and physical hazards can be reduced or controlled by a variety of these control methods, or by a combination of methods.

## Engineering Controls

Engineering controls are methods of environmental control whereby the hazard is "engineered out," either by initial design specifications, or by applying methods of substitution (e.g., replacing toxic chlorine used in disinfection processes with relatively nontoxic sodium hypochlorite). Engineering control may entail utilization of isolation methods. For example, a diesel generator that, when operating, produces noise levels in excess of 120 decibels (120 dBA) could be controlled by housing it inside a soundproofed enclosure—effectively isolating the noise hazard. Another example of hazard isolation can be seen in the use of tight enclosures that isolate an abrasive blasting operation. This method of isolation is typically used in conjunction with local exhaust ventilation. Ventilation is one of the most widely used and effective engineering controls (because it is so crucial in controlling workplace atmospheric hazards).

## Work Practice Controls

Work practice controls alter the manner in which a task is performed. Some fundamental and easily implemented work practice controls include (1) following proper procedures that minimize exposures while operating production and control equipment; (2) inspecting and maintaining process and control equipment on a regular basis; (3) implementing good housekeeping procedures; (4) providing good supervision; and (5) mandating that eating, drinking, smoking, chewing tobacco or gum, and applying cosmetics in regulated areas be prohibited.

## Administrative Controls

After the design, construction, and installation phase has been completed, adding engineering controls to a workplace hazard often becomes difficult and expensive. An ongoing question industrial hygienists face is "If I can't engineer out the hazard, what can I do?"

This question would not arise, of course, if the industrial hygienist had been invited to participate in the design, construction, and installation phases. This constitutes "good engineering practice," however, based on experience, this is the exception to the rule.

As a remedial action (a third line of defense)—after determining that engineering controls and work practice controls can't be accomplished for technological, budgetary, or other reasons—administrative controls might be called for.

What are administrative controls? Simply, administrative controls include controlling workers' exposure by scheduling production and workers' tasks, or both, in ways that minimize exposure levels. The employer might schedule operations with the highest exposure potential during periods when the fewest workers are present. For example, a worker who is required to work in an extremely high noise area where engineering controls and work practice controls are not possible would be rotated from the high noise area to a quiet area when the daily permissible noise exposure limit is reached.

It should be noted that reducing exposures by limiting the duration of exposure (basically by modifying the work schedule) must be carefully managed (most managers soon find that attempting to properly manage this procedure takes a considerable amount of time, effort, and "imagination"). When practiced, reducing worker exposure is based on limiting the amount of time a worker is exposed, ensuring that OSHA Permissible Exposure Limits (PELs) are not exceeded.

Because administrative controls are not easy to implement and manage, many practicing industrial hygienists don't particularly like the practice; furthermore, they feel that such a strategy merely spreads the exposure out, and does nothing to control the source. Experience has shown that in many instances this view is correct. Nevertheless, work schedule modification is commonly used for exposures to such stressors as noise and lead.

Though mentioned under work practice controls, good housekeeping practices are also an administrative control. Housekeeping practices? Absolutely. If dust and spilled chemicals are allowed to accumulate in the work area, workers will be exposed to these substances. This is of particular importance for flammable and toxic materials (in the prevention of fires, explosions, or poisoning). Housekeeping practices that prevent toxic or hazardous materials from being dispersed into the air are also an important concern.

Administrative controls implemented at work can also reach beyond the workplace. For example, if workers abate asbestos all day, they should only wear approved protective suits and other required personal protective equipment (PPE). After the work assignment is completed each day, the workers must be decontaminated, following the standard protocol. Moreover, these workers should be prohibited from wearing personal clothing while removing asbestos. They should be required to decontaminate and remove their contaminated protective clothing and change into uncontaminated personal clothing before leaving the jobsite. The proper procedure is to leave any contaminated clothing at work and not to take it into the household where family members could be exposed. The bottom line: leave asbestos (and any other contaminant) at the worksite.

Implementation of standardized *materials handling or transferring procedures* is another administrative control often used to protect workers. In handling chemicals, any transfer operation taken should be closed-system (i.e., the transfer of a chemical directly from its storage container to the application point), or should have adequate exhaust systems to prevent worker exposure to contamination of the workplace air. This practice should also include the use of spill trays to collect overfill spills or leaking materials between transfer points.

Administrative controls that involve visual inspection and automatic sensor services (leak detection programs) allow not only for quick detection, but also for quick repair and minimal exposure. When automatic system sensors and alarms are deployed as administrative controls, tying the alarm system into an automatic shutdown system (close a valve, open an electrical circuit, etc.) allows the sensor to detect a leak, sound the alarm, and initiate corrective action (e.g., immediate shutdown of the system).

Two other administrative control practices include training and personal hygiene. For workers to best protect themselves from workplace hazards (to reduce the risk of injury or illness), they must be made aware of the hazards; they must be trained. OSHA puts well-placed emphasis on the worker training requirement. No worker can be expected to know about every workplace process or equipment hazard unless he or she has been properly trained on the hazards and/or potential hazards. Thus, an important part of the training process is worker awareness. Legally (and morally) workers have the right to know what they are working with, what they are exposed to while on the job; they must be made aware of the hazards. They must also be trained on what actions to take when they are exposed to specific hazards.

Personal hygiene practices are an important part of worker protection. The industrial hygienist must ensure that appropriate cleaning agents and facilities such as soap, sinks, showers, and toilets are available to workers. Changing rooms, in addition to such equipment as emergency eyewashes and deluge showers, must be made available and conveniently located for worker use.

## Personal Protective Equipment (PPE)

Personal protective equipment is the workers' last line of defense against injury on the job. Industrial hygienists prefer to incorporate engineering, work practice, and administrative controls whenever possible. However, when the work environment can't be made safe by any other method, PPE is used as the last resort. PPE imposes a barrier between the work and the hazard, but does nothing to reduce or eliminate the hazard. Typical PPE includes safety goggles, helmets, face shields, gloves, safety shoes, hearing protection devices, full-body protective clothing, barrier creams, and respirators (Olishifski, 1988). To be effective,

PPE must be individually selected; properly fitted and periodically refitted; conscientiously and properly worn; regularly maintained; and replaced as necessary.

## REFERENCES

Goetsch, D. L., 1996. *Occupational Safety and Health: In the Age of High Technology for Technologists, Engineers, and Managers*, 2nd ed. Englewood Cliffs, N.J.: Prentice Hall.

Olishifski, J. B., 1998. "Overview of Industrial Hygiene," *Fundamentals of Industrial Hygiene*, 3rd ed. Chicago: National Safety Council.

OSHA, 1998. *Informational Booklet on Industrial Hygiene*, 3143. Washington, D.C.: U.S. Department of Labor. www.osha.gov (accessed June 2005).

Spellman, F. R., 1998. *Surviving an OSHA Audit*. Boca Raton, Fla.: CRC Press.

# 2

# Industrial Hygiene/Safety Terminology

Every branch of science, every profession, and every engineering process has its own language for communication. Industrial hygiene is no different. To work even at the edge of industrial hygiene, you must acquire a fundamental vocabulary of the components that make up the process of administering industrial hygiene practice.

## INTRODUCTION

As Voltaire said, "If you wish to converse with me, define your terms." In this chapter, we define many of the terms or "tools" (concepts and ideas) used by industrial hygienists in applying their skills to make our dynamic technological world safer. These concepts are presented early in the text rather than later (as is traditionally done in an end-of-book glossary), so you can become familiar with the terms early, before the text approaches the issues those terms describe. The practicing industrial hygienist and/or student of industrial hygiene should know these concepts—without them it is difficult (if not impossible) to practice industrial hygiene.

Industrial hygiene has extensive and unique terminology, most with well-defined meanings, but a few terms (especially *safety, accident, injuries,* and *engineering*, as used in the safety context) often are poorly defined, or are defined from different and conflicting points of view. For our purpose, we present the definitions of key terms, highlighting and explaining those poorly defined terms, discussing differing views where necessary.

Not every industrial hygiene and safety term is defined herein, only those terms and concepts necessary to understand the technical jargon presented in this text. For those practicing industrial hygienists and safety engineers and students of industrial hygiene who want a complete, up-to-date, and accurate dictionary of terms used in the industrial hygiene and safety professions, we recommend the American Society of Safety Engineers (ASSE) text, *The Dictionary of Terms Used in the Safety Profession*. This concise, informative, and valuable industrial hygiene/safety asset can be obtained from the ASSE at 1800 East Oakton Street, Des Plaines, Illinois 60018 (1-312-692-4121).

Finally, anytime we look to a definition for meaning, we are wise to remember the words of Yogi Berra, that great philosopher who provides meanings in terms we can all relate to and understand: "95% of baseball is pitching, the other 50% is hitting."

## TERMS

**abatement period**: The amount of time given an employer to correct a hazardous condition that has been cited.

**abrasive blasting**: A process for cleaning surfaces by means of materials such as sand, alumina, or steel grit in a stream of high-pressure air.

**absorption**: The taking up of one substance by another, such as a liquid by a solid or a gas by a liquid.

**accident**: This term is often misunderstood and is often mistakenly used interchangeably with *injury*. The dictionary defines an accident as "a happening or event that is not expected, foreseen or intended." The legal definition is: "an unexpected happening causing loss or injury which is not due to any fault or misconduct on the part of the person injured, yet entitles some kind of legal relief."

Consider the confusion in the following definition of the term *accident*:

> With rare exception, an accident is defined, explicitly or implicitly, by the unexpected occurrence of physical or chemical change to an animate or inanimate structure. It is important to note that the term covers only damage of certain types. Thus, if a person is injured by inadvertently ingesting poison, an accident is said to have taken place; but if the same individual is injured by inadvertently ingesting poliovirus, the result is but rarely considered accidental. This illustrates a curious inconsistency in the approach to accidents as opposed to other sources of morbidity, one which continues to delay progress in the field [of accident research]. In addition, although accidents are defined by the unexpected occurrence of damage, it is the unexpectedness, rather than the production and prevention of that damage per se, that has been emphasized by much of accident research. (Haddon, Suchm, and Klein, 1964, 28)

However, another definition—the one we will adopt, the one that is more applicable to our context—is provided by safety experts, the authors of *The Dictionary of Terms Used in the Safety Profession* (ASSE, 1988). Let's see how they define *accident*.

An accident is an unplanned and sometimes injurious or damaging event which interrupts the normal progress of an activity and is invariably preceded by an unsafe act or unsafe condition thereof. An accident may be seen as resulting from a failure to identify a hazard or from some inadequacy in an existing system of hazard controls. Based on applications in casualty insurance, an accident is an event that is definite in point of time and place but unexpected as to either its occurrence or its results.

**accident analysis:** A comprehensive, detailed review of the data and information compiled from an accident investigation. An accident analysis should be used to determine causal factors only, and not to point the finger of blame at any one. Once the causal factors have been determined, corrective measures should be prescribed to prevent recurrence.

**accident prevention:** The act of prevention of a happening which may cause loss or injury to a person.

**accommodation:** The ability of the eye to become adjusted after viewing the visual display terminal (VDT) so as to be able to focus on other objects, particularly objects at a distance.

**accuracy:** The exactness of an observation obtained from an instrument or analytical technique with the true value.

**ACGIH:** American Conference of Governmental Industrial Hygienists.

**acid:** Any chemical with a low pH that in water solution can burn the skin or eyes. Acids turn litmus paper red and have pH values of 0 to 6.

**acoustics:** In general, the experimental and theoretical science of sound and its transmission; in particular, that branch of the science that has to do with the phenomena of sound in a particular space such as a room or theater. Industrial hygiene is concerned with the technical control of sound, and, to engineer out the noise hazard, involves architecture and construction, studying control of vibration, soundproofing, and the elimination of noise.

**action level:** Term used by OSHA and NIOSH (National Institute for Occupational Safety and Health—a federal agency that conducts research on safety and health concerns) and is defined in the Code of Federal Regulations (CFR), title 40, *Protection of Environment.* Under OSHA, action level is the level of toxicant, which requires medical surveillance, usually 50% of the Personal Exposure Level (PEL). Note that OSHA also uses the action level in other ways besides setting the level of "toxicant." For example, in its hearing conservation standard, 29 CFR 1910.95, OSHA defines the action level as an 8-hour time-weighted average (TWA) of 85 decibels measured on the A-scale, slow response, or equivalently, a dose of 50%. Under 40 CFR 763.121, action level means an airborne concentration of asbestos of 0.1 fiber per cubic centimeter (f/cc) of air calculated as an 8-hour TWA.

**activated charcoal:** Charcoal is an amorphous form of carbon formed by burning wood, nutshells, animal bones, and other carbonaceous material. Charcoal becomes activated by heating it with steam to 800–900°C. During this treatment, a porous, submicroscopic internal structure is formed which gives it an extensive internal surface area. Activated charcoal is commonly used as a gas or vapor adsorbent in air-purifying respirators and as a solid sorbent in air-sampling.

**acute**: Health effects, which show up a short length of time after exposure. An acute exposure runs a comparatively short course and its effects are easier to reverse than those of a chronic exposure.

**acute toxicity**: The discernible adverse effects induced in an organism with a short period of time (days) of exposure to an agent.

**administrative controls**: Methods of controlling employee exposures by job rotation, work assignment, time periods away from the hazard, or training in specific work practices designed to reduce exposure.

**adsorption**: The taking up of a gas or liquid at the surface of another substance, usually a solid (e.g., activated charcoal adsorbs gases).

**aerosols**: Liquid or solid particles that are so small they can remain suspended in air long enough to be transported over a distance.

**air**: The mixture of gases that surrounds the earth; its major components are as follows: 78.08% nitrogen, 20.95% oxygen, 0.03% carbon dioxide, and 0.93% argon. Water vapor (humidity) varies.

**air cleaner**: A device designed to remove atmospheric airborne impurities, such as dusts, gases, vapors, fumes, and smoke.

**air contamination**: The result of introducing foreign substances into the air so as to make the air contaminated.

**air-line respirator**: A respirator that is connected to a compressed breathing air source by a hose of small inside diameter. The air is delivered continuously or intermittently in a sufficient volume to meet the wearer's breathing.

**air monitoring**: The sampling for and measurement of pollutants in the atmosphere.

**air pollution**: Contamination of the atmosphere (indoor or outdoor) caused by the discharge (accidental or deliberate) of a wide range of toxic airborne substances.

**air-purifying respirator**: A respirator that uses chemicals to remove specific gases and vapors from the air or that uses a mechanical filter to remove particulate matter. An air-purifying respirator must only be used when there is sufficient oxygen to sustain life and the air contaminant level is below the concentration limits of the device.

**air sampling**: Industrial hygienists are interested in knowing what contaminants workers are exposed to, and at what concentrations. Determining the quantities and types of atmospheric contaminants is accomplished by measuring and evaluating a representative sample of air. The types of air contaminants that occur in the workplace depend upon the raw materials used and the processes employed. Air contaminants can be divided into two broad groups, depending upon physical characteristics: (1) gases and vapors, and (2) particulates.

**air-supplied respirator**: Respirator that provides a supply of breathable air from a clean source outside of the contaminated work area.

**allergic reactions**: Due to the presence of allergens on spores, all molds studied to date have the potential to cause allergic reaction in susceptible humans. Allergic reactions are believed to be the most common exposure reaction to molds (Rose, 1999).

**alpha particle**: A small, positively charged particle made up of two neutrons and two protons of very high velocity, generated by many radioactive materials, including uranium and radium.

**alveoli**: Tiny air sacs in the lungs, located at the ends of bronchioles. Through the thin walls of the alveoli, blood takes in oxygen and gives up carbon dioxide in respiration.

**ambient**: Descriptive of any condition of the surrounding environment at a given point. For example, ambient air means that portion of the atmosphere, external to buildings, to which the general public has access. Ambient sound is the sound generated by the environment.

**amorphous**: Noncrystalline.

**ANSI**: The American National Standards Institute is a voluntary membership organization (run with private funding) that develops consensus standards nationally for a wide variety of devices and procedures.

**aromatic**: Applied to a group of hydrocarbons and their derivatives characterized by the presence of the benzene nucleus.

**asbestosis**: A disease of the lungs caused by inhalation of fine airborne asbestos fibers.

**asphyxiant**: A vapor gas which can cause unconsciousness or death by suffocation (lack of oxygen).

**asphyxiation**: Suffocation from lack of oxygen. A substance (e.g., carbon monoxide), that combines with hemoglobin to reduce the blood's capacity to transport oxygen produces chemical asphyxiation. Simple asphyxiation is the result of exposure to a substance (such as methane) that displaces oxygen. Asphyxiation is one the principal potential hazards of working in confined spaces.

**ASTM**: American Society for Testing and Materials.

**atmosphere**: In physics, a unit of pressure whereby 1 atmosphere (atm) equals 14.7 pounds per square inch (psi).

**atmosphere-supplying respirator**: A respirator that provides breathing air from a source independent of the surrounding atmosphere. There are two types: air-line and self-contained breathing apparatus.

**atmospheric pressure:** The pressure exerted in all directions by the atmosphere. At sea level, mean atmospheric pressure is 29.92 in. H, 14.7 psi, or 407 in. wg.

**atomic weight:** The atomic weight is approximately the sum of the number of protons and neutrons found in the nucleus of an atom. This sum is also called the mass number. The atomic weight of oxygen is approximately 16, with most oxygen atoms containing 8 neutrons and 8 protons.

**attenuate:** To reduce in amount. Used to refer to noise or ionizing radiation.

**attenuation:** The reduction of the intensity at a designated first location as compared with intensity at a second location, which is farther from the source (reducing the level of noise by increasing distance from the source is a good example).

**audible range:** The frequency range over which normal hearing occurs—approximately 20 Hz through 20,000 Hz. Above the range of 20,000 Hz, the term ultrasonic is used. Below 20 Hz, the term subsonic is used.

**audiogram:** A record of hearing loss or hearing level measured at several different frequencies—usually 500 to 6000 Hz. The audiogram may be presented graphically or numerically. Hearing level is shown as a function of frequency.

**audiometric testing:** Objective measuring of a person's hearing sensitivity. By recording the response to a measured signal, a person's level of hearing sensitivity can be expressed in decibels, as related to an audiometric zero, or no-sound base.

**authorized person:** A person designated or assigned by an employer or supervisor to perform a specific type of duty or duties, to use specified equipment, and/or to be present in a given location at specified times (for example, an authorized or qualified person is used in confined space entry).

**auto-ignition temperature:** The lowest temperature at which a vapor-producing substance or a flammable gas will ignite even without the presence of a spark or flame.

**Avogadro's Number:** The number of molecules in a mole of any substance ($6.02217 \times 10^3$). Named after Italian physicist Amedeo Avogadro (1776–1856). At 0°C and 29.92 in. HG, one mole occupies 22.4 liters of volume.

**background noise:** The radiation coming from sources other than the particular noise sources being monitored.

**baghouse:** Term commonly used for the housing containing bag filters for recovery of fumes from arsenic, lead, sulfa, etc. It has many different trade meanings, however.

**base:** A compound that reacts with an acid to form a salt. It is another term for alkali.

**baseline data:** Data collected prior to a project for later use in describing conditions before the project began. Also commonly used to describe the first audiogram given (within six months) to a worker after he or she has been exposed to the action level (85 dBA)—to establish his or her baseline for comparison to subsequent audiograms for comparison.

**bel**: A unit equal to 10 decibels (see decibel).

**behavior-based management models**: A management theory, based on the work of B. F. Skinner, that explains behavior in terms of stimulus, response, and consequences.

**benchmarking**: A process for rigorously measuring company performance versus "best-in-class" companies, and using analysis to meet and exceed the best in class.

**benign**: Not malignant. A benign tumor is one which does not metastasize or invade tissue. Benign tumors may still be lethal, due to pressure on vital organs.

**beta particle**: Beta radiation. A small electrically charged particle thrown off by many radioactive materials, identical to the electron. Beta particles emerge from radioactive material at high speeds.

**bioaerosols**: Mold spores, pollen, viruses, bacteria, insect parts, animal dander, etc.

**biohazard**: Biological hazard—organisms or products of organisms that present a risk to humans.

**boiler code**: ANSI/ASME Pressure Vessel Code. A set of standards prescribing requirements for the design, construction, testing, and installation of boilers and unfired pressure vessels.

**boiling point**: The temperature at which the vapor pressure of a liquid equals atmospheric pressure.

**Boyle's Law**: States that the product of a given pressure and volume is constant with a constant temperature.

**breathing zone**: A hemisphere-shaped area from the shoulders to the top of the head.

**cancer**: A cellular tumor formed by mutated cells.

**capture velocity**: Air velocity at any point in front of an exhaust hood necessary to overcome opposing air currents and to capture the contaminated air by causing it to flow into the hood.

**carbon monoxide**: A colorless, odorless toxic gas produced by any process that involves the incomplete combustion of carbon-containing substances. It is emitted through the exhaust of gasoline powered vehicles.

**carcinogen**: A substance or agent capable of causing or producing cancer in mammals, including humans. A chemical is considered to be a carcinogen if: (a) it has been evaluated by the International Agency for Research on Cancer (IARC) and found to be a carcinogen or potential carcinogen; (b) it is listed as a carcinogen or potential carcinogen in the Annual Report on Carcinogens published by the National Toxicology Program (NTP) (latest edition); or (c) it is regulated by OSHA as a carcinogen.

**carpal tunnel syndrome**: An injury to the median nerve inside the wrist.

**CAS**: Chemical Abstracts Service, an organization under the American Chemical Society. CAS abstracts and indexes chemical literature from all over the world in *Chemical Abstracts*. "CAS Numbers" are used to identify specific chemicals or mixtures.

**catalyst**: A substance that alters the speed of, or makes possible, a chemical or biochemical reaction, but remains unchanged at the end of the reaction.

**catastrophe**: A loss of extraordinary large dimensions in terms of injury, death, damage, and destruction.

**causal factor**: A person, thing, or condition that contributes significantly to an accident or to a project outcome.

**ceiling limit (C)**: An airborne concentration of a toxic substance in the work environment, which should never be exceeded.

**CERCLA**: Comprehensive Environmental Response, Compensation and Liability Act of 1980. Commonly known as "Superfund."

**CFR**: Code of Federal Regulations. A collection of the regulations that have been promulgated under U.S. law.

**Charles's Law**: Law stating that the volume of a given mass of gas at constant pressure is directly proportional to its absolute temperature (Kelvin).

**Chemical cartridge respirator**: A respirator that uses various chemical substances to purify inhaled air of certain gases and vapors. This type respirator is effective for concentrations no more than ten times the TLV of the contaminant, if the contaminant has warning properties (odor or irritation) below the TLV.

**chemical change**: Change that occurs when two or more substances (reactants) interact with each other, resulting in the production of different substances (products) with different chemical compositions. A simple example of chemical change is the burning of carbon in oxygen to produce carbon dioxide.

**chemical hazards**: Include mist, vapors, gases, dusts, and fumes.

**chemical spill**: An accidental dumping, leakage, or splashing of a harmful or potentially harmful substance.

**CHEMTREC**: Chemical Transportation Emergency Center. Public service of the Chemical Manufacturers Association that provides immediate advice for those at the scene of hazardous materials emergencies. CHEMTREC has a 24-hour toll-free telephone number (800-424-9300) to help respond to chemical transportation emergencies.

**chromatograph**: An instrument that separates and analyzes mixtures of chemical substances.

**chronic:** Persistent, prolonged, repeated. Chronic exposure occurs when repeated exposure to or contact with a toxic substance occurs over a period of time, the effects of which become evident only after multiple exposures.

**coefficient of friction**: A numerical correlation of the resistance of one surface against another surface.

**colorimetry**: A term used for all chemical analysis involving reactions in which a color is developed when a particular contaminant is present in the sample and reacts with the collection medium. The resultant stain length or color intensity is measured to determine the actual concentration.

**combustible gas indicator**: An instrument which samples air and indicates whether an explosive mixture is present.

**combustible liquid**: Liquids having a flash point at or above 37.8°C (100°F).

**combustion**: Burning, defined in chemical terms as the rapid combination of a substance with oxygen, accompanied by the evolution of heat and usually light.

**competent person**: As defined by OSHA, one who is capable of recognizing and evaluating employee exposure to hazardous substances or to unsafe conditions, and who is capable of specifying protective and precautionary measures to be taken to ensure the safety of employees as required by particular OSHA regulations under the conditions to which such regulations apply.

**concentration**: The amount of a given substance in a stated unit of measure. Common methods of stating concentration are percent by weight or by volume, weight per unit volume, normality, etc.

**conductive hearing loss**: Type of hearing loss usually caused by a disorder affecting the middle or external ear.

**confined space**: A vessel, compartment, or any area having limited access and (usually) no alternate escape route, having severely limited natural ventilation or an atmosphere containing less that 19.5% oxygen, and having the capability of accumulating a toxic, flammable, or explosive atmosphere, or of being flooded (engulfing a victim).

**contact dermatitis**: Dermatitis caused by contact with a substance.

**containment**: In fire terminology, restricting the spread of fire. For chemicals, restricting chemicals to an area that is diked or walled off to protect personnel and the environment.

**contingency plan**: (commonly called the emergency response plan) Under 40 CFR 260.10, a document that sets forth an organized, planned, and coordinated course of action to be followed in the event of an emergency that could threaten human health or the environment.

**convection**: The transfer of heat from one location to another by way of a moving medium.

**corrosive material**: Any material that dissolves metals or other materials, or that burns the skin.

**coulometer**: A chemical analysis instrument that determines the amount of a substance released in electrolysis by measurement of the quantity of electricity used. The number of electrons transferred in terms of coulombs is an indication of the contaminant concentration.

**cumulative injury**: A term used to describe any physical or psychological disability that results from the combined effects of related injuries or illnesses in the workplace.

**cumulative trauma disorder (CTD)**: A disorder of a musculoskeletal or nervous system component, which in some cases can result in moderate to total disability. It is caused by highly repetitive and/or forceful movements required of one or more parts of the worker's musculoskeletal system.

**cutaneous**: Pertaining to or affecting the skin.

**cyclone device**: A dust-collecting instrument that has the ability to separate particles by size.

**Dalton's Law of Partial Pressures**: States that in a mixture of theoretically ideal gases, the pressure exerted by the mixture is the sum of the pressure exerted by each component gas of the mixture.

**decibel (dB)**: A unit of measure used originally to compare sound intensities and subsequently electrical or electronic power outputs; now also used to compare voltages. In hearing conservation, a logarithmic unit used to express the magnitude of a change in level of sound intensity.

**decontamination**: The process of reducing or eliminating the presence of harmful substances such as infectious agents, to reduce the likelihood of disease transmission from those substances.

**degrees Celsius (Centigrade)**: The temperature on a scale in which the freezing point of water is 0°C and the boiling point is 100°C. To convert to degrees Fahrenheit, use the following formula: °F = (°C × 1.8) + 32.

**degrees Fahrenheit**: The temperature on a scale in which the boiling point of water is 212°F and the freezing point is 32°F.

**density**: A measure of the compactness of a substance; it is equal to its mass per unit volume and is measured in kg per cubic meter/LB per cubic foot (D = mass/Volume).

**dermatitis**: Inflammation or irritation of the skin from any cause. Industrial dermatitis is an occupational skin disease.

**dermatosis**: A broader term than dermatitis; it includes any cutaneous abnormality, thus encompassing folliculitis, acne, pigmentary changes, and nodules and tumors.

**design load**: The weight that can be safely supported by a floor, equipment, or structure, as defined by its design characteristics.

**dike**: An embankment or ridge of either natural or man-made materials used to prevent the movement of liquids, sludges, solids, or other materials.

**dilute**: Adding material to a chemical by the user or manufacturer to reduce the concentration of active ingredient in the mixture.

**direct reading instruments**: Devices that provide an immediate indication of the concentration of aerosols, gases, and vapors by means of a color change in colorimetric devices, or a register on a meter or instrument.

**dose**: An exposure level. Exposure is expressed as weight or volume of test substance per volume of air (mg/l), or as parts per million (ppm).

**dose-response relationship**: Correlation between the amount of exposure to an agent or toxic chemical and the resulting effect on the body.

**dosimeter:** Provides a time-weighted average over a period of time such as one complete work shift.

**dry bulb temperature:** The temperature of a gas or mixture of gases indicated on an ordinary thermometer after correction for radiation and moisture.

**duct:** A conduit used for moving air at low pressures.

**duct velocity:** Air velocity through the duct cross section.

**dusts:** Solid particles generated by handling, crushing, grinding, rapid impact, detonation, and decrepitation of organic or inorganic materials, such as rock ore, metal, coal, wood and grain. Dusts do not tend to flocculate, except under electrostatic forces; they do not diffuse in air but settle under the influence of gravity.

**dyspnea:** Shortness of breath, difficult or labored breathing.

**electrical grounding:** Precautionary measures designed into an electrical installation to eliminate dangerous voltages in and around the installation, and to operate protective devices in case of current leakage from energized conductors to their enclosures.

**emergency plan:** *See* contingency plan.

**emergency response:** The response made by firefighters, police, health care personnel, and/or other emergency service upon notification of a fire, chemical spill, explosion, or other incident in which human life and/or property may be in jeopardy.

**energized ("live"):** The conductors of an electrical circuit. Having voltage applied to such conductors and to surfaces that a person might touch; having voltage between such surfaces and other surfaces that might complete a circuit and allow current to flow.

**energy:** The capacity for doing work. Potential energy (PE) is energy deriving from position; thus a stretched spring has elastic PE, and an object raised to a height above the Earth's surface, or the water in an elevated reservoir has gravitational PE. A lump of coal and a tank of oil, together with oxygen needed for their combustion, have chemical energy. Other sorts of energy include electrical and nuclear energy, light, and sound. Moving bodies possess kinetic energy (KE). Energy can be converted from one form to another, but the total quantity stays the same (in accordance with the conservation of energy principle). For example, as an orange falls, it loses gravitational PE, but gains KE.

**engineering:** The application of scientific principles to the design and construction of structures, machines, apparatus, manufacturing processes, and power generation and utilization, for the purpose of satisfying human needs. Industrial hygiene is concerned with control of environment and humankind's interface with it, especially safety interaction with machines, hazardous materials, and radiation.

**engineering controls:** Methods of controlling employee exposures by modifying the source or reducing the quantity of contaminants released into the workplace environment.

**Epidemiological Theory:** Holds that the models used for studying and determining epidemiological relationships can also be used to study causal relationships between environmental factors and accidents or diseases.

**ergonomics:** A multidisciplinary activity dealing with interactions between man and his total working environment, plus stresses related to such environmental elements as atmosphere, heat, light, and sound, as well as all tools and equipment of the workplace.

**etiology:** The study or knowledge of the causes of disease.

**evaporation:** The process by which a liquid is changed into the vapor state.

**evaporation rate:** The ratio of the time required to evaporate a measured volume of a liquid to the time required to evaporate the same volume of a reference liquid (butyl acetate, ethyl ether) under ideal test conditions. The higher the ratio, the slower the evaporation rate. The evaporation rate can be useful in evaluating the health and fire hazards of a material.

**exhaust ventilation:** A mechanical device used to remove air from any space.

**exposure:** Contact with a chemical, biological, or physical hazard.

**exposure ceiling:** Refers to the concentration level of a given substance that should not be exceeded at any point during an exposure period.

**face velocity:** Average air velocity into an exhaust system measured at the opening of the hood or booth.

**fall arresting system:** A system consisting of a body harness, a lanyard or lifeline, and an arresting mechanism with built-in shock absorber, designed for use by workers performing tasks in locations from which falls would be injurious or fatal, or where other kinds of protection are not practical.

**Federal Register:** Publication of U.S. government documents officially promulgated under the law, documents whose validity depends upon such publication. It is published on each day following a government working day. It is, in effect, the daily supplement to the Code of Federal Regulations (CFR).

**fiber:** Particle with an aspect ratio greater than 3:1 (NIOSH 5:1).

**fire:** A chemical reaction between oxygen and a combustible fuel.

**fire point:** The lowest temperature at which a material can evolve vapors fast enough to support continuous combustion.

**flame ionization detector:** A direct reading instrument that ionizes gases and vapors with an oxyhydrogen flame and measures the differing electrical currents thus generated.

**flammable liquid:** Any liquid having a flash point below 37.8°C (100°F), except any mixture having components with flashpoints of 100°F or higher, the total of which make up 99% or more of the total volume of the mixture.

**flammable range**: The difference between the lower and upper flammable limits, expressed in terms of percentage of vapor or gas in air by volume, and is also often referred to as the "explosive range."

**flammable solid**: A nonexplosive solid liable to cause fire through friction, absorption of moisture, spontaneous chemical change, or heat retained from a manufacturing process, or that can be ignited readily and when ignited, burns so vigorously and persistently as to create a serious hazard.

**flash point**: The lowest temperature at which a liquid gives off enough vapor to form ignitable moisture with air, and produce a flame when a source of ignition is present. Two tests are used—open cup and closed cup.

**foot-candle**: A unit of illumination. The illumination at a point on a surface which is one foot from, and perpendicular to, a uniform point source of one candle.

**fume**: Airborne particulate matter formed by the evaporation of solid materials, e.g., metal fume emitted during welding. Usually less than one micron in diameter.

**gage pressure**: Pressure measured with respect to atmospheric pressure.

**gamma rays**: High energy x-rays.

**gas**: A state of matter in which the material has very low density and viscosity, can expand and contract greatly in response to changes in temperature and pressure, easily diffuses into other gases, and readily and uniformly distributes itself throughout any container. Examples include sulfur dioxide, ozone, and carbon monoxide.

**gas chromatography**: A detection technique that separates a gaseous mixture by passing it through a column, enabling the components to be released at various times depending on their molecular structure. Used as an analytical tool for air sampling gases and vapors.

**Geiger-Muller counter**: A gas-filled electrical instrument that indicates the presence of an atomic particle or ray by detecting the ions produced.

**general ventilation**: A ventilation system using natural or mechanically generated make-up air to mix and dilute contaminants in the workplace.

**globe thermometer**: A thermometer set in the center of a black metal sphere to measure radiant heat.

**grab sample**: A sample taken within a short duration to quantify or identify air contaminants.

**gram (g)**: A metric unit of weight. One ounce equals 28.4 grams.

**grounded system**: A system of conductors in which at least one conductor or point is intentionally grounded, either solidly or through a current-limiting (current transformer) device.

**ground-fault circuit interrupter (GFCI)**: A sensitive device intended for shock protection, which functions to de-energize an electrical circuit or portion thereof within a fraction of a second, in case of leakage to ground of current sufficient to be dangerous to persons but less than that required to operate the overcurrent protective device of the circuit.

**half-life:** For a single radioactive decay process, the time required for the activity to decrease to half its value by that process.

**hazard:** The potential for an activity, condition, circumstance, or changing conditions or circumstances to produce harmful effects. Also an unsafe condition.

**hazard analysis:** A systematic process for identifying hazards and recommending corrective action.

**Hazard and Operability (HAZOP) Analysis:** A systematic method in which process hazards and potential operating problems are identified, using a series of guide words to investigate process deviations.

**hazard assessment:** A qualitative evaluation of potential hazards in the interrelationships between and among the elements of a system, upon the basis of which the occurrence probability of each identified hazard is rated.

**Hazard Communication Standard (HazCom):** An OSHA workplace standard found in 29 CFR1910.1200 (OSHA, 1995) that requires all employers to become aware of the chemical hazards in their workplace and relay that information to their employees. In addition, a contractor conducting work at a client's site must provide chemical information to the client regarding the chemicals that are brought onto the work site.

**hazard control:** A means of reducing the risk from exposure to a hazard.

**hazard identification:** The pinpointing of material, system, process, and plant characteristics that can produce undesirable consequences through the occurrence of an accident.

**hazardous material:** Any material possessing a relatively high potential for harmful effects upon persons.

**hazardous substance:** Any substance which has the potential for causing injury by reason of its being explosive, flammable, toxic, corrosive, oxidizing, irritating, or otherwise harmful to personnel.

**hazardous waste:** A solid, liquid, or gaseous waste that may cause or significantly contribute to serious illness or death, or that poses a substantial threat to human health or the environment when the waste is improperly managed. When a hazardous material is spilled, it becomes a hazardous waste.

**hearing conservation:** The prevention of, or minimizing of noise-induced deafness through the use of hearing protection devices, the control of noise through engineering controls, annual audiometric tests and employee training.

**heat cramps:** A type of heat stress resulting in muscle spasms. Exposure to excessive heat results in salt and potassium depletion.

**heat exhaustion:** A condition usually caused by loss of body water from exposure to excessive heat. Symptoms include headache, tiredness, nausea, and sometimes fainting.

**heat rash**: A rash caused by sweating and inadequate hygiene practices.

**heat stroke**: A serious disorder resulting from exposure to excessive heat. Caused by sweat suppression and increased storage of body heat. Symptoms include hot, dry skin, high temperature, mental confusion, collapse, and sometimes convulsions or coma.

**HEPA (High Efficiency Particulate Air) filter**: A disposable, extended medium, dry type filter with a particle removal efficiency of no less than 99.97% for 0.3m particles.

**Homeland Security**: Federal cabinet level department created to protect America as a result of 9/11. The new Department of Homeland Security (DHS) has three primary missions: prevent terrorist attacks within the United States, reduce America's vulnerability to terrorism, and minimize the damage from potential attacks and natural disasters.

**hood entry loss**: The pressure loss from turbulence and friction as air enters a ventilation system.

**hot work**: Work involving electric or gas welding, cutting, brazing, or similar flame or spark-producing operations.

**human factor engineering** (used in the United States) or **ergonomics** (used in Europe): For practical purposes, the terms are synonymous, and focus on human beings and their interaction with products, equipment, facilities, procedures, and environments used in work and everyday living. The emphasis is on human beings (as opposed to engineering, where the emphasis is more strictly on technical engineering considerations) and how the design of things influences people. Human factors, then, seek to change the things people use and the environments in which they use these things to better match the capabilities, limitations, and needs of people (Sanders & McCormick, 1993).

**IDLH**: Immediately Dangerous to Life and Health. An atmospheric concentration of any toxic, corrosive, or asphyxiant substance that poses an immediate threat to life, would cause irreversible or delayed adverse health effects, or would interfere with an individual's ability to escape from a dangerous atmosphere.

**ignition source**: Anything that provides heat, spark, or flame sufficient to cause combustion or an explosion.

**ignition temperature**: The temperature at which a given fuel can burst into flame.

**illumination**: The amount of light flux a surface receives per unit area. May be expressed in lumens per square foot or in foot-candles.

**impaction**: Forcibly lodging particles into matter.

**impervious**: A material that does not allow another substance to pass through or penetrate it. Frequently used to describe gloves or chemical protective clothing.

**impingement**: The process of collecting particles by pulling contaminated air through a device filled with water or reagent (particles remain in the liquid).

**impulse noise**: A noise characterized by rapid rise time, high peak value, and rapid decay.

**inches of mercury column**: A unit used in measuring pressures. One inch of mercury column equals a pressure of 1.66 kPa (0.491 psi).

**inches of water column**. A unit used in measuring pressures. One inch of water column equals a pressure of 0.25 kPa (0.036 psi).

**incident**: An undesired event that, under slightly different circumstances, could have resulted in personal harm or property damage; any undesired loss of resources.

**incompatible**: Materials that could cause dangerous reactions from direct contact with one another.

**Indoor Air Quality (IAQ)**: Refers to the effect, good or bad, of the contents of the air inside a structure, on its occupants. Usually, temperature (too hot and cold), humidity (too dry or too damp), and air velocity (draftiness or motionless) are considered "comfort" rather than IAQ issues. Unless those are extreme, they may make someone unhappy, but they won't make a person ill. Nevertheless, most IAQ professionals will take these factors into account in investigating air quality situations.

**industrial hygiene**: The American Industrial Hygiene Association (AIHA) defines industrial hygiene as "that science and art devoted to the anticipation, recognition, evaluation, and control of those environmental factors or stresses—arising in the workplace—which may cause sickness, impaired health and well-being, or significant discomfort and inefficiency among workers or among citizens of the community."

**ingestion**: Taking in by the mouth.

**inhalation**: Breathing of a substance in the form of a gas, vapor, fume, mist, or dust.

**injury**: A wound or other specific damage.

**insoluble**: Incapable of being dissolved in a liquid.

**interlock**: A device that interacts with another device or mechanism to govern succeeding operations. For example, an interlock on an elevator door prevents the car from moving unless the door is properly closed.

**ionizing radiation**: Radiation that becomes electrically charged or changed into ions.

**irritant**: A substance that produces an irritating effect when it contacts skin, eyes, nose, or respiratory system.

**job hazard analysis**: The breaking down into its component parts of any method or procedure, to determine the hazards connected therewith and the requirements for performing it safely. Also called job safety analysis.

**Kelvin**: A temperature scale, also called absolute temperature, where the temperature is measured on the average kinetic energy per molecule of a perfect gas.

**kinetic energy**: The energy resulting from a moving object.

**Laboratory Safety Standard**: A specific hazard communication program for laboratories, found in 29 CFR 1910.1450 (OHSA, 1995). These regulations are essentially a blend of hazard communication and emergency response for laboratories. The cornerstone of the Lab Safety Standard is the requirement for a written Chemical Hygiene Plan.

**laser**: Light amplification by stimulated emission of radiation.

**latent period**: The time that elapses between exposure and the first manifestation of damage.

**$LC_{50}$**: Lethal concentration that will kill 50% of the test animals within a specified time.

**$LD_{50}$**: The dose required to produce death in 50% of the exposed species within a specified time.

**Liter (L)**: A measure of capacity. One quart equals 0.9L.

**local exhaust ventilation**: A ventilation system that captures and removes contaminants at the point of generation before escaping into the workplace.

**Lockout/Tagout Procedure**: An OSHA procedure found in 29 CFR 1910.147 (OHSA, 1995). A tag or lock is used to tag out or lock out a device, so that no one can inadvertently actuate the circuit, system, or equipment that is temporarily out of service.

**Log and Summary of Occupational Injuries and Illnesses (OSHA-200 Log)**: A cumulative record that employers (generally of more than 10 employees) are required to maintain, showing essential facts of all reportable occupational injuries and illnesses.

**loss**: The degradation of a system or component. Loss is best understood when related to dollars lost. Examples include death or injury to a worker, destruction or impairment of facilities or machines, destruction or spoiling of raw materials, and creation of delay. In the insurance business, loss connotes dollar loss, and we have seen underwriters who write it as LO$$ to make that point.

**lower explosive limit (LEL)**: The minimum concentration of a flammable gas in air required for ignition in the presence of an ignition source. Listed as a percent by volume in air.

**makeup air**: Clean, tempered outdoor air supplied to a workplace to replace air removed by exhaust ventilation.

**malignant**: As applied to a tumor. Cancerous and capable of undergoing metastasis, or invasion of surrounding tissue.

**Material Safety Data Sheet (MSDS)**: Chemical information sheet provided by the chemical manufacturer that includes information such as chemical and physical characteristics; long and short term health hazards; spill control procedures; personal protective equipment (PPE) to be used when handling the chemical; reactivity with other chemicals; incompatibility with other chemicals; and manufacturer's name, address, and phone number. Employee access to and understanding of MSDS are important parts of the HazCom Program.

**medical monitoring**: The initial medical exam of a worker, followed by periodic exams. The purpose of medical monitoring is to assess workers' health, determine their fitness to wear personal protective equipment, and maintain records of their health.

**mesothelioma**: Cancer of the membranes that line the chest and abdomen, almost exclusively associated with asbestos exposure.

**metabolic heat**: Produced within a body as a result of activity that burns energy.

**metastasis**: Transfer of the causal agent (cell or microorganisms) of a disease from a primary focus to a distant one through the blood or lymphatic vessels. Also, spread of malignancy from site of primary cancer to secondary sites.

**meter**: A metric unit of length, equal to about 39 inches.

**micron (micrometer, m)**: A unit of length equal to one millionth of a meter, approximately 1/25,000 of an inch.

**milligram (mg)**: A unit of weight in the metric system. One thousand milligrams equals one gram.

**milligrams per cubic meter (mg/m3)**: Unit used to measure air concentrations of dusts, gases, mists, and fumes.

**milliliter (mL)**: A metric unit used to measure volume. One milliliter equals one cubic centimeter.

**millimeter of mercury (mmHg)**: The unit of pressure equal to the pressure exerted by a column of liquid mercury one millimeter high at a standard temperature.

**mist**: Tiny liquid droplets suspended in air.

**molds**: The most typical form of fungus found on earth, comprising approximately 25% of the earth's biomass (McNeel and Kreutzer, 1996).

**monitoring**: Periodic or continuous surveillance or testing to determine the level of compliance with statutory requirements and/or pollutant levels, in various media or in humans, animals, or other living things.

**mucous membranes**: Lining of the hollow organs of the body, notably the nose, mouth, stomach, intestines, bronchial tubes, and urinary tract.

**mutagen**: A substance or material that causes change in the genetic material of a cell.

**mycotoxins**: Some molds are capable of producing mycotoxins, natural organic compounds that are capable of initiating a toxic response in vertebrates (McNeel and Kreutzer, 1996).

**NFPA**: National Fire Protection Association. A voluntary membership organization whose aim is to promote and improve fire protection and prevention. The NFPA publishes 16 volumes of codes known as the National Fire Codes.

**NIOSH**: The National Institute for Occupational Safety and Health. A federal agency that conducts research on health and safety concerns, tests and certifies respirators, and trains occupational health and safety professionals.

**nonionizing radiation**: That radiation on the electromagnet spectrum that has a frequency of $10^{15}$ or less and a wavelength in meters of $3 \times 10^{-7}$.

**NTP**: National Toxicology Program. The NTP publishes an annual report on carcinogens.

**nuisance dust**: Has a long history of little adverse effect on the lungs and does not produce significant organic disease or toxic effect when exposures are kept under reasonable control.

**Occupational Safety and Health Act (OSH Act):** A federal law passed in 1970 to assure, so far as possible, every working man and woman in the nation safe and healthful working conditions. To achieve this goal, the Act authorizes several functions, such as encouraging safety and health programs in the workplace and encouraging labor-management cooperation in health and safety issues.

**OSHA Form 300:** Log and Summary of Occupational Injuries and Illnesses.

**oxidation:** When a substance either gains oxygen, or loses hydrogen or electrons in a chemical reaction. One of the chemical treatment methods.

**oxidizer:** Also known as an oxidizing agent, a substance that oxidizes another substance. Oxidizers are a category of hazardous materials that may assist in the production of fire by readily yielding oxygen.

**oxygen deficient atmosphere:** The legal definition of an atmosphere where the oxygen concentration is less than 19.5% by volume of air.

**oxygen-enriched atmosphere:** An atmosphere containing more than 23.5% oxygen by volume.

**particulate matter:** Substances (such as diesel soot and combustion products resulting from the burning of wood) released directly into the air; any minute, separate particle of liquid or solid material.

**performance standards:** A form of OSHA regulation standards that lists the ultimate goal of compliance, but does not explain exactly how compliance is to be accomplished. Compliance is usually based on accomplishing the act or process in the safest manner possible, based on experience (past performance).

**permissible exposure limit (PEL):** The time-weighted average concentration of an airborne contaminant that a healthy worker may be exposed to 8 hours per day or 40 hours per week without suffering any adverse health effects. Established by legal means and enforceable by OSHA.

**personal protective equipment (PPE):** Any material or device worn to protect a worker from exposure to or contact with any harmful substance or force.

**pH:** Means used to express the degree of acidity or alkalinity of a solution with neutrality indicated as seven.

**pitot tube:** A device used for measuring static pressure within ventilation ducts.

**polymerization:** A chemical reaction in which two or more small molecules (monomers) combine to form larger molecules (polymers) that contain repeating structural units of the original molecules. A hazardous polymerization is the above reaction, with an uncontrolled release of energy.

**ppm:** Parts per million parts of air by volume of vapor or gas or other contaminant. Used to measure air concentrations of vapors and gases.

**precision:** The degree of exactness of repeated measurements.

**preliminary assessment:** A quick analysis to determine how serious the situation is, and to identify all potentially responsible parties. The preliminary assessment uses readily available information, such as forms, records, aerial photographs, and personnel interviews.

**pressure:** The force exerted against an opposing fluid or thrust distributed over a surface.

**psi:** Pounds per square inch. For MSDS purposes, the pressure a material exerts on the walls of a confining vessel or enclosure. For technical accuracy, pressure must be expressed as psig (pounds per square inch gauge) or psia (pounds per square absolute; that is, gauge pressure plus sea level atmospheric pressure, or psig plus approximately 14.7 pounds per square inch).

**radiant heat:** The result of electromagnetic nonionizing energy that is transmitted through space without the movement of matter within that space.

**radiation:** Consists of energetic nuclear particles and includes alpha rays, beta rays, gamma rays, x-rays, neutrons, high-speed electrons, and high-speed protons.

**Raynaud's syndrome:** Caused by vibrating hand tools, an abnormal constriction of blood vessels in the fingers when exposed to cold temperature.

**RCRA:** Resource Conservation and Recovery Act of 1976.

**reactivity (chemical):** A substance that reacts violently by catching on fire, exploding, or giving off fumes when exposed to water, air, or low heat.

**reactivity hazard:** The ability of a material to release energy when in contact with water. Also, the tendency of a material, when in its pure state or as a commercially produced product, to vigorously polymerize, decompose, condense, or otherwise self-react and undergo violent chemical change.

**reportable quantity (RQ):** The minimum amount of a hazardous material that, if spilled while in transport, must be reported immediately to the National Response Center. Minimum reportable quantities range from 1 pound to 5,000 pounds per 24-hour day.

**respirable size particulates:** Particulates in the size range that permits them to penetrate deep into the lungs upon inhalation.

**respirator (approved):** A device which has met the requirements of National Institute of Safety and Health (NIOSH) 42 CFR Part 84 and is designed to protect the wearer for inhalation of harmful atmospheres and has been approved by the National Institute for Occupational Safety and Health (NIOSH) and the Mine Safety and Health Administration (MSHA).

**respiratory system:** Consists of (in descending order)—the nose, mouth, nasal passages, nasal pharynx, pharynx, larynx, trachea, bronchi, bronchioles, air sacs (alveoli) of the lungs, and muscles of respiration.

**risk:** The combination of the expected frequency (event/year) and consequence (effects/event) of a single accident or a group of accidents; the result of a loss-probability occurrence and the acceptability of that loss.

**risk assessment**: A process that uses scientific principles to determine the level of risk that actually exists in a contaminated area.

**risk characterization**: The final step in the risk assessment process, it involves determining a numerical risk factor. This step ensures that exposed populations are not at significant risk.

**risk management**: The professional assessment of all loss potentials in an organization's structure and operations, leading to the establishment and administration of a comprehensive loss control program.

**rotameters**: A small tapered tube with a solid ball (float) inside used to measure flow rate of air sampling equipment.

**safety**: A general term denoting an acceptable level of risk of, relative freedom from, and low probability of harm.

**safety factor**: Based on experimental data, the amount added (e.g., 1000-fold) to ensure worker health and safety.

**safety standard**: A set of criteria specifically designed to define a safe product, practice, mechanism, arrangement, process, or environment, produced by a body representative of all concerned interests, and based upon currently available scientific and empirical knowledge concerning the subject or scope of the standard.

**SARA**: Superfund Amendments and Reauthorization Act of 1986.

**SCBA**: Self-contained breathing apparatus.

**secondary containment**: A method using two containment systems so that if the first is breached, the second will contain all of the fluid in the first. For underground storage tanks (USTs), secondary containment consists of either a double-walled tank or a liner system.

**security assessment**: An intensified security test in scope and effort, the purpose of which is to obtain an advanced and very accurate idea of how well the organization has implemented security mechanisms and, to some degree, policy.

**sensitizers**: Chemicals that in very low dose trigger an allergic response.

**short term exposure limit (STEL)**: The time weighted average concentration to which workers can be exposed continuously for a short period of time (typically 15 minutes) without suffering irritation, chronic or irreversible tissue damage, or impairment for self-rescue.

**silica**: Crystalline silica ($SiO_2$) is a major component of the earth's crust and is responsible for causing silicosis.

**"skin"**: A notation (sometimes used with PEL or TLV exposure data) which indicates that the stated substance may be absorbed by the skin, mucous membranes, and eyes—either airborne or by direct contact—and that this additional exposure must be considered part of the total exposure to avoid exceeding the PEL or TLV for that substance.

**solubility in water**: A term expressing the percentage of a material (by weight) that will dissolve in water at ambient temperature. Solubility information can be useful in determining spill cleanup methods and re-extinguishing agents and methods for a material.

**solvent**: A substance, usually a liquid, in which other substances are dissolved. The most common solvent is water.

**sorbent**: (1) A material that removes toxic gases and vapors from air inhaled through a canister or cartridge. (2) Material used to collect gases and vapors during air-sampling.

**specific gravity**: The ratio of the densities of a substance to water.

**stability**: An expression of the ability of a material to remain unchanged. For MSDS purposes, a material is stable if it remains in the same form under expected and reasonable conditions of storage or use. Conditions which may cause instability (dangerous change) are stated, for example, temperatures above 150°F or shock from dropping.

**synergism**: Cooperative action of substances whose total effect is greater than the sum of their separate effects.

**systemic**: Spread throughout the body, affecting all body systems and organs, rather than localized in one spot or area.

**temporary threshold shift (TTS)**: Temporary hearing loss due to noise exposure. May be partially or completely recovered when exposure ends.

**tendonitis**: Inflammation of a tendon.

**tenosynovitis**: Inflammation of the connective tissue sheath of a tendon.

**threshold**: The lowest dose or exposure to a chemical at which a specific effect is observed.

**threshold limit value (TLV)**: The same concept as PEL except that TLVs do not have the force of governmental regulations behind them, but are based on recommended limits established and promoted by the American Conference of Governmental Industrial Hygienists.

**time-weighted average (TWA)**: A mathematical average of exposure concentration over a specific time.

**total quality management (TQM)**: A way of managing a company that revolves around a total and willing commitment of all personnel at all levels to quality.

**toxicity**: The relative property of a chemical agent with reference to a harmful effect on some biologic mechanism and the condition under which this effect occurs. In other words, the quality of being poisonous.

**toxicology**: The study of poisons, which are substances that can cause harmful effects to living things.

**toxin**: A poison.

**unsafe condition**: Any physical state that deviates from that which is acceptable, normal, or correct in terms of past production or potential future production of personal injury and/or damage to property; thus, any physical state that results in a reduction in the degree of safety normally present.

**upper explosive limit (UEL):** The maximum concentration of a flammable gas in air required for ignition in the presence of an ignition source.

**vapor:** The gaseous form of substances that are normally solid or liquid at room temperature.

**vapor pressure:** Pressure—measured in pounds per square inch absolute (psia)—exerted by a vapor. If a vapor is kept in confinement over its liquid so that the vapor can accumulate above the liquid (the temperature being held constant), the vapor pressure approaches a fixed limit called the maximum (or saturated) vapor pressure, dependent only on the temperature and the liquid.

**vapors:** The gaseous form of substances that are normally in the solid or liquid state (at room temperature and pressure). The vapor can be changed back to the solid or liquid state either by increasing the pressure or decreasing the temperature alone. Vapors also diffuse. Evaporation is the process by which a liquid is changed into the vapor state and mixed with the surrounding air. Solvents with low boiling points will volatize readily. Examples include benzene, methyl alcohol, mercury, and toluene.

**velometer:** A device used in ventilation to measure air velocity.

**viscosity:** The property of a fluid that resists internal flow by releasing counteracting forces.

**volatility:** The tendency or ability of a liquid to vaporize. Liquids such as alcohol and gasoline, because of their well-known tendency to evaporate rapidly, are called volatile liquids.

**vulnerability assessment:** A very regulated, controlled, cooperative, and documented evaluation of an organization's security posture from outside-in and inside-out, for the purpose of defining or greatly enhancing security policy.

**water column:** A unit used in measuring pressure.

**wet bulb globe temperature:** Temperature as determined by the wet bulb thermometer (a bulb covered with a cloth saturated with water) or a standard sling psychrometer or its equivalent. Influenced by the evaporation rate of water which, in turn, depends on relative air humidity.

**workers' compensation:** A system of insurance required by state law and financed by employers, which provides payments to employees and their families for occupational illnesses, injuries, or fatalities incurred while at work and resulting in loss of wage income, usually regardless of the employer's or employee's negligence.

**zero energy state:** The state of equipment in which every power source that can produce movement of a part of the equipment, or the release of energy, has been rendered inactive.

**zoonoses:** Diseases communicable from animals to humans under natural conditions.

## REFERENCES

ASSE, 1988. *Dictionary of Terms Used in the Safety Profession*, 3rd ed. Des Plaines, Ill.: American Society of Safety Engineers.

Haddon, W., Jr., E. A. Suchm, and D. Klein, 1964. *Accident Research*. New York: Harper & Row.

McNeel, S., and Kreutzer, R., 1996. "Fungi & Indoor Air Quality." *Health & Environment Digest* 10, no. 2, (May/June): 9–12. http://www.dhs.cahwet,gov/org.ps/deodc/ehib/EHIB2/topics/fungi_inorro.html.

OSHA, 1987. *Protection of Environment*. CFR 40. Washington, D.C.: U.S. Department of Labor.

OSHA, 1995. *Occupational Safety and Health Standards for General Industry*. CFR 29. Washington, D.C.: U.S. Department of Labor.

Rose, C. F., 1999. "Antigens ." *ACGIH Bioaerosols Assessment and Control*. Cincinnati: American Conference of Governmental Industrial Hygienists. 25.1–25.11.

Sanders, M. S. and E. J. McCormick, 1993. *Human Factors in Engineering and Design*. 7th ed., New York: McGraw-Hill.

## SUGGESTED READING

Bird, F. E., and G. L. Germain. *Damage Control*. New York: American Management Association, 1966.

Boyce, A. *Introduction to Environmental Technology*. New York: Van Nostrand Reinhold, 1997.

Center for Chemical Process Safety (CCPS), 1985. *Guidelines for Hazard Evaluation Procedures*, 2nd ed. New York: American Institute of Chemical Engineers.

Fletcher, J. A. *The Industrial Environment—Total Loss Control*. Ontario, Canada: National Profile Limited, 1972.

National Institute of Safety and Health (NIOSH), 1995. Respiratory Protective Devices. 42 CFR Part 84.

Plog, B. A., ed. *Fundamentals of Industrial Hygiene*, current edition. Chicago: National Safety Council.

# 3

# Hazard Communication, Occupational Environmental Limits, and Air Monitoring and Sampling

**HAZARD COMMUNICATION**

The 1984 release of methyl isocyanate (MIC) by Union Carbide in Bhopal, India, and the resulting deaths and injuries are well known. However, not all of the repercussions—the lessons learned—from this incident are as well known. After Bhopal arose a worldwide outcry: How could such an incident occur? Why wasn't something done to protect the inhabitants? Weren't there safety measures taken or in place to prevent such a disaster from occurring?

In the U.S., these questions and others were bandied around and about by the press and Congress. Congress took the first major step to prevent such incidents from occurring in the U.S. by directing OSHA to take a close look at chemical manufacturing to see if a Bhopal-type incident could occur in this country. After conducting a study, OSHA reported to Congress that a Bhopal-type incident in the U.S. was very unlikely. Within a few months of the report, however, a chemical spill occurred in Institute, West Virginia, similar to Bhopal, but fortunately not deadly (over 100 people became ill).

Needless to say, Congress was upset. Because of the Bhopal and Institute fiascoes, OSHA mandated its Hazard Communication Standard (29 CFR 1910.1200) in 1984. Later, other programs—like SARA (Superfund) Title III reporting requirements for all chemical users, producers, suppliers, and storage entities—were mandated by the U.S. Environmental Protection Agency (USEPA).

There is no all-inclusive list of chemicals covered by the Hazard Communication Standard; however, the regulation refers to "any chemical which is a physical or health hazard." Those specifically deemed hazardous include:

- Chemicals regulated by OSHA in 29 CFR Part 1910, Subpart Z, Toxic and Hazardous Substances
- Chemicals included in the American Conference of Governmental Industrial Hygienists' (ACGIH) latest edition of Threshold Limit Values (TLVs) for *Chemical Substances and Physical Agents in the Work Environment*
- Chemicals found to be suspected or confirmed carcinogens by the National Toxicology Program in the *Registry of Toxic Effects of Chemicals Substances* published by NIOSH or appearing in the latest edition of the *Annual Report on Carcinogens*, or by the International Agency for Research on Cancer in the latest editions of its IARC *Monographs*

Congress decided that those personnel working with or around hazardous materials "had a right to know" about those hazards. Thus, OSHA's Hazard Communication Standard was created—without a doubt, the regulation most important to the communication of hazards to employees.

Under its Hazard Communication Standard (more commonly known as "HazCom" or the "Right to Know Law"), OSHA (1995) requires employers who use or produce chemicals on the worksite to inform all employees of the hazards that might be involved with those chemicals. HazCom says that employees have the right to know what chemicals they are handling, or could be exposed to. HazCom's intent is to make the workplace safer. Under HazCom, the employer is required to fully evaluate all chemicals on the worksite for possible physical and health hazards. All information relating to these hazards must be made available to the employee 24 hours a day. The standard is written in a performance manner, meaning that the specifics are left to the employer to develop.

HazCom also requires the employer to ensure proper labeling of each chemical, including chemicals that might be produced by a *process* (process hazards). For example, in the wastewater industry, deadly methane gas is generated in the waste stream. Another common wastewater hazard (off gas) is the generation of hydrogen sulfide (which produces the characteristic rotten-egg odor) during degradation of organic substances in the wastestream, and can kill quickly. OSHA's HazCom requires the employer to label methane and hydrogen sulfide hazards so that workers are warned, and safety precautions are followed.

Labels must be designed to be clearly understood by all workers. Employers are required to provide both training and written materials to make workers aware of what they are working with and what hazards they might be exposed to. Employers are also required to make Material Safety Data Sheets (MSDS) available to all employees. An MSDS is a fact sheet for a chemical posing a physical or health hazard at work. The MSDS must be in English and contain the following information:

# HAZARD COMMUNICATION

- Identity of the chemical (label name)
- Physical hazards
- Control measures
- Health hazards
- Whether it is a carcinogen
- Emergency and first aid procedures
- Date of preparation of latest revision
- Name, address, and telephone number of manufacturer, importer, or other responsible party

Blank spaces are not permitted on an MSDS. If relevant information in any one of the categories is unavailable at the time of preparation, the MSDS must indicate that no information was available. Every facility must have an MSDS for each hazardous chemical it uses. Copies must be made available to other companies working on the worksite (outside contractors, for example), and they must do the same for the host site. The facility hazard communication program must be in writing and, along with MSDS, made available to all workers at all times during all shifts.

## Material Safety Data Sheets (MSDS)

The following information is required on an MSDS, which, as previously stated, must be in English.

1. The specific identity of each hazardous chemical or mixture ingredient and common product names.
2. The physical and chemical characteristics of the hazardous material including: (a) density or specific gravity of liquid or solid; (b) density of gas or vapor relative to air; (c) boiling point; (d) melting point; (e) flash point; (f) flammability range; and (g) vapor pressure.
3. Physical hazard data such as stability, reactivity, flammability, corrosivity, explosivity.
4. Health hazard data including acute and chronic health effects, and target organ effects.
5. Exposure limits such as OSHA Permissible Exposure Limits (PELs).
6. Carcinogenicity of material.
7. Precautions to be taken, including use of PPE.
8. Emergency and first aid procedures. This includes spill cleanup information and EPA spill reportability information.

9. Supplier or manufacturer data including: name, address, telephone number, and date prepared.

## HazCom and the Industrial Hygienist

The industrial hygienist must take a personal interest in ensuring that his or her facility is in full compliance with the Hazard Communication Standard for three major reasons:

1. It is the law.
2. It is consistently the number one cause of citations issued by OSHA for noncompliance.
3. Compliance with the standard goes a long way toward protecting workers.

Major elements of the HazCom Standard, which the industrial hygienist must ensure are part of the organization's written hazard communication program, include hazard determination, labels and other forms of warning, material safety data sheets, and employee training. Another required element is trade secrets. For the sake of brevity in describing HazCom requirements, and because we are presenting a hazard communication program that is typically required in the general manufacturing type workplace, we don't discuss the trade secrets element beyond the brief mention that it allows the chemical manufacturer, importer, or employer to withhold the specific chemical identity—including the chemical name and other specific identification of a hazardous chemical—from the MSDS under certain conditions. For more information on the trade secrets, review 29 CFR 1910.1200 (Hazard Communication Standard).

*Hazard determination* primarily affects the chemical manufacturer and importers (for manufactured products), not the facility employer and safety/industrial hygiene official, unless they choose not to rely on the evaluation performed by the chemical's manufacturer or importer to satisfy this requirement. However, this is not the case with site process produced hazards, such as dangerous off gases. In this case, the industrial hygienist must play a major role in assessing the hazards or potential hazards generated, evaluating their possible impact on employees, and determining proper mitigation procedures.

The chemical manufacturers and importers must supply the purchaser with *Material Safety Data Sheets (MSDS)*. Purchasers (employers) are required to have an MSDS in the workplace for each hazardous chemical they use or produce. As stated previously the MSDS must be in English and readily available 24/7 to all employees, contractors, and/or visitors.

# HAZARD COMMUNICATION

*Note:* I cannot overemphasize the need for the industrial hygienist to ensure that an MSDS is present and available in the workplace for every chemical in the workplace. It is absolutely essential that all workers know where these MSDS are located.

The employer must provide *employee training* on the hazard communication program. Training on the hazardous chemicals in work areas must be provided to employees upon their initial assignment. Moreover, whenever a new physical or health hazard is introduced into the workplace (one for which training has not previously been provided), the employer must provide the training. Specifically, employee training must include:

1. the methods and observation that may be used to detect the presence or release of a hazardous chemical in the work area;
2. the physical and health hazards of the chemicals in the work area;
3. the measures employees can take to protect themselves from these hazards, including specific procedures the employer has implemented to protect employees from exposure to hazardous chemicals, such as appropriate work practices, emergency procedures, and personal protective equipment to be used; and
4. the details of the hazard communication program developed by the employer, including an explanation of the labeling system and the material safety data sheet, and how employees can obtain and use the appropriate hazard information.

*Note:* As with all OSHA-required training, you must not only ensure that the training is conducted, you must also ensure that it has been properly documented. Our legal system views training not properly documented as training not done.

*Labels and other forms of warning* are elements of HazCom that the industrial hygienist must pay particular attention to. Specifically, the chemical manufacturer, importer, or distributor must ensure that each container of hazardous chemicals being shipped is labeled, tagged, or marked with the following information:

- Identity of the hazardous chemical(s)
- Appropriate hazard warnings
- Name and address of the chemical manufacturer, importer, or other responsible party

The employer's (thus, the industrial hygienist's) responsibilities include: signs, placards, process sheets, batch tickets, operating procedures, or other such written materials in lieu of affixing labels to individual stationary process containers—as long as the alternative method identifies the containers to which it is applicable and conveys the information required on the label. The written materials must be readily accessible to the employees in their work area throughout each shift.

The employer must not remove or deface existing labels on incoming containers of hazardous chemicals unless the container is immediately marked with the required information.

The industrial hygienist must ensure that labels or warnings in his or her workplace are legible, in English, and are prominently displayed on the container or readily available in the work area throughout each work shift. It is highly recommended that employers with employees who speak other languages *also* add the information in their languages.

If existing labels already convey the required information, the employee need not affix new labels.

If the employer becomes newly aware of any significant information regarding the hazards of a chemical, the employer must revise the labels for the chemical within three months of becoming aware of the new information. (**Note:** We recommend immediate corrective action on this item. Knowledge of a problem or hazard entails a certain amount of liability if anything untoward occurs.) Labels on containers of hazardous chemicals shipped after that time shall contain the new information.

*Note:* The facility industrial hygienist, supervisors, and employees must be constantly vigilant to ensure that hazard warnings and/or labels are correct, in place, and legible.

The employer is required to develop a *written hazard communication program*. This particular requirement is often cited as the most common noncompliance (with OSHA) violation found in industry today. The written hazard communication program must be present and maintained in each workplace. The written program must include a list of hazardous chemicals known to be present—using an identity that is referenced on the appropriate material safety data sheet, the methods the employer uses to inform employees of the hazards of nonroutine tasks, and the hazards associated with chemicals contained in unlabeled pipes in their work areas.

## Sample Written Hazard Communication Program

I. Introduction
The OSHA Hazard Communication Standard was promulgated to ensure that all chemicals in the workplace would be evaluated, and that information regarding the hazards of these chemicals would be communicated to employers and employees. The goal of the Standard is to reduce the number of chemically related occupational illnesses and injuries.

# HAZARD COMMUNICATION

To comply with the Hazard Communication Standard, this written program has been established for the Company. All departments and work centers of the Company are included within this program. Copies of this written program will be available (for review by any employee) in the following locations:

[LIST LOCATIONS]

Department Standard Operating Procedures (SOPs) work in conjunction with this basic document in providing the safest possible environment to all employees.

II. Responsibilities

   A. Department Directors will be responsible for implementing and ensuring the compliance of their departmental personnel with the Company's Hazard Communication Program. Additionally, they will assign appropriate supervisors with the responsibility of ensuring compliance.

   B. The Company's Industrial Hygiene Director has the following responsibilities under the Company's Hazard Communication Program:

   1. Develop and modify as necessary the Company's Hazard Communication Program.
   2. Annually check and review the effectiveness of the overall program and all work center programs.
   3. Inspect quarterly each work center's Hazardous Chemical Inventory List and corresponding Material Safety Data Sheet to ensure they are current and complete.
   4. Receive and review all incoming or updated editions of MSDS and distribute them to pertinent work centers.
   5. Maintain a current master MSDS file.
   6. Train supervisors in requirements for the Hazard Communication Program and assist in training personnel as required.

   C. Assigned supervisors have the following responsibilities under the Company's Hazard Communication Program:

   1. Reporting receipt of all new chemicals to Safety/Industrial Hygiene Division.
   2. Ensuring that no chemical is used at the work center until it is listed on the Hazardous Chemical Inventory List, the corresponding MSDS is inserted into the "Right To Know" Station Binder, and each employee has received the appropriate HazCom training.
   3. Maintaining current work center "Right To Know" work stations, which include ensuring the MSDS and Chemical Inventory List are current and accurately reflect chemicals used on site.
   4. Ensuring proper labeling practices for all hazardous chemicals in accordance with this program.

5. Forwarding to the Industrial Hygiene Division all MSDS received from sources other than the Industrial Hygiene Division.
6. Ensuring that every employee is trained on this program and the hazards involved with chemicals used in the work center.
7. Ensuring that all on-site contractors receive copies of the Hazard Communication Program and all work center MSDS. In turn, the work center supervisor is responsible for obtaining MSDS from the contractor about chemicals that he or she may bring to the site that Company employees may be exposed to.

D. Company personnel will be responsible for familiarizing themselves with the Hazard Communication Program, and complying with instructions contained within the Hazard Communication Program.

III. Definition of Terms

The Hazard Communication Program defines various terms [these terms either appear in the company's Hazard Communication Program or are definitions appropriate to MSDS] as follows:

**chemical**: Any element, compound, or mixture of elements and/or compounds.

**chemical name**: The scientific designation of a chemical in accordance with the nomenclature system developed by the International Union of Pure and Applied chemistry (IUPAC) or the Chemical Abstracts Service (CAS) Rules of Nomenclature, or a name that will clearly identify the chemical for the purpose of conducting a hazard evaluation.

**combustible liquid**: Any liquid having a flashpoint at or above 100°F (37.8°C), but below 200°F (93.3°C).

**common name**: Any designation or identification, such as code name, code number, trade name, brand name, or generic name used to identify a chemical other than its chemical name.

**compressed gas**: A compressed gas is (1) a gas or mixture of gases in a container having an absolute pressure exceeding 40 psi at 70°F (21.1°C); (2) a gas or mixture of gases in a container having an absolute pressure exceeding 104 psi at 130°F (54.4°C) regardless of the pressure at 70°F (21.1°C); or (3) a liquid having a vapor pressure exceeding 10 psi at 100°F (37.8°C), as determined by ASTM D-323-72.

**container**: Any bag, barrel, bottle, box, can, cylinder, drum, reaction vessel, storage tank, or the like that contains a hazardous chemical.

**explosive**: A chemical that causes a sudden, almost instantaneous release of pressure, gas, and heat when subjected to sudden shock, pressure, or high temperature.

# HAZARD COMMUNICATION

**exposure**: The actual or potential subjection of an employee to a hazardous chemical through any route of entry, in the course of employment.

**flammable aerosol**: An aerosol that, when tested by the method described in 16 CFR 1500.45, yields a flame projection exceeding 18 inches at full valve opening, or a flashback (flame extending back to the valve) at any degree of valve opening.

**flammable gas**: A gas that at ambient temperature and pressure forms a flammable mixture with air at a concentration of 13% by volume or less, or a gas that at ambient temperature and pressure forms a range of flammable mixtures with air wider than 12% by volume regardless of the lower limit.

**flammable liquid**: A liquid having flashpoint 100°F (37.8°C).

**flammable solid**: A solid, other than a blasting agent or explosive as defined in 29 CFR 1910.109 (a), that is likely to cause fire through friction, absorption of moisture, spontaneous chemical change, or retained heat from manufacturing or processing, or which can be ignited, and that when ignited, burns so vigorously and persistently as to create a serious hazard. A chemical shall be considered to be a flammable solid if, when tested by the method described in 16 CFR 1500.44, it ignites and burns with a self-sustained flame at a rate greater than one-tenth of an inch per second along its major axis.

**flashpoint**: The minimum temperature at which a liquid gives off a vapor in sufficient concentration to ignite.

**hazard warning**: Any words, pictures, symbols, or combination thereof appearing on a label or other appropriate form of warning which convey the hazards of the chemical(s) in the container.

**hazardous chemical**: Any chemical which is a health or physical hazard.

**hazardous chemical inventory list**: An inventory list of all hazardous chemicals used at the site, and containing the date of each chemical's MSDS insertion.

**health hazard**: A chemical for which there is statistically significant evidence based on at least one study conducted in accordance with established scientific principles that acute or chronic health effects may occur in exposed employees.

**immediate use**: The use under the control of the person who transfers the hazardous chemical from a labeled container, and only within the work shift in which it is transferred.

**label**: Any written, printed, or graphic material displayed on or affixed to containers or hazardous chemicals.

**Material Safety Data Sheet**: The written or printed material concerning a hazardous chemical, developed in accordance with 29 CFR 1910.

**mixture**: Any combination of two or more chemicals if the combination is not, in whole or in part, the result of a chemical reaction.

**NFPA hazardous chemical label**: A color-coded labeling system developed by the National Fire Protection Association (NFPA) which rates the severity of the health hazard, fire hazard, reactivity hazard, and special hazard of the chemical.

**organic peroxide**: An organic compound that contains the bivalent 0–0 structure, and which may be considered to be a structural derivative of hydrogen peroxide, where one or both of the hydrogen atoms has been replaced by an organic radical.

**oxidizer**: A chemical (other than a blasting agent or explosive as defined in 29 CFR 1910.198 (a)), that initiates or promotes combustion in other materials thereby causing fire either of itself or through the release of oxygen of other gases.

**physical hazard**: A chemical for which there is scientifically valid evidence that it is a combustible liquid, a compressed gas explosive, flammable, an organic peroxide, an oxidizer, pyrophoric, unstable (reactive), or water reactive.

**portable container**: A storage vessel which is mobile, such as a drum, side-mounted tank, tank truck, or vehicle fuel tank.

**primary route of entry**: The primary means (such as inhalation, ingestion, skin contact, etc.) whereby an employee is subjected to a hazardous chemical.

**"Right To Know" workstation**: Provides employees with a central information work station where they can have access to site MSDS sheets, Hazardous Chemical Inventory List, and the company's written Hazard Communication Program.

**"Right To Know" station binder**: a station binder located in the "Right To Know" work station that contains the company's Hazard Communication Program, the Hazardous Chemicals Inventory List and corresponding MSDS, and the Hazard Communication Program Review and Signature Form.

**pyrophoric**: A chemical that will ignite spontaneously in air at a temperature of 130°F (54.4°C) or below.

**stationary container**: A permanently mounted chemical storage tank.

**unstable (reactive) chemical**: A chemical which in its pure state or as produced or transported will vigorously polymerize, decompose, condense, or will become self-reactive under conditions or shock, pressure, or temperature.

# HAZARD COMMUNICATION

**water reactive chemical**: A chemical that reacts with water to release a gas that is either flammable or presents a health hazard.

**work center**: Any convenient or logical grouping of designated unit processes or related maintenance actions.

IV. "Right To Know" Workstations

Each work center will establish an employee "Right To Know" workstation. This "Right To Know" workstation will be accessible to employees during their work hours. The "Right To Know" workstation will contain a "Right To Know" Station Binder. This Station Binder will contain Company's Hazard Communication Program, Hazardous Chemical Inventory List and corresponding MSDS, and the Hazard Communication Program Review and Signature Form.

V. Hazardous Chemical Inventory List

A list of all hazardous chemicals or fuels used or produced at each work center will be maintained in each work center's Hazardous Chemical Inventory List. This Hazardous Chemical Inventory List is to be filed in the front of each work center's "Right To Know" Station Binder, maintained within its "Right To Know" workstation.

The Hazardous Chemical Inventory List will also show the date of the most recent MSDS insertion for each chemical. Only the Hazardous Chemicals or fuels actually used within each work center will be listed in that work center's Hazardous Chemical Inventory List. Company's master Hazardous Chemical Inventory List for all chemicals used within the company will be maintained by the Industrial Hygiene Office.

Each work center supervisor is to ensure that its Hazardous Chemical Inventory List is accurate, updated, and available for employee use.

Work centers receiving new chemicals or chemicals not on their current Hazardous Chemical Inventory List shall follow these procedures:

- Add the hazardous chemical to the Hazardous Chemical Inventory List.
- Procure and insert the chemical MSDS into the "Right To Know" Station Binder.
- Train employees on the hazards associated with the chemical.

The Hazardous Chemical Inventory List for each work center will be reviewed by each work center at least quarterly; a verifying signature for this quarterly review will be made on the Hazardous Chemical Inventory List. The Hazardous Chemical Inventory List for each work center will be reviewed quarterly by the Industrial Hygiene Division to ensure it is current and that corresponding MSDS are available.

Hazardous Chemical Inventory Form

[NAME OF WORK CENTER]

Chemical             Date of MSDS Insertion

_____           _____

_____           _____

_____           _____

_____           _____

_____           _____

_____           _____

VI. Material Safety Data Sheet (MSDS)

The MSDS are a set of individual data sheets providing related safety information for each hazardous chemical utilized or produced at the work center. Material Safety Data Sheets are filed in each work center's "Right To Know" Station Binder, located in the "Right To Know" workstations. Each chemical listed on the Hazardous Chemical Inventory List must have a corresponding MSDS. MSDS are provided to work centers by the Industrial Hygiene Division any time manufacturers forward new copies or new editions. Work center supervisors are responsible for insuring that MSDS are current and available for all chemicals listed on their work center's Hazardous Chemical Inventory List, and that chemicals are not used unless this information is available.

The Material Safety Data Sheets should contain information as follows:
    a. Identity of hazardous chemical.
    b. Identity of hazardous ingredients in a hazardous chemical mixture.
    c. Chemical and physical characteristics of the hazardous chemical.
    d. Chemical and physical hazards of the hazardous chemical.
    e. Acute and chronic health hazards, including signs and symptoms of exposure and medical conditions, which are generally aggravated by exposure to the hazardous chemical.
    f. Primary route of entry.
    g. Personal exposure limits in terms of maximum duration and concentration.
    h. Protective measures and special precautions.
    i. Emergency procedures and first aid procedures.
    j. Date of preparation of the Material Safety Data Sheet.

# HAZARD COMMUNICATION

k. Identification of person or agency responsible for the information contained on the Material Safety Data Sheet.

The Material Safety Data Sheet shall not contain any blank spaces. "Not applicable" or "unknown" should be indicated as appropriate.

The Purchasing Department shall specify that MSDS will be required with all orders. Supervisors using Local Supply Orders (LSOs) to obtain chemicals should ensure that an MSDS is available for the product prior to or at receipt of the product.

Any MSDS received by the work center should be forwarded to the Industrial Hygiene Office with notations on which work center forwarded the sheet, and an indication as to whether the chemical is in use or not. If the work center is using a chemical and does not have an existing MSDS, the Industrial Hygiene Division should be notified immediately. The Industrial Hygiene Division will procure the needed MSDS or will generate a generic form. The Industrial Hygiene Division will review all incoming or self-generated MSDS for completeness before forwarding copies to pertinent work centers.

VII. Hazard Warnings and Labeling

Hazard warnings are individual warnings on hazardous chemical containers, which provide related safety information for each respective hazardous chemical utilized or produced within Company.

A. Hazard warnings should be displayed on or affixed to all hazardous chemical containers, providing the information as follows:

1. *Each portable container* should be labeled, tagged or otherwise marked as follows:
    a. Chemical or common name of the hazardous chemical
    b. Hazard warnings
    c. Name and address or the chemical manufacturer

*Note:* Company personnel should not remove or deface existing hazard warnings or labels on hazardous chemical containers received or used at the work center, unless the container is immediately marked with the required information.

2. *Each stationary container* shall be labeled as follows:
    a. Chemical or common name of the hazardous chemical stenciled with six (6) inch block letters
    b. NFPA Hazardous Chemical Label

*Note:* Company supervisors should verify that all containers received for use are appropriately labeled.

3. Where applicable, all *chemical piping* should be labeled in accordance with the Piping Identification Code or with standard industry color codes. [HazCom

does not require the labeling of pipes, but when they contain hazardous chemicals, labeling is recommended.]
- B. NFPA Hazardous Chemical Labels should be affixed to hazardous chemical containers wherever appropriate and/or informative.
- C. NFPA "0–4" number rating system for each chemical can be obtained from the Industrial Hygiene Division.

VIII. Training

The Hazard Communication Program requires periodic training of Company personnel to ensure that the program's required safety precautions are properly conducted. Supervisors should consider personnel training as a primary responsibility.

- A. The Hazard Communication Program training duties and responsibilities for company personnel are as follows:
  1. The Industrial Hygiene Division shall be responsible for training supervisors in the program requirements, and will be available by appointment to present to work center personnel the required Hazard Communication Program brief.
  2. Company supervisors shall be responsible for training company personnel in the program requirements.
  3. Company personnel shall be responsible for familiarizing themselves with the program requirements.
- B. The Hazard Communication Program training should be planned so that:
  1. Training is given both in the program requirements and in related safety information, including protective measures, special precautions, and emergency procedures.
  2. Training is given both in the classroom and by on-the-job training.
  3. Training is given, utilizing group participation during both discussion and question-and-answer periods.
  4. Training is ongoing, to preserve the continuity and integrity of program and work center safety.
- C. The Hazard Communication Program training should be provided as follows:
  1. Training is conducted prior to assignment of a new employee to his work duties.
  2. Training is conducted whenever a new hazardous chemical is utilized or produced at the work center.
  3. Training is conducted prior to starting work on nonroutine tasks involving hazardous chemicals.
- D. The Hazard Communication Program training is to be conducted by the supervisor, according to the recommended outline below:

# HAZARD COMMUNICATION

1. Locate and identify the Hazard Communication Program, Hazardous Chemical Inventory List, and Material Safety Data Sheets, contained in the "Right To Know" workstations.
2. Discuss the objective and content of the Hazard Communication Program.
3. Explain that the Hazardous Chemical Inventory List lists *all* hazardous chemicals at the work center concerned.
4. Discuss the physical and health effects of each hazardous chemical at the individual work center as contained in the MSDS.
5. Discuss the methods and techniques used to determine the presence or release of hazardous chemicals at the individual work center.
6. Discuss the protective measures and special precautions used to lessen or prevent exposure to hazardous chemicals at the individual work center.
7. Discuss emergency procedures for each hazardous chemical at the individual work center as contained in the MSDS.
8. Discuss the hazardous warning and labeling system.

*Note*: Hazard Communication Program training shall be recorded in each employee training record; these records will be examined during Industrial Hygiene/Safety Division's quarterly safety inspections.

[Company]'s Employee Training Record

Hazard Communication Training

| Date | Employee Name | Work center |
|------|---------------|-------------|
| _____ | _____ | _____ |
| _____ | _____ | _____ |
| _____ | _____ | _____ |

IX. On-Site Contractors/Visitors
   A. Company work center supervisors and the Engineering Division should notify all contractors, vendors, etc. performing work within Company work centers of the Hazard Communication Program as follows:
      1. The Industrial Hygiene Division, when directed by work center supervisor/Engineering Division, will provide a copy of Company's Hazard Communication Program to regulatory agencies, consulting engineers, contractors, etc., as appropriate upon their *initial visit* on site to perform each specific project.

*Note:* A copy of the Hazard Communication Program should be provided upon the initial visit for each specific project. It is not necessary to provide additional copies for ongoing visits to accomplish the same specific project.

    2. All written requests for proposals or quotations, all written specifications, and all written contracts and work orders should include a written notification of the Hazard Communication Program as follows:

> "Company is required in accordance with 29 CFR 1910.1200 to inform Company and contractor personnel that work centers within Company have hazardous chemicals on site. Company and contract personnel may be exposed to these hazardous chemicals while working at Company work centers. A written Hazard Communication Program has been developed to inform personnel of the specific hazardous chemicals at the work center, and the related safety information including protective measures, special precautions and emergency procedures to be observed. The Hazard Communication Program including Material Safety Data Sheets for each hazardous chemical at the work center will be made available to contractors. Contractors are responsible for communicating the information contained in the Material Safety Data Sheets (MSDS) to their personnel working at the work center."

B. It is the responsibility of work center supervisors to provide a copy of the Hazard Communication Program to other supervisors or personnel outside of the work center. It is the responsibility of other supervisors or personnel outside the work center to familiarize themselves and their personnel with the objective and content of the Hazard Communication Program. Whenever Company personnel visit another work center to perform maintenance, provide assistance, or other activity, they should acquaint themselves with the information contained in the work center's "Right To Know" work station.

The work center supervisor should provide a copy of the Hazard Communication Program and training to the other Company departments and personnel as appropriate.

## Hazard Communication Program Audit Items

If your facility has a written hazard communication program similar to the one above, you are well on the road toward compliance. If your program is audited by OSHA, the goal, of course, is for any auditor who might visit your facility to readily "see" that you are in compliance. Often an auditor will not even review your written hazard communication program if he or she can plainly see you are in compliance.

# HAZARD COMMUNICATION

Let's take a look at some of the HazCom items OSHA will be looking at. You must be able to answer yes to each of the following items, if site-applicable.

- Are all chemical containers marked with contents' name and hazards?
- Are storage cabinets used to hold flammable liquids labeled "Flammable—Keep Fire Away"?
- For a fixed extinguishing system, is a sign posted warning of the hazards presented by the extinguishing medium?
- Are all aboveground storage tanks properly labeled?
- If you store hazardous materials (including gasoline) in aboveground storage tanks, are tanks or other containers holding hazardous materials appropriately labeled with chemical name and hazard warning?
- Are all chemicals used in spray painting operations correctly labeled?
- If you store chemicals, are all containers properly labeled with chemical name and hazard warning?

Along with checking these items, the OSHA auditor will make notes on the chemicals he or she finds in the workplace. When the walk-around is completed, the auditor will ask you to provide a copy of the MSDS for each chemical in his or her notes.

To avoid a citation, you must not fail this major test. If the auditor, for example, noticed during the walk-around that employees were using some type of solvent or cleaning agent in the performance of their work, he or she will want to see a copy of the MSDS for that particular chemical. If you can't produce a copy, you are in violation and will be cited. Be careful on this item—it is one of the most commonly cited offenses. Obviously, the only solution to this problem is to ensure that your facility has an MSDS for each chemical used, stored, or produced, and that your Chemical Inventory List is current and accurate. Save yourself a big hassle—make MSDS available to employees for each chemical used on site.

Keep in mind that the OSHA auditor will look at each work center within your company, and that each different work center will present its own specialized requirements. If your company has an environmental laboratory, for example, the auditor will spend considerable time in the lab, ensuring you are in compliance with OSHA's Laboratory Standard, and that you have a written Chemical Hygiene Plan.

## Frequently Asked Questions on HazCom (OSHA, 2005)

**Q** What are temporary agency employers required to do to meet HazCom requirements?

**A** In meeting the requirements of OSHA's HazCom Standard, the temporary agency employer would, for example, be expected to provide generic hazard training and information concerning categories of chemicals employees may potentially encounter. Host employers would then be responsible for providing site-specific hazard training pursuant to 29 CFR 1910.1200(h)(1) and 1926.59 (i.e., OSHA's Construction Standard).

**Q** Can MSDS be stored on a computer to meet the accessibility requirements of HazCom?

**A** If the employee's work area includes the area where the MSDS can be obtained, then maintaining MSDS on a computer would be in compliance. If the MSDS can be accessed only out of the employee's work area(s), then the employer would be out of compliance with paragraphs (g)(8) or (g)(9) of the Hazard Communication Standard.

**Q** What are the container labeling requirements under HazCom?

**A** Under HazCom, the manufacturer, importer, or distributor is required to label each container of hazardous chemicals. If the hazardous chemicals are transferred into unmarked containers, these containers must be labeled with the required information, unless the container into which the chemical is transferred is intended for the immediate use of the employee who performed the transfer.

**Q** How does HazCom apply to pharmaceutical drugs?

**A** HazCom only applies to pharmaceuticals that the drug manufacturer has determined to be hazardous and that are known to be present in the workplace in such a manner that employees are exposed under normal conditions of use or in a foreseeable emergency. The pharmaceutical manufacturer and the importer have the primary duty for the evaluation of chemical hazards. The employer may rely upon the hazard determination performed by the pharmaceutical manufacturer or importer.

**Q** When is the chemical manufacturer required to distribute MSDS?

**A** Hazard information must be transmitted on MSDS that must be distributed to the customer at the time of first shipment of the product. The Hazard Communication Standard also requires that MSDS be updated by the chemical manufacturer or importer within three months of learning of "new or significant information" regarding the chemical's hazard potential.

**Q** What is considered proper training under the HazCom standard?

**A** Employees are to be trained at the time they are assigned to work with a hazardous chemical. The intent of this provision (1910.1200(h)) is to have information prior to exposure to prevent the occurrence of adverse health effects. This purpose cannot be met if training is delayed until a later date.

The training provisions of the Hazard Communication Standard are not satisfied solely by giving employee the data sheets to read. An employer's training program is to be a forum for explaining to employees not only the hazards of the chemicals in their work area, but also how to use the information generated in the hazard communication program. This can be accomplished in many ways (audiovisuals, classroom instruction, interactive video), and should include an opportunity for employees to ask questions to ensure that they understand the information presented to them.

Training need not be conducted on each specific chemical found in the workplace, but may be conducted by categories of hazard (e.g., carcinogens, sensitizers, acutely toxic agents) that are or may be encountered by an employee during the course of his duties.

Furthermore, the training must be comprehensible. If the employees receive job instructions in a language other than English, then the training and information to be conveyed under the Hazard Communication Standard will also need to be conducted in a foreign language.

**Q** What are the requirements for refresher training or retaining a new hire?

**A** Additional training is to be done whenever a new physical or health hazard is introduced into the work area, not a new chemical. For example, if a new solvent is brought into the workplace, and it has hazards similar to existing chemicals for which training has already been conducted, then no new training is required. As with initial training, and in keeping with the intent of the standard, the employer must make employees specifically aware which hazard category (i.e., corrosive, irritant, etc.) the solvent falls within. The substance-specific data sheet must still be available, and the product must be properly labeled. If the newly introduced solvent is a suspect carcinogen, and there has never been a carcinogenic hazard in the workplace before, then new training for carcinogenic hazards must be conducted for employees in those work areas where employees will be exposed.

It is not necessary that the employer retrain each new hire if that employee has received prior training by a past employer, an employee union, or any other entity. General information, such as the rudiments of the Hazard Communications Standard could be expected to remain with an employee from one position to another. The employer, however, maintains the responsibility to ensure that their employees are adequately trained and are equipped with the knowledge and information necessary to conduct their jobs safely. It is likely that additional training will be needed since employees must know the specifics of their new employers' programs such as where the MSDS are located, details of the employer's in-plant labeling system, and the hazards of new chemicals to which they will be exposed. For example, (h)(3)(iii) requires that employees be trained on the measures they can take to protect themselves from hazards, including specific procedures the employer has implemented such as work practices, emergency procedures, and personal protective equipment to be used. An employer, therefore, has a responsibility to evaluate an employee's level of knowledge with regard to the hazards in the workplace, their familiarity with the requirements of the standard, and the employer's hazard communication program.

**Q** Do you need to keep MSDS for commercial products such as Windex and Wite-Out?

**A** OSHA does not require that MSDS be provided to purchasers of household consumer products when the products are used in the workplace in the same manner that a consumer would use them (i.e., where the duration and frequency of use—and therefore exposure—is not greater than what the typical consumer would experience). This exemption in OSHA's regulation is based, however, not upon the chemical manufacturer's intended use of the product, but upon the chemicals and the manner in which they are used resulting in a duration and frequency of exposure greater than what a normal consumer would experience. The employees have a right to know about the properties of those hazardous chemicals.

**Q** What are the requirements and limits to using generic MSDS?

**A** [Regarding] the suitability of a generic material safety data sheet (MSDS). As you are probably aware, the requirements for MSDS are found in paragraph (g) of 29 CFR 1910.1200. MSDS must be developed for hazardous chemicals used in the workplace, and must list the hazardous chemicals that are found in a product in quantities of 1% of greater, or 0.1% or greater if the chemical is a carcinogen. The MSDS does not have to list the amount that the hazardous chemical occurs in the product.

Therefore, a single MSDS can be developed for the various combinations of . . . [chemicals], as long as the hazards of the various . . . mixtures are the same. This "generic" MSDS must meet all of the minimum requirements found in OSHA's Hazard Communication Standard (29 CFR 1910.1200(g)), including the name, address and telephone number of the responsible party preparing or distributing the MSDS who can provide additional information.

**Q** What is the application of HazCom to an office environment?

**A** Office workers who encounter hazardous chemicals only in isolated instances are not covered by the rule. The Occupational Safety and Health Administration (OSHA) considers most office products (such as pens, pencils, adhesive tape) to be exempt under the provisions of the rule, either as articles or as consumer products. How about copy toner? Is it exempt? OSHA has previously stated that intermittent or occasional use of a copying machine does not result in coverage under the rule. However, if an employee handles the chemicals to service the machine, or operates it for long periods of time, then the program would have to be applied.

**Q** Is a material safety data sheet (MSDS) required for a nonhazardous chemical?

**A** MSDS that represent nonhazardous chemicals are not covered by the Hazard Communication Standard. Paragraph 29 CFR 1910.1200(g)(8) of the standard requires that "the employer shall maintain in the workplace copies of the required MSDS for each hazardous chemical, and shall ensure that they are readily accessible during each work shift to employees when they are in their workarea(s)." OSHA does not require nor encourage employers to maintain MSDS for nonhazardous chemicals. Consequently, an employer is free to discard MSDS for nonhazardous chemicals.

## OCCUPATIONAL ENVIRONMENTAL LIMITS

Many processes and procedures generate hazardous air contaminants that can get into the air people breathe. Normally, the body can take in limited amounts of hazardous air contaminants, metabolize them and eliminate them from the body without producing harmful effects. Safe levels of exposure to many hazardous materials have been established by governmental agencies after much research in their short tem (acute) and cumulative (chronic) health effects using available human exposure data (usually from industrial sources) and animal testing. When the average air concentrations repeatedly exceed certain thresholds, called *exposure limits*, adverse health effects are more likely to occur. Exposure limits do change with time as more research is conducted and more occupational data is collected.

A fairly standard terminology has come to be used with regard to occupational environmental limits (OELs). A workplace exposure level, such as a permissible exposure limit (PEL) or threshold limit value (TLV), is expressed as the concentration of the air contaminant in a volume of air. It is important to know what they are and what significance they play in the industrial hygienist's daily activities.

*Threshold limit values* (TLVs) are published by the American Conference of Governmental Industrial Hygienists (ACGIH) (an organization made up of physicians, toxicologists, chemists, epidemiologists, and industrial hygienists) in its *Threshold Limit Values for Chemical Substances and Physical Agents in the Work Environment.* These values are used in assessing the risk of a worker exposed to a hazardous chemical vapor; concentrations in the workplace can often be maintained below these levels with proper controls. The substances listed by ACGIH are evaluated annually, limits are revised as needed, and new substances are added to the list as information becomes available. The values are established from the experience of many groups in industry, academia, and medicine, and from laboratory research.

The chemical substance exposure limits listed under both ACGIH and OSHA are based strictly on airborne concentrations of chemical substances in terms of milligrams per cubic meter ($mg/m^3$), parts per million (ppm; the number of "parts" of air contaminant per million parts of air), and fibers per cubic centimeters (fibers/$cm^3$). The smaller the concentration number, the more toxic the substance is by inhalation. The ACGIH has established some "rules of thumb" in regards to exposure limits. Substances with exposure limits below 100 ppm are considered highly toxic by inhalation. Those substances with exposure limits of 100–500 ppm are considered moderately toxic by inhalation. Substances with exposure limits greater than 500 ppm are considered slightly toxic by inhalation. Allowable limits are based on three different time periods of average exposure: (1) 8-hour work shifts known as TWA (time-weighted average), (2) short terms of 15 minutes or STEL (short-term exposure limit), and (3) instantaneous exposure of "C" (ceiling). Unlike OSHA's PELs, TLVs are recommended levels only, and do not have the force of regulation to back them up.

OSHA has promulgated limits for personnel exposure in workplace air for approximately 400 chemicals listed in tables Z1, Z2, and Z3 in Part 1910.1000 of the Federal Occupational Safety and Health Standard. These limits are defined as *permissible exposure limits* and like TLVs are based on 8-hour time-weighted averages or ceiling limits when preceded by a "C." (Exposure limits expressed in terms other than ppm must be converted to ppm before comparing to the guidelines.) Keeping within the limits in the Subpart Z tables is the only requirement specified by OSHA for these chemicals. The significance of OSHA's PELs is that they have the force of regulatory law behind them—compliance with OSHA's PELs is the law.

Evaluation of personnel exposure to physical and chemical stresses in the industrial workplace requires the use of the guidelines provided by TLVs and the regulatory guidelines of PELs. For the industrial hygienist to carry out the goals of recognizing, measuring, and effecting controls (of any type) for workplace stresses, such limits are a necessity, and have become the ultimate guidelines in the science of industrial hygiene. A word of caution, however. These values are set only as guides for the best practice and are not to be considered absolute values. What does that mean to the industrial hygienist? These values provide reasonable assurance that occupational disease will not occur if exposures are kept below these levels. On the other hand, occupational disease is likely to develop in some people—if the recommended levels are exceeded on a consistent basis.

*Time weighted average* (TWA) is the fundamental concept of most OELs. It is usually presented as the average concentration over an 8-hour workday for a 40-hour workweek.

Eight-hour *threshold limit values—time weighted averages* (TLV-TWA) exist for some four hundred plus chemical agents commonly found in the workplace. NIOSH lists sampling and analytical methods for most of these agents.

*Short-term exposure limits* (STELs) are recommended when exposures of even short duration to high concentrations of a chemical are known to produce acute toxicity.

The STEL is the concentration to which workers can be exposed continuously for a short period of time without suffering from (1) irritation, (2) chronic or irreversible tissue damage, or (3) narcosis of sufficient degree to increase the likelihood of accidental injury, impaired self rescue, or reduced work efficiency.

# HAZARD COMMUNICATION

The STEL is defined as a 15-minute TWA exposure that should not be exceeded at any time during a workday, even if the overall 8-hour TWA is within limits, and it should not occur more than four times per day. **Note:** There should be at least 60 minutes between successive exposures in this range. If warranted, an averaging period other than 15 minutes can also be used. STELS are not available for all substances.

*Ceiling* (C) is the concentration that should not be exceeded during any part of the working exposure, assessed over a 15-minute period.

The *action level* is concentration or level of an agent at which it is deemed that some specific action should be taken. Action levels are found only in certain substance specific standards by OSHA. In practice the action level is usually set at one-half of the TLV.

*Skin notation* denotes the possibility that dermal absorption may be a significant contribution to the overall body burden of the chemical. A "SKIN" notation that follows the exposure limit indicates that a significant exposure can be received if the skin is in contact with the chemical in the gas, vapor, or solid form.

*Airborne particulate matter* is divided into three classes based on its likely deposition within the respiratory tract. While past practice was to provide TLVs in terms of total particulate mass, the recent approach is to take into account the aerodynamic diameter of the particle and its site of action. The three classes of airborne particulate matter are described below.

1. *Inhalable particulate mass (IPM): TLVs* are designated for compounds that are toxic if deposited at any site within the respiratory tract. The typical size for these particles can range from submicron size to approximately 100 microns.
2. *Thoracic particulate mass (TPM): TLVs* are designated for compounds that are toxic if deposited within either the airways of the lung or the gas-exchange region. The typical size for these particles can range from approximately 5 to 15 microns.
3. *Respirable particulate mass (RPM): TLVs* are designated for those compounds that are toxic if deposited within the gas-exchange region of the lung. The typical size for these particles is approximately 5 microns or less.

*Nuisance dust* is no longer used since all dusts have biological effects at some dose. The term *particulates* not otherwise classified is now being used in place of nuisance dusts.

*Biological exposure limits* (BELs) covers nearly 40 chemicals. A BEL has been defined as a level of a determinant that is likely to be observed in a specimen (such as blood, urine, or air) collected from a worker who was exposed to a chemical and who has similar levels of the determinant as if he or she had been exposed to the chemical at the TLV.

## AIR MONITORING AND SAMPLING

Air monitoring is widely used to measure human exposure and to characterize emission sources. It is often employed within the context of the general survey, investigating a specific complaint, or simply for regulatory compliance. It is also used for more fundamental purposes, such as in confined space entry operations. Just about any confined space entry team member can be trained to properly calibrate and operate air monitors for safe confined space entry, but a higher level of knowledge and training is often required in the actual evaluation of confined spaces for possible oxygen deficiency and/or air contaminant problems. After an overview of the basics of air monitoring and sampling, we discuss air monitoring requirements for permit required confined space entry.

### Air Sample Volume

OSHA's *Analytical Method* and NIOSH's *Manual of Analytical Methods* list a range of air sample volumes, minimum volume (VOL-MIN) to maximum volume (VOL-MAX) that should be collected for an exposure assessment. The volume is based on the sampler's sorptive capacity and assumes that the measured exposure is the OSHA PEL.

The range of volumes listed may not collect a sufficient mass for accurate laboratory analysis if the actual contaminant concentration is less than the PEL or the TWA-TLV.

If a collection method recommends sampling 10 liters of air, the industrial hygienist may not be sure if an 8.5-liter air sample will collect sufficient mass for the lab to quantify. Finding that an insufficient air volume (i.e., too little mass) was sampled after the lab results are returned can turn out to be an enormous waste of resources. To avoid this situation, it is essential to understand the restrictions faced by analytical labs: limit of detection (LOD) and limit of quantification (LOQ).

To compute a minimum air sample volume to provide useful information for evaluation of airborne contaminant concentration in the workplace, the industrial hygienist must understand how to correctly manipulate the LOD and the LOQ. Knowledge of these limits will provide increased flexibility in sampling.

### Limit of Detection (LOD)

The *limit of detection* has many definitions in the literature. For example, the American Chemical Society (ACS) Committee on Environmental Analytical Chemistry (2002) defines the LOD as the lowest concentration level than can be determined to be statistically different from a blank sample (a sample of a carrying agent—a gas, liquid, or solids—that is normally used to selectively capture a material of interest, and that is subjected to the usual analytical or measurement process to establish a zero baseline or background value, which is used to adjust or correct routine analytical results). The ACS definition of LOD is used in this text.

The analytical instrument output signal produced by the sample must be three to five times the instrument's background noise level to be at the limit of detection; that is, the signal-to-noise ratio (S/N) is three-to-one (3/1). S/N ratio is the analytical method's lower limit of detection.

## Limit of Quantification (LOQ)

The *limit of quantification* is the concentration level above which quantitative results may be obtained with a certain degree of confidence. That is, LOQ is the minimum mass of the analyte above which the precision of the reported result is better than a specified level. The recommended value of the LOQ is the amount of analyte that will give rise to a signal that is 10 times the standard deviation of the signal from a series of media blanks.

## Precision, Accuracy, and Bias

Sample results are only as good as the sampling technique and equipment used. Thus, in any type of air monitoring operation (air, water, or soil), it is important that the industrial hygienist factor in the precision, accuracy, and any possible bias involved in the monitoring process.

*Precision* is the reproducibility of replicate analyses of the same sample (mass or concentration). For example, how close to each other is a target shooter able to place a set of shots anywhere on the target?

*Accuracy* is the degree of agreement between measured values and the accepted reference value. The investigator must carefully design his sampling program and use certain statistical tools to evaluate his data before making any inferences from the data. In our target shooting analogy, accuracy can be equated to how close does a target shooter come to the bull's-eye?

*Bias* is the error introduced into sampling that causes estimates of parameters to be inaccurate. More specifically, bias is the difference between the average measured mass or concentration and a reference mass or concentration, expressed as a fraction of the reference mass or concentration. For example, how far from the bull's-eye is the target shooter able to place a cluster of shots?

## Calibration Requirements

The American National Standards Institute (1994) defines calibration as the set of operations which establishes, under specified conditions (i.e., instrument manufacturer's guidelines or regulator's protocols), the relationship between values indicated by a measuring instrument or measuring system, and the corresponding standard or known values derived from the standard. *Note:* Before any air-monitoring device can be relied on as accurate, it must be calibrated. Calibration procedures can be found in OSHA's Personal Sampling for Air Contaminants [www.osha.gov].

## Types of Air Sampling

Although the information provided in the following discussion is area specific (*area sampling*), it is important to point out that one of the most important air sampling operations is personal sampling. *Personal sampling* puts the sample detection device on the worker. This is done to obtain samples that represent the worker's exposure while working. As the preferred method of evaluating worker exposure to airborne contaminants, personal sampling, as mentioned, requires the worker to wear the detection device on their person in the breathing zone area. A small air pump and associated tubing connected to the detector is also worn by the worker. Personal sampling allows the industrial hygienist to define a potential hazard, check compliance with specific regulations, and determine the worker's daily TWA exposure.

## Analytical Methods for Gases and Vapors

In routine practice the industrial hygienist will collect air samples to determine the concentration of a known contaminant or group of contaminants and will request or conduct analyses for these compounds. The first step in this procedure is to develop a sampling plan. When developing a sampling plan or strategy, the sampler should review the specific sampling and analytical methods available for the contaminants of interest. Several organizations have compiled and published collections of sampling and analytical methods for gases and vapors (see Table 3.1).

For discussion purposes, analytical methods for gases and vapors (vapors are the gaseous phase of a substance that is liquid or solid at normal temperature and pressure; vapors diffuse) are grouped into chromatographic, volumetric, and optical methods. In the following we briefly discuss chromatography.

Table 3.1. Available Publications on Sampling and Analytical Methods for Gases and Vapors

*NIOSH Manual of Analytical Methods*, 4th ed. Centers for Disease Control and Prevention National Institute for Occupational Safety and Health
4676 Columbia Parkway
Cincinnati, OH 45226

*Annual Book of ASTM Standards*
American Society for Testing and Materials
100 Barr Harbor Dr.
West Conshohocken, PA 19428

*OSHA Analytical Methods Manual*
Occupational Safety and Health Administration
OSHA Salt Lake Technical Center
P.O Box 65200
1781 South 300 West
Salt Lake City, UT 84165

*Methods of Air Sampling and Analysis*
Lewis Publishers/CRC Press Inc.
2000 Corporate Blvd. NW
Boca Raton, FL 33431

According to Breysse and Lees (1999), the primary type of analytical equipment used in *chromatography* for the analysis of gases and vapors in air samples is the *gas chromatograph* (GC). The GC is a powerful tool for the analysis of low-concentration air contaminants. It is generally a reliable analytical instrument. GC analysis is applicable to compounds with sufficient vapor pressure and thermal stability to dissolve in the carrier gas and pass through the chromatographic column in sufficient quantity to be detectable. Air samples to be analyzed by GC are typically collected on sorbent tubes and desorbed into a liquid for analysis. It should be noted that the GC instrument can not be used for reliable identification of specific substances. Because of this limitation, the GS is often married to the mass spectrometer (MS) instrument to provide specific results.

As mentioned, the GC is limited in its analysis for contaminant specificity. To correct this limitation, the GC/MS combination is used. When the industrial hygienist or analyst uses the GC instrument to separate compounds before analysis with an MS instrument, a complementary relationship exists. The technician has access to both the retention times and mass spectral data. Many industrial hygienists consider GC/MS analysis as a tool for conclusive proof of identity— the "gold standard" in scientific analysis.

Some common applications of GC/MS include:

1. Evaluation of complex mixtures
2. Identification of pyrolysis and combustion products from fires
3. Analysis of insecticides and herbicides—conventional analytical methods frequently cannot resolve or identify the wide variety of industrial pesticides currently in use, but GC/MS can both identify and quantify these compounds.

## Air Monitoring versus Air Sampling

In the practice of industrial hygiene, the terms air monitoring and air sampling are often used interchangeably to mean the same thing. But are they the same? This depends on your choice and use of the vernacular.

In reality, they are different; that is, air monitoring and air sampling are separate functions. The difference is related to time: real time versus time-integration.

*Air monitoring* is real-time monitoring and generally includes monitoring with handheld, direct-reading units such as portable gas chromatographs (GC), photoionization detectors (PIDs), flame ionization detectors (FIDs), dust monitors, and colorimetric tubes. Real-time air monitoring instrumentation is generally easily portable and allows the user to collect multiple samples in a relatively short sample period—ranging from a few seconds to a few minutes. Most portable real-time instruments measure low parts per million (ppm) of total volatile organics.

Real-time monitoring methods have higher detection limits than time-integrated sampling methods, react with entire classes of compounds and, unless real-time monitoring is conducted continuously, provide only a "snapshot" of the monitored ambient air concentration. Air monitoring instruments and methods provide results that are generally used for evaluation of short-term exposure limits and can be useful in providing timely information to those engaged in various activities such as confined space entry operations. That is, in confined space operations, proper air monitoring can detect the presence or absence of life-threatening contaminants and/or insufficient oxygen levels within the confined space, alerting the entrants to make the space safe (e.g., by using forced air ventilation) before entry.

On the other hand, time-integrated *air sampling* is intended to document actual exposure for comparison to long-term exposure limits. Air sampling data is collected at "fixed" locations along the perimeter of the sample area (work area) and at locations adjacent to other sensitive receptors. Because most contaminants are (or will be) present in ambient air at relatively low levels, some type of sample concentrating is necessary to meet detection limits normally required in evaluating long-term health risks. Air sampling is accomplished using air-monitoring instrumentation designed to continuously sample large volumes of air over extended periods of time (typically 8–24 hours).

Air sampling methods involve collecting air samples on sampling media designed specifically for collection of the compounds of interest or as whole air samples. Upon completion of the sampling period the sampling media is collected, packaged, and transported for subsequent analysis. Analysis of air samples usually requires a minimum of 48 hours to complete.

Both air monitoring and air sampling are important and significant tools in the industrial hygienist's toolbox.

To effectively evaluate a potentially hazardous worksite, an industrial hygienist (IH) must obtain objective and quantitative data. To do this, the IH must perform some form of air sampling, dependent upon, of course, the airborne contaminant in question. Moreover, sampling operations involve the use of instruments to measure the concentration of the particulate, gas, or vapor of interest. Many instruments perform both sampling and analysis. The instrument of choice in conducting sampling and analysis typically is a direct-reading-type instrument. The IH must be familiar with the uses, advantages, and limitations of such instruments. In addition, the IH must use math calculations to calculate sample volumes, sample times, TLVs, air concentrations from vapor pressures, and determine the additive effects of chemicals when multiple agents are used in the workplace. These calculations must take into account changing conditions, such as temperature and pressure change in the workplace. Finally, the IH must understand how particulates, gases, and vapors are generated; how they enter the human body; how they impact worker health; and how to evaluate particulate-, gas-, and vapor-laden workplaces.

Because air sampling is integral to just about everything the industrial hygienist does and is about, in the following we include important sections focusing on air sampling principles, dealing with airborne particulates, airborne gases and vapors, direct-reading instruments, and basic air sampling calculations.

## AIR SAMPLING FOR AIRBORNE PARTICULATES

As discussed in chapter 1, *airborne particulates* (or particulate matter, PM) include solid and liquid matter such as dusts, fumes, mists, smokes, and bioaerosols. Inhalation of particulates is a major cause of occupational illness and disease. Pneumoconiosis ("dusty lung"), is a lung disease cause by inhalation.

There are four critical factors that influence the health impact of airborne particulates. Each of these four factors is interrelated in such a way that no one factor can be considered independently of the others:

- The size of the particles
- The duration of exposure time
- The nature of the dust in question
- The airborne concentration of the dust, in the breathing zone of the exposure person. (The *breathing zone* of the worker is described by a hemisphere bordering the shoulders to the top of the head.)

## Dusts

Dusts are generated by mechanical processes such as grinding or crushing. Dusts range in size from 0.5–50µm. Note that dust is a relatively new term used to describe dust that is hazardous when deposited anywhere in the respiratory tree including the nose and mouth. It has a 50% cut-point of 10 microns and includes the big and the small particles. The cut-point describes the performance of cyclones and other particle size selective devices. For personal sampling, the 50% cut-point is the size of the dust that the device collects with 50% efficiency.

Alpaugh (1988) points out that common workplace dusts are either inorganic or organic. Inorganic dusts are derived from metallic and nonmetallic sources. Nonmetallic dusts can be silica bearing—that is, in combined or free silica as crystalline or amorphous form. Organic dusts are either synthetic or natural. Natural organic dust can be animal or vegetable derived.

Examples of organic and inorganic dust are:

- Sand (inorganic, nonmetallic, silica bearing, free silica, crystalline)
- Beryllium (inorganic, metallic)
- Cotton (organic, naturally occurring, vegetable)

Elsewhere, Grantham (1992) has classified dusts on the basis of their health effects. He identifies dusts as being:

- Innocuous (e.g., iron oxide, limestone; may also be considered as nuisance dusts)
- Acute respiratory hazards (e.g., cadmium fume)

- Chronic respiratory hazards (e.g., airborne asbestos fibers)
- Sensitizers (e.g., many hardwood dusts)

## Duration of Exposure

The duration of exposure may be *acute* (short-term) or it may be *chronic* (long-term).

Some airborne particulates, for example beryllium, may exert a toxic effect after a single acute exposure, or metal fume fever may occur following acute exposure to metal fumes.

Other particulates, such as lead or manganese, may exert a toxic effect following a longer period of exposure, maybe several days to several weeks. Such exposures could be termed *sub-chronic*. Chronic lung conditions, such as pneumoconiosis or mesothelioma, may follow prolonged exposure to silica dusts or asbestos (crocidolite or blue asbestos), respectively.

## Particle Size

Particle size is critical in determining where particulates will settle in the lung. Smaller particles outnumber larger ones but vary widely in size. Larger particles will settle in the upper respiratory tract in the bronchi and the bronchioles, and will not tend to penetrate the smaller airways found in the alveolar (air sac) region. These are termed *inspirable* particles. Those smaller sized particles that can penetrate the alveolar (the gas exchange) region of the lungs are termed *respirable* particles.

Particle size is expressed as "aerodynamic" or "equivalent" diameter. This is equal to the diameter of spherical particles of unit density that have the same falling velocity (*terminal*, or settling, *velocity*) in air as the particle in question. The terminal velocity is proportional to the specific gravity of the particle, p, and the square of its diameter, d.

Particles with an aerodynamic diameter greater than approximately 20 microns (μm) will be trapped in the nose and upper airways.

Particles in the region of 7–20 microns will penetrate to the bronchioles and are inspirable, while particles in the size range 0.5–7 microns are respirable. Particles smaller than this will not settle out as their terminal velocity is so small that there is insufficient time for them to be deposited in the alveolus before they are exhaled.

An understanding of aerodynamic diameter is important when calculating terminal settling rates of particulates. Constants for these calculations include:

$$1 \text{ gm/cm}^3 = \text{unit density}$$

$$\text{Gravity} = 32.2 \text{ ft/sec}^2 \text{ or } 98 \text{ cm/sec}^2$$

## Stoke's Law

Stoke's Law is the relationship that relates the "settling rate" to a particle's density and diameter. Stoke's Law applies to the fate of particulates in the atmosphere. Stoke's Law is given as:

$$u = \frac{gd^2 \, \rho_1 - \rho_2}{18\eta} \quad (3.1)$$

where:

- u = settling velocity in cm/s (settling velocity = 0.006 ft/min(specific gravity)$d^2$)
- g = acceleration due to gravity in cm/s$^2$
- $d^2$ = diameter of particle squared in cm$^2$
- $\rho_1$ = particle density in g/cm$^3$
- $\rho_2$ = air density in g/cm$^3$
- $\eta$ = air viscosity in poise g/cm – s

What Stoke's Law tells us is, all other things being constant, that dense particles settle faster, that larger particles settle faster, and that denser, more viscous air causes particles to settle slower.

With Stoke's Law, we can predict settling rate for a given particle if its diameter and density are known. Stoke's Law is also used to estimate particle diameters (called "Stoke's diameter") from observed settling rates.

## Airborne Dust Concentration

The concentration of dust is a critical factor in the impact on the health of the worker exposed. This is measured in the breathing zone of the worker.

Airborne concentrations of dust are usually assessed by collecting dust on a pre-weighed filter. A known volume of air is drawn through the filter, which is then re-weighed. The difference in weight is the mass of dust, usually in milligrams (mg) or micrograms (µg), and the volume is expressed as cubic meters of air (m$^3$). Hence, the overall concentration of dust in air is measured in mg/m$^3$ or µg/m$^3$.

## Particulate Collection

In order to evaluate workplace atmosphere for particulates a sample must be taken and analyzed. To obtain a sample the following collection mechanisms are used:

- Impaction
- Sedimentation
- Diffusion
- Direct interception
- Electrostatic attraction

The *filter* is the most common particulate collection device. The mixed cellulose ester membrane filter is the most commonly used type of filter. This type filter is used for collecting asbestos and metals. Other filter types include polyvinyl chloride, silver, glass fiber, and Teflon filters.

## Analysis of Particulates

There are several methods for analysis of particulates, including:

- Gravimetric (e.g., for coal dust, free silica, total dust)
- Instrumental (e.g., atomic absorption [AA] for metals)
- Optical microscopy (e.g., for fibers, asbestos, dust)
- Direct-reading instruments (e.g., aerosol photometers, piezo-electric instruments)
- Wet chemical (e.g., for lead, isocyanates, free silica)

## Health and Environmental Impacts of Particulates

Particulates cause a wide variety of health and environmental impacts. Many scientific studies have linked breathing particulates to a series of significant health problems, including aggravated asthma, increases in respiratory symptoms like coughing and difficult or painful breathing, chronic bronchitis, decreased lung function, and premature death.

Particulate matter is the major cause of reduced visibility (haze) in parts of the United States, including many of our national parks. Soot particulate both stains and damages stone and other materials, including culturally important objects such as monuments and statues.

Particles can be carried over long distances by wind and then settle on ground or water. This settling can make lakes and streams acidic, change the nutrient balance in coastal waters and large river basins, deplete nutrients in the soil, damage sensitive forests and farm crops, and affect the diversity of ecosystems.

## Control of Particulates

As with all other methods of industrial hygiene hazard control, control of particulates is most often accomplished through the use of engineering controls, administrative controls, and PPE. The best method for controlling particulates is the use of ventilation—an engineering control. To be most effective, the particulate generating process should be totally enclosed with a negative pressure exhaust ventilation system in place. Administrative controls include using wet methods of housekeeping and wet cleanup methods to minimize dust regeneration and prohibiting the use of compressed air to clean work surfaces. PPE, used as a last resort, includes equipping workers with proper respiratory protection to prevention inhalation of particulates, and protective clothing to protect worker contact with particulates.

## AIR SAMPLING FOR GASES AND VAPORS

*Gases* and *vapors* are "elastic fluids," so-called because they take the shape and volume of their containers.

A *fluid* is generally termed as gas if its temperature is very far removed from that required for liquefaction; it is called a vapor if its temperature is close to that of liquefaction.

In the industrial hygiene field, a substance is considered a gas if this is its normal physical state at room temperature and atmospheric pressure. It is considered a vapor if, under the existing environmental conditions, conversion of its liquid or solid form to the gaseous state results from its vapor pressure affecting its volatilization or sublimation into the atmosphere of the container, which may be the process equipment or the worksite. Our chief interest in distinguishing between gases and vapors lies in our need to assess the potential occupational hazards associated with the use of specific chemical agents, an assessment which requires knowledge of the physical and chemical properties of these substances (NIOSH, 2005).

The type of air sampling for gases and vapors employed depends on the purpose of sampling, environmental conditions, equipment available, and nature of the contaminant.

## The Gas Laws

In sampling gases and vapors, it is important to have a basic understanding of the physics of air. Specifically, it is important to have an understanding of the gas laws (Spellman, 1999a).

Gases can be contaminants as well as conveyors of contaminants. Air (which is mainly nitrogen) is usually the main gas stream.

# HAZARD COMMUNICATION 83

Gas conditions are usually described in two ways: *Standard Temperature and Pressure* (STP) and *Standard Conditions* (SC). STP represents 0°C (32°F) and 1 atm. SC is more commonly used and represents typical room conditions of 20°C (70°F) and 1 atm; SC is usually measured in cubic meters, $Nm^3$, or standard cubic feet (scf).

To understand the physics of air, it is imperative that you understand the various physical laws that govern the behavior of pressurized gases.

**Pascal's Law** states that a confined gas (fluid) transmits externally applied pressure uniformly in all directions without change in magnitude. If the container is flexible, it will assume a spherical (balloon) shape.

**Boyle's Law** states the absolute pressure of a confined quantity of gas varies inversely with its volume at a given temperature. For example, if the pressure of a gas doubles, its volume will be reduced by half, and vice versa.

Boyle's Law is expressed in equation form as

$$P_1 \times V_1 = P_2 \times V_2 \tag{3.2}$$

where:

$P_1$ and $V_1$ are the initial pressure and volume
$P_2$ and $V_2$ are the final pressure and volume

**Charles's Law** states the volume of a given mass of gas at constant pressure is directly proportional to its absolute temperature.

The temperature should be in Kelvin (273 + °C) or Rankine (absolute zero = –460°F or 0°R).

$$P_2 = P_1 \times T_2/T_1 \tag{3.3}$$

Charles's Law also states: If the pressure of a confined quantity of gas remains the same, the change in the volume (V) of the gas varies directly with a change in the temperature of the gas, as given in the equation

$$V_2 = V_1 \times T_2/T_1 \tag{3.4}$$

The **Ideal Gas Law** combines Boyle's and Charles's Laws because air cannot be compressed without its temperature changing. The Ideal Gas Law is expressed by the equation

$$P_1 \times V_1/T_1 = P_2 \times V_2/T_2 \tag{3.5}$$

Note that the Ideal Gas Law is used as a design equation even though the equation shows that the pressure, volume, and temperature of the second state of a gas are equal to the pressure, volume, and temperature of the first state. In actual practice, however, other factors (humidity, heat of friction, and efficiency losses, for example) affect the gas. This equation uses absolute pressure (psia) and absolute temperatures (°R).

The Ideal Gas Law can also be expressed as

$$PV = nRT \qquad (3.6)$$

where:

n = number of moles of gas (or mass in grams divided by the molecular weight)
R = molar gas constant
P = Pressure
V = volume
T = temperature, always in Kelvin

Note that this relationship explains why one-gram molecular weight (GMW) of gas occupies 22.4 liters of space at STP. This number represents the number of molecules in one mole of any compound. Equation 3.6 is based on Avogadro's assumption that equal volumes of all gases under the same conditions of temperature and pressure contain the same number of molecules.

## Types of Air Samples

No matter the type of air sampling used, workplace samples must be obtained that represent the worker's exposure (i.e., a representative sample).

In taking a *representative sample*, a sampling plan should be used that specifies

- where to sample,
- whom to sample,
- how long to sample,
- how many samples to take, and
- when to sample.

Generally, as mentioned previously, two methods of sampling for airborne contaminants are used: personal air sampling and area sampling. *Personal air sampling* (the worker wears a sampling device that collects an air sample) is the preferred method of evaluating worker exposure to airborne contaminants. *Area monitoring* (e.g., in confined spaces) is used to identify high exposure areas.

# HAZARD COMMUNICATION

## Methods of Sampling

Standardized sampling methods provide the information needed to sample air for specific contaminants. Standard air sampling methods specify procedures, collection media, sample volume, flowrate, and chemical analysis to be used. For example, NIOSH's *Manual of Analytical Methods* and OSHA's *Chemical Information Manual* provide the information necessary to sample air for specific contaminants.

Generally, two methods of sampling are used in sampling for airborne contaminants: grab sampling and continuous (or integrated) sampling.

*Grab sampling* (i.e., instantaneous sampling) is conducted using a heavy walled evacuated (air removed) flask. The flask is placed in the work area and a valve is opened to allow air to fill the flask. The sample represents a "snapshot" of an environmental concentration at a particular point in time. The sample is analyzed either in the laboratory or with suitable field instruments.

*Continuous sampling* is the preferred method for determining TWA exposures. The sample is taken for a sample air stream.

## Air Sampling Collection Processes

Airborne contaminants are collected on media or in liquid media through absorption or adsorption processes.

*Absorption* is the process of collecting gas or vapor in a liquid (dissolving gas/vapor in a liquid). Absorption theory states that gases and vapors will go into solution up to an equilibrium concentration.

Samplers include gas washing bottles (impingers), fritted bubblers, spiral and helical absorbers, and glass-bead columns.

*Gas adsorbents* (gas onto a solid) typically use activated charcoal, silica gel, or other materials to collect gases.

*Diffusive samplers* (passive samplers) depend on flow of contaminant across a quiescent (uses no pump to draw air across adsorbent) layer of air, or a membrane. Diffusion depends on well-established rules from physical chemistry, known as Fick's Law.

## Calibration of Air Sampling Equipment

In order to gather accurate sampling data, the equipment used must be properly calibrated. The calibration of any instrument is an absolute necessity if the data are to have any meaning. Various devices are used to calibrate air sampling equipment. Calibration is based on primary or secondary calibrations standards.

Primary calibrations standards include

- the soap bubble meter (or frictionless piston meter),
- the spirometer (measures displaced air),
- the Mariotti bottle (measures displaced water), and
- electronic calibrators (provide instantaneous airflow readings)

Secondary calibration standards include

- the wet test meter,
- the dry gas meter, and
- rotameters.

## DIRECT READING INSTRUMENTS FOR AIR SAMPLING

Direct reading instruments are used for on-site evaluations for a number of reasons:

- To find the sources of emission of hazardous contaminants on the spot.
- To ascertain if select OSHA air standards are being exceeded.
- To check the performance of control equipment.
- As continuous monitors at fixed locations to trigger an alarm system in the event of breakdown in a process control, which could result in the accidental release of copious amounts of harmful substances to the workroom atmosphere.
- As continuous monitors at fixed locations to obtain permanent recorded documentation of the concentrations of a contaminant in the atmospheric environment for future use in epidemiological and other types of occupational studies, in legal actions, to inform employees as to their exposure, and for information required for improved design of control measures.

Such on-site evaluations of the atmospheric concentrations of hazardous substances make possible the immediate assessment of undesirable exposures and enable the industrial hygienist to make an immediate correction of an operation, in accordance with his or her judgment of the seriousness of a situation, without permitting further risk of injury to the workers (NIOSH, 2005).

### Direct Reading Physical Instruments

One type of direct reading instruments used in air sampling are *direct reading physical instruments*. The physical properties of gases, aerosols, and vapors are used in the design of direct reading physical instruments for quantitative estimations of these types of contaminants in the atmosphere. The various types of these instruments, the principle of operation, and a brief description of application are presented here.

**aerosol photometry**: Measures, records, and controls particulates continuously in areas requiring sensitive detection of aerosol levels; detection of 0.05 to 40 µm diameter particles. Computer interface equipment is available.

**chemiluminescence**: Measurement of NO in ambient air selectivity and $NO_x$ after conversion to NO by hot catalyst. Specific measurement of $O_2$. No atmospheric interferences.

**colorimetry**: Measure and separate recording of $NO_2$, $NO_x$, $SO_2$, total oxidants, $H_2S$, HF, $NH_3$, $Cl_2$, and aldehydes in ambient air.

**combustion**: Detects and analyzes combustible gases in terms of percent LEL (lower explosive limit) on graduated scale. Available with alarm set at 1/3 LEL.

**conductivity, electrical**: Records $SO_2$ concentrations in ambient air. Some operate off a 12-volt car battery. Operate unattended for periods up to 30 days.

**coulometry**: Continuous monitoring of NO, $NO_2$, $O_x$, and $SO_2$ in ambient air. Provided with strip chart recorders. Some require attention only once a month.

**flame ionization (with gas chromatograph)**: Continuous determination and recording of methane, total hydrocarbons, and carbon monoxide. Catalytic conversion of SO to $CH_4$. Operates up to 3 days unattended. Separate model for continuous monitoring of $SO_2$, $H_2S$, and total sulfur in air. Unattended operation up to 3 days.

**flame ionization (hydrocarbon analyzer)**: Continuous monitoring of total hydrocarbons in ambient air; potentiometric or optional current outputs compatible with any recorded. Electronic stability from 32°F to 110°F.

**gas chromatograph, portable**: On-site determination of fixed gases, solvent vapors, nitro and halogenated compounds, and light hydrocarbons. Instruments available with choice of flame ionization, electron capture, or thermal conductivity detectors and appropriate columns for desired analyses. Rechargeable batteries.

**infrared analyzer (photometry)**: Continuous determination of a given component in a gaseous or liquid stream by measuring amount of infrared energy absorbed by component of interest using pressure sensor technique. Wide variety of applications include CO, $CO_2$, Freons, hydrocarbons, nitrous oxide, $NH_3$, $SO_2$ and water vapor.

**photometry, ultraviolet (tuned to 253.7 mµ)**: Direct readout of mercury vapor; calibration filter is built into the meter. Other gases or vapors which interfere include acetone, aniline, benzene, ozone, and others with absorb radiation at 253.7 mµ.

**photometry, visible (narrow-centered 394 mµ band pass)**: Continuous monitoring of $SO_2$, $SO_3$,

$H_2S$, mercaptans, and total sulfur compounds in ambient air. Operates more than 3 days unattended.

**particle counting (near forward scattering)**: Reads and prints particle concentrations at 1 of 3 preset time intervals of 100, 1000 or10,000 seconds, corresponding to 0.01, 0.1 and 1 cubic foot of sampled air.

**polarography**: Monitor gaseous oxygen in flue gases, auto exhausts, hazardous environments, and in food storage atmospheres and dissolved oxygen in wastewater samples. Battery operated, portable, sample temperature 32° to 110°F, up to 95% relative humidity. Potentiometric recorder output. Maximum distance between sensor and amplifier is 1000 feet.

**radioactivity**: Continuous monitoring of ambient gamma and x-radiation by measurement of ion chamber currents, averaging or integrating over a constant recycling time interval, sample temperature limits 32°F to 120°F; 0 to 95% relative humidity (weatherproof detector); up to 1,000 feet remote sensing capability. Recorder and computer outputs. Complete with alert, scram, and failure alarm systems. All solid-state circuitry.

**radioactivity**: Continuous monitoring of beta or gamma emitting radioactive materials within gaseous or liquid effluents; either a thin wall Geiger-Mueller tube or a gamma scintillation crystal detector is selected depending on the isotope of interest; gaseous effluent flow, 4 cfm; effluent sample temperature limits 32°F to 120°F using scintillation detector and –65°F to 165°F using G-M detector. Complete with high radiation, alert and failure alarms.

**radioactivity**: Continuous monitoring of radioactive airborne particulates collected on a filter tape transport system; rate or air flow, 10 SCFM; scintillation and G-M detectors, optional but a beta-sensitive plastic scintillator is provided to reduce shielding requirements and offer greater sensitivity. Air sample temperature limits 32°F to 120°F; weight 550 pounds. Complete with high and low flow alarm and a filter failure alarm.

## Direct-Reading Colorimetric Devices

The second type of direct reading instruments used in air sampling are *direct-reading colorimetric devices*, which are widely used, easy to operate, and inexpensive. They utilize the chemical properties of an atmospheric contaminant for the reaction of that substance with a color-producing reagent, revealing stain length or color intensity. Stain lengths or color intensities can be read directly to provide an instantaneous value of the concentration accurate within ± 25%. Reagents used in detector kits may be in either a liquid or a solid phase or provided in the form of chemical treated papers. The liquid and solid reagents are generally supported in sampling devices through which a measured amount of contaminated air is drawn. On the other hand, chemically treated papers are usually exposed to the atmosphere and the reaction time noted for a color change to occur (NIOSH, 2005).

# HAZARD COMMUNICATION

## Calibration of Direct-Reading Instruments

Two common methods used for calibrating direct-reading instruments are the *static method* and the *dynamic method*. The static method is easy to use and efficient; a known volume of gas is introduced into the instrument and sampling is performed for a limited period of time. With the dynamic method, the instrument is used to monitor a known concentration of the contaminant to test its accuracy.

## AIR SAMPLING CALCULATIONS

### Gram Molecular Volumes

1 mole = 22.4 liters at STP

1 mole = 24.45 liters at NTP

Standard Pressure is 760 torr (760 mm of mercury (Hg) at sea or atmosphere level)

Standard Temperature is 0°C or 273°K (Kelvin = 273 + x°C).

### Sample Conversions and Calculations Using Boyle's Law

If 500 ml of oxygen is collected at a pressure of 780 mm Hg, what volume will the gas occupy if the pressure is changed to 740 mm Hg?

New volume ($V_2$) = 500 ml × 780 mm Hg/740 mm Hg = 527 ml

What is the volume of a gas at a pressure of 90 cm Hg if 300 ml of the gas was collected at a pressure of 86 cm Hg?

New volume ($V_2$) = 300 ml × 86 cm Hg/90 cm Hg = 287 ml

Calculate the volume of a gas that occupies a volume of 110 ml, if it occupies a volume of 300 ml at a pressure of 80 cm Hg.

New pressure ($P_2$) = 80 cm Hg × 300 ml/110 ml = 218 cm Hg

### Sample Conversions and Calculations Using Charles's Law

What volume will an amount of gas occupy at 25°C if the gas occupies a volume of 500 ml at a temperature of 0°C? Assume that the pressure remains constant.

$$°K = 273° + °C$$

$$\text{New volume } (V_2) = 500 \text{ ml} \times 298°K/273°K = 546 \text{ ml}$$

What is the volume of a gas at −25°C if the gas occupied 48 ml at a temperature of 0°C?

$$\text{New volume } (V_2) = 48 \text{ ml} \times 248°K/273°K = 43.6 \text{ ml}$$

If a gas occupies a volume of 800 ml at 15°C, at what temperature will it occupy a volume of 1000 ml if the pressure remains constant?

$$\text{New absolute pressure } (T_2) = 288°K \times 1000 \text{ ml}/800 \text{ ml} = 360°K$$

## Sample Conversions and Calculations Using Boyle's Law and Charles's Law Combined

Calculate the volume of a gas at STP if 600 ml of the gas is collected at 25°C and 80 cm Hg.

$$\text{New volume } (V_2) = 600 \text{ ml} \times 80 \text{ cm Hg}/76 \text{ cm Hg} \times 273°K/298°K = 578 \text{ ml}$$

If a gas occupies a volume of 100 ml at a pressure of 76 cm Hg and 25°C, what volume will the gas occupy at 900 mm Hg and 40°C?

$$\text{New volume } (V_2) = 100 \text{ ml} \times 760 \text{ mm Hg}/900 \text{ mm Hg} \times 313°K/298°K = 88.7 \text{ ml}$$

If 500 ml of oxygen is collected at 20°C, and the atmospheric pressure is 725.0 mm Hg, what is the volume of the dry oxygen at STP?

$$\text{New volume of dry gas} = 500 \text{ ml} \times 725.0 \text{ mm Hg}/760 \text{ mm Hg} \times 273°K/293°K$$

$$= 443.5 \text{ ml}$$

2.50 g of a gas occupy 240 ml at 20°C and 740 torr. What is the gram molecular weight of the gas?

Step 1: $240 \text{ ml} \times 273°K/293°K \times 740 \text{ torr}/760 \text{ torr} = 217 \text{ ml}$

Step 2: $2.50 \text{ g}/217 \text{ ml} \times 1000 \text{ml}/1L \times 22.4 \text{ L}/1 \text{ mole} = 258 \text{ g/mole}$

## AIR MONITORING: CONFINED SPACE ENTRY

When a confined space is to be entered, it is important to remember that one can never rely on his or her senses to determine if the air in the confined space is safe. You cannot see or smell many toxic gases and vapors, nor can you determine the level of oxygen present (Spellman, 1999b).

As mentioned, one of the most common and important functions that an industrial hygienist is called upon to perform is in regards to evaluation of confined spaces for safe entry. A *confined space* is defined as a space large enough and so configured that an employee can bodily enter and perform assigned work; has limited or restricted means for entry or exit; and is not designed for continuous employee occupancy. A *permit-required confined space* (a permit is a written or printed document provided by the employer to allow and control entry into a specified space) has one or more of the following characteristics: (1) it contains or has the potential to contain a hazardous atmosphere; (2) it contains a material that has the potential for engulfing an entrant; (3) it has a configuration such that an entrant could be trapped or asphyxiated by inwardly converging walls or by a floor that slopes downward and tapers to a smaller cross section; or (4) it contains any recognized serious safety or health hazard.

## Monitoring Procedures

Atmospheric monitoring (testing) is required for two distinct purposes: evaluation of the hazards of the permit space and verification that acceptable entry conditions for entry into that space exist.

For *evaluation testing*, the atmosphere of a confined space is analyzed using equipment of sufficient sensitivity and specificity to identify and evaluate any hazardous atmospheres that may exist or arise, so that appropriate permit entry procedures can be developed and acceptable entry conditions stipulated for that space. Evaluation and interpretation of these data, and development of entry procedure, should be done by, or reviewed by, a technically qualified industrial hygienist based on evaluation of all serious hazards.

For *verification testing*, the atmosphere of a permit space that may contain a hazardous substance is tested for residues of all contaminants identified by evaluation testing using permit-specified equipment to determine that residual concentrations at the time of testing and entry are within the range of acceptable entry conditions. Results of the testing (i.e., actual concentrations, etc.) should be recorded on the permit in the space provided adjacent to the acceptable entry condition.

*Note:* Measurement of values for each atmospheric parameter should be made for at least the minimum response time of the test instrument specified by the manufacturer.

When testing *stratified atmospheres*—that is, monitoring for entries involving a descent into atmosphere that may be stratified—the atmospheric envelope should be tested a distance of approximately 4 feet in the direction of travel and to each side. If a sampling probe is used, the entrant's rate of progress should be slowed to accommodate the sampling speed and detector response.

## REFERENCES

Alpaugh, E. L., 1988. "Particulates." *Fundamentals of Industrial Hygiene*, 3rd ed. Ed. Barbara A. Plog. Publ. National Safety Council, USA.

American Chemical Society (ACS), 2002. *Principles of Environmental Sampling and Analysis—Two Decades Later*. Boston, Mass.: American Chemical Society Meeting.

American National Standards Institute (ANSI), 1994. *Calibration Laboratories and Measuring and Text Equipment General Requirements*. Washington, D.C.

Breysse, P. N., and P. S. J. Lees, 1999. "Analysis of gases and vapors." *The Occupational Environment—Its Evaluation and Control.* Ed. S. R. DiNardi. Fairfax, Va.: American Industrial Hygiene Association.

Grantham, D., 1992. "Dusts In the Workplace." *Occupational Health and Hygiene Guidebook for the WHSO*. Brisbane: D.L. Grantham.

NIOSH, 2005. *Direct Reading Instruments for Determining Concentrations of Aerosols, Gases and Vapors*. By Robert G. Keenan. www.cdc.gov/niosh/pdgs/74-177-h.pdf (accessed July 2005).

OSHA, 1995. *Occupational Safety and Health Standards for General Industry*. CFR 29 1910.147. Washington, D.C.: U.S. Department of Labor.

OSHA, 2005. *Frequently Asked Questions: Hazard Communication (HazCom)*. U.S. Department of Labor.www.osha.gov.htmol/gaq-hazcom.html (accessed June 2005).

Spellman, F. R., 1999a. *The Science of Air: Concepts & Applications*. Boca Raton, Fla.: CRC Press.

Spellman, F. R., 1999b. *Confined Space Entry*. Boca Raton, Fla.: CRC Press.

## SUGGESTED READING

*Code of Federal Regulations*. Title 29 Parts 1900–1910 (.146). Washington, D.C.: Office of the Federal Register, 1995.

*OSHA's Hazard Communication Standard*. Rockville, Md.: Government Institutes, 1996.

# 4

# Indoor Air Quality and Mold Control

For those familiar with *Star Trek*, and for those who are not, consider a quotable quote: "The air is the air." However, in regards to the air we breathe, according to the U.S. Environmental Protection Agency (USEPA, 2001), few of us realize that we all face a variety of risks to our health as we go about our day-to-day lives. Driving our cars, flying in planes, engaging in recreational activities and being exposed to environmental pollutants all pose varying degrees of risk. Some risks are simply unavoidable. Some we choose to accept because to do otherwise would restrict our ability to lead our lives the way we want. And some are risks we might decide to avoid if we had the opportunity to make informed choices. Indoor air pollution is one risk that we can do something about.

In the last several years, a growing body of scientific evidence has indicated that the air within homes and other buildings can be more seriously polluted than the outdoor air in even the largest and most industrialized cities. Other research indicates that people spend approximately 90 percent of their time indoors. Thus, for many people, the risks to health may be greater due to exposure to air pollution indoors than outdoors (USEPA, 2001).

In addition, people who may be exposed to indoor air pollutants for the longest periods of time are often those most susceptible to the effects of indoor air pollution. Such groups include the young, the elderly, and the chronically ill, especially those suffering from respiratory or cardiovascular disease.

The impact of energy conservation on inside environments may be substantial, particularly with respect to decreases in ventilation rates (Hollowell et al., 1979a) and "tight" buildings constructed to minimize infiltration of outdoor air (Woods, 1980; Hollowell et al., 1979b). The purpose of constructing tight buildings is to save energy—to keep the heat or air conditioning inside the structure. The problem is indoor air contaminants within these tight structures are not only trapped within but also can be concentrated, exposing inhabitants to even more exposure.

What about indoor air quality problems in the workplace? In this chapter we discuss this pervasive but often overlooked problem. We discuss the basics of indoor air quality (as related to the workplace environment) and the major contaminants that currently contribute to this problem. Moreover, mold and mold remediation, although not new to the workplace, are the new buzzwords attracting the industrial hygienist's attention these days. We also briefly discuss exposure to contaminants such as asbestos, silica, lead, and formaldehyde contamination. Various related remediation practices are also discussed.

## WHAT IS INDOOR AIR QUALITY?

According to Byrd (2003), indoor air quality (IAQ) refers to the effect, good or bad, of the contents of the air inside a structure, on its occupants. For our purposes, IAQ refers to the quality of the air inside workplaces as represented by concentrations of pollutants and thermal conditions (temperature and relative humidity) that affect the health, comfort, and performance of employees. Usually, temperature (too hot or too cold), humidity (too dry or too damp), and air velocity (draftiness or motionlessness) are considered "comfort" rather than indoor air quality issues. Unless they are extreme, they may make someone uncomfortable, but they won't make a person ill. Other factors affecting employees, such as light and noise, are important indoor environmental quality considerations, but are not treated as core elements of indoor air quality problems. Nevertheless, most industrial hygienists must take these factors into account in investigating environmental quality situations.

Byrd (2003) further points out that "good IAQ is the quality of air, which has no unwanted gases or particles in it at concentrations which will adversely affect someone. Poor IAQ occurs when gases or particles are present at an excessive concentration so as to affect the health of occupants."

In the workplace, poor IAQ may only be annoying to one person, or, at the extreme, could be fatal to all the occupants in the workplace. The concentration of the contaminant is what is crucial. Potentially infectious, toxic, allergenic, or irritating substances are always present in the air; there is nearly always a threshold level below which no effect occurs.

### Why is IAQ Important to Workplace Owners?

Workplace structures (buildings) exist to protect workers from the elements and to otherwise support worker activity. Workplace buildings should not make workers sick, cause

# INDOOR AIR QUALITY AND MOLD CONTROL

them discomfort, or otherwise inhibit their ability to perform. How effectively a workplace building functions to support its workers and how efficiently the workplace building operates to keep costs manageable is a measure of the workplace building's performance.

The growing proliferation of chemical pollutants in industrial and consumer products, the tendency toward tighter building envelopes and reduced ventilation to save energy, and pressures to defer maintenance and other building services to reduce costs have fostered indoor air quality problems in many workplace buildings. Employee complaints of odors, stale and stuffy air, and symptoms of illness or discomfort breed undesirable conflicts between workplace occupants and workplace managers. Lawsuits sometimes follow.

If IAQ is not well managed on a daily basis, remediation of ensuing problems and/or resolution in court can be extremely costly. Moreover, air quality problems in the workplace can lead to reduced worker performance. So it helps to understand the causes and consequences of indoor air quality and to manage your workplace buildings to avoid these problems.

## WORKER SYMPTOMS ASSOCIATED WITH POOR AIR QUALITY

Worker responses to pollutants, climatic factors, and other stressors such as noise and light are generally categorized according to the type and degree of responses and the time frame in which they occur. Workplace managers should be generally familiar with these categories, leaving detailed knowledge to industrial hygienists.

*Acute effects* are those that occur immediately (e.g., within 24 hours) after exposure. Chemcials released from building materials may cause headaches, or mold spores may result in itchy eyes and runny nose in sensitive individuals shortly after exposure. Generally, these effects are not long lasting and disappear shortly after exposure ends. However, exposure to some biocontaminants (fungi, bacteria, viruses) resulting from moisture problems, poor maintenance, or inadequate ventilation have been known to cause serious, sometimes life threatening respiratory diseases which themselves can lead to chronic respiratory conditions.

*Chronic effects* are long-lasting responses to long-term or frequently repeated exposures. Long-term exposures to even low concentrations of some chemicals may induce chronic effects. Cancer is the most commonly associated long-term health consequence of exposure to indoor air contaminants. For example, long-term exposures to environmental tobacco smoke, radon, asbestos, and benzene increase cancer risk.

*Discomfort* is typically associated with climatic conditions but workplace building contaminants may also be implicated. Workers complain of being too hot or too cold, or experience eye, nose, or throat irritation because of low humidity. However, reported symptoms can be difficult to interpret. Complaints that the air is "too dry" may result from irritation from particles on the mucous membranes rather than low humidity, or "stuffy air" may mean that the temperature is too warm or there is lack of air movement, or "stale air" may mean that there is a mild but difficult to identify odor. These conditions may be unpleasant and cause discomfort among workers, but there is usually no serious health implication involved. Absenteeism, work performance, and employee morale, however, can be seriously affected when building managers fail to resolve these complaints.

*Performance effects*—significant measurable changes in worker's ability to concentrate or perform mental or physical tasks—have been shown to result from modest changes in temperature and relative humidity. In addition, recent studies suggest that similar effects are associated with indoor pollution due to lack of ventilation or the presence of pollution sources. Estimates of performance losses from poor indoor air quality for all buildings suggest a 2–4% loss on average. Future research should further document and quantify these effects.

## Workplace Building Associated Illnesses

The rapid emergence of IAQ problems and associated occupant complaints have led to terms which describe illnesses or effects particularly associated with buildings. These include sick building syndrome, building related illness, and multiple chemical sensitivity.

Sick building syndrome (SBS) is a catch-all term that refers to a series of acute complaints for which there is no obvious cause and where medical tests reveal no particular abnormalities. It describes situations in which more than 20% of the building occupants experience acute health and comfort effects that appear to be linked to time spent in a building because all other probable causes have been ruled out. The 20% figure is arbitrarily set, as there will always be some workers complaining about adverse health effects associated with occupancy of a building. However, if the figure is 20% or more, it is considered that there must be some determinable cause that can be remedied. Symptoms include headaches; eye, nose, and throat irritation; dry cough; dry or itchy skin; dizziness and nausea; difficulty in concentration; fatigue; and sensitivity to odors (USEPA, 2001).

SBS is attributed to inadequate ventilation, chemical contaminants from indoor and outdoor sources, and biological contaminants such as molds, bacteria, pollens, and viruses. Passon et al. explain in "Sick-Building Syndrome and Building-Related Illnesses" how increased air tightness of buildings in the 1970s, as a means of reducing energy consumption, has created environmental conditions conducive to the "proliferation of microorganisms [including mold] in indoor environments" (1996, 33). Once growth has occurred, harmful organisms can be spread by improperly designed and maintained ventilation systems (USEPA, 2001).

A single causative agent (e.g., a contaminant) is seldom identified and complaints may be resolved when building operational problems and/or occupant activities identified by investigators are corrected.

Increased absenteeism, reduced work efficiency, and deteriorating employee morale are the likely outcomes of SBS problems which are not quickly resolved.

*Building related illness* (BRI) refers to a defined illness with a known causative agent resulting from exposure to the building air. While the causative agent can be chemical (e.g., formaldehyde), it is often biological. Typical sources of biological contaminants are humidification systems, cooling towers, drain pans or filters, other wet surfaces, or water damaged building material. Symptoms may be specific or may mimic symptoms commonly associated with the flu, including fever, chills, and cough. Serious lung and respiratory conditions can occur. Legionnaires' disease, hypersensitivity pneumonitis, and humidifier fever are common examples of BRI.

It is generally recognized that some workers can be sensitive to particular agents at levels which do not have an observable affect in the general population. In addition, it is recognized that certain chemicals can be sensitizers in that exposure to the chemical at high levels can result in sensitivity to that chemical at much lower levels.

Some evidence suggests that a subset of the worker population may be especially sensitive to low levels of a broad range of chemicals at levels common in today's home and working environments. This apparent condition has come to be known as *multiple chemical sensitivity* (MCS).

Workers reported to have MCS apparently have difficulty being in most buildings. There is significant professional disagreement concerning whether MCS actually exists and what the underlying mechanism might be. Building managers may encounter occupants who have been diagnosed with MCS. Resolution of complaints in such circumstances may or not be possible. Responsibility to accommodate such workers is subject to negotiation and may involve arrangements to work at home or in a different location.

## Building Factors Affecting Indoor Air Quality

Building factors affecting IAQ can be grouped into two factors: factors affecting indoor climate and factors affecting indoor air pollution.

*Factors affecting indoor climate*—the thermal environment (temperature, relative humidity and airflow)—are important dimensions of indoor air quality for several reasons. First, many complaints of poor indoor air may be resolved by simply altering the temperature or relative humidity. Second, people that are thermally uncomfortable will have a lower tolerance to other building discomforts. Third, the rate at which chemicals are released from building material is usually higher at higher building temperatures. Thus, if occupants are too warm, it is also likely that they are being exposed to higher pollutant levels.

*Factors affecting indoor air pollution* include much of the building fabric, its furnishings and equipment, and its occupants and their activities. In a well functioning building, some of these pollutants will be directly exhausted to the outdoors and some will be removed as outdoor air enters that building and replaces the air inside. The air outside may also contain contaminants which will be brought inside in this process. This air exchange is brought about by the mechanical introduction of outdoor air (outdoor air ventilation rate), the mechanical exhaust of indoor air, and the air exchanged through the building envelope (infiltration and exfiltration).

Pollutants inside can travel through the building as air flows from areas of higher atmospheric pressure to areas of lower atmospheric pressure. Some of these pathways are planned and deliberate so as to draw pollutants away from occupants, but problems arise when unintended flows draw contaminants into occupied areas. In addition, some contaminants may be removed from the air through natural processes, as with the adsorption of chemicals by surfaces or the settling of particles onto surfaces. Removal processes may also be deliberately incorporated into the building systems. Air filtration devices, for example, are commonly incorporated into building ventilation systems.

Thus, the factors most important to understanding indoor pollution are (a) indoor sources of pollution, (b) outdoor sources of pollution, (c) ventilation parameters, (d) airflow patterns and pressure relationships, and (e) air filtration systems.

## Types of Pollutants

Common pollutants or pollutant classes of concern in commercial buildings along with common sources of these pollutants are provided in table 4.1.

## Sources of Indoor Air Pollutants

Air quality is affected by the presence of various types of contaminants in the air. Some are in the form of gases. These would be generally classified as toxic chemicals. The types of interest are combustion products (carbon monoxide, nitrogen dioxide), volatile organic compounds (formaldehyde, solvents, perfumes and fragrances, etc.), and semi-volatile organic compounds (pesticides). Other pollutants are in the form of animal dander; soot; particles from buildings, furnishings and occupants such as fiberglass, gypsum powder, paper dust, lint from clothing, carpet fibers, etc.; dirt (sandy and earthy material); etc.

# INDOOR AIR QUALITY AND MOLD CONTROL

**Table 4.1. Indoor Pollutants and Potential Sources**

| Pollutant or Pollutant Class | Potential Sources |
|---|---|
| Environmental Tobacco Smoke | Lighted cigarettes, cigars, pipes |
| Combustion Contaminants | Furnaces, generators, gas or kerosene space heaters, tobacco products, outdoor air, vehicles |
| Biological Contaminants | Wet or damp materials, cooling towers, humidifiers, cooling coils or drain pans, damp duct insulation or filters, condensation, re-entrained sanitary exhausts, bird droppings, cockroaches or rodents, dustmites on upholstered furniture or carpeting, body odors |
| Volatile Organic Compounds (VOCs) | Paints, stains, varnishes, solvents, pesticides, adhesives, wood preservatives, waxes, polishes, cleansers, lubricants, sealants, dyes, air fresheners, fuels, plastics, copy machines, printers, tobacco products, perfumes, dry cleaned clothing |
| Formaldehyde | Particle board, plywood, cabinetry, furniture, fabrics |
| Soil gases (radon, sewer gas, VOCs, methane) | Soil and rock (radon), sewer drain leak, dry drain traps, leaking underground storage tanks, land fill |
| Pesticides | Termiticides, insecticides, rodenticides, fungicides, disinfectants, herbicides |
| Particles and Fibers | Printing, paper handling, smoking and other combustion, outdoor sources, deterioration of materials, construction/renovation, vacuuming, insulation. |

Burge and Hoyer (1998) point out many specific sources for contaminants that result in adverse health effects in the workplace, including the workers (as carriers of contagious diseases, allergens, and other agents on clothing); building compounds (VOCs, particles, fibers); contamination of building components (allergens, microbial agents, pesticides); and outdoor air (microorganisms, allergens, and chemical air pollutants).

When workers complain of IAQ problems, the industrial hygienist is called upon to determine if the problem really is an IAQ problem. If he or she determines that some form of contaminant is present in the workplace, proper remedial action is required. This usually includes removing the source of the contamination.

Tables 4.2 and 4.3 identify indoor and outdoor sources (respectively) of contaminants commonly found in the workplace and offer some measures for maintaining control of these contaminants.

Table 4.3 identifies common sources of contaminants that are introduced from outside buildings. These contaminants frequently find their way inside through the building shell, openings, or other pathways to the inside.

**Table 4.2. Indoor Sources of Contaminants**

| Category/Common Sources | Mitigation and Control |
|---|---|
| *Housekeeping and Maintenance* | |
| cleanser | Use low-emitting products |
| waxes and polishes | Avoid aerosols and sprays |
| disinfectants | Dilute to proper strength |
| air fresheners | Do not overuse; use during unoccupied hours |
| adhesives | |
| janitor's/storage closets | Use proper protocol when diluting and mixing |
| wet mops | |
| drain cleaners | Store properly with containers closed and lid tight |
| vacuuming | |
| paints and coatings | Use exhaust ventilation for storage spaces (eliminate return air) |
| solvents | |
| pesticides | Clean mops, store mop top up to dry |
| lubricants | Avoid "air fresheners"—clean and exhaust instead |
| | Use high efficiency vacuum bags/filters |
| | Use integrated Pest Management |
| *Occupant-Related Sources* | |
| tobacco products | Smoking policy |
| office equipment (printers/copiers) | Use exhaust ventilation with pressure control for major local sources |
| cooking/microwave | |
| art supplies | Low emitting art supplies/marking pens |
| marking pens | Avoid paper clutter |
| paper products | Education material for occupants and staff |
| personal products (e.g., perfume) | |
| tracked in dirt/pollen | |
| *Building Uses as Major Sources* | |
| print/photocopy shop | Use exhaust ventilation and pressure control |
| dry cleaning | |
| science laboratory | Use exhaust hoods where appropriate; check hood airflows |
| medical office | |
| hair/nail salon | |
| cafeteria | |
| pet store | |
| *Building-related Sources* | |
| plywood/compressed wood | Use low emitting sources |
| construction adhesives | Air out in an open/ventilated area before installing |
| asbestos products | |
| insulation | Increase ventilation rates during and after installing |
| wall/floor coverings (vinyl/plastic) | |
| carpets/carpet adhesives | Keep material dry prior to enclosing |
| wet building products | |

*(continued)*

# INDOOR AIR QUALITY AND MOLD CONTROL

Table 4.2. **Indoor Sources of Contaminants** *(continued)*

| Category/Common Sources | Mitigation and Control |
|---|---|
| *Building-related Sources (continued)* | |
| transformers<br>upholstered furniture<br>renovation/remodeling | |
| *HVAC system* | |
| contaminated filters<br>contaminated duct lining<br>dirty drain pans<br>humidifiers<br>lubricants<br>refrigerants<br>mechanical room<br>maintenance activities<br>combustion appliances<br>boilers/furnaces/stoves/generators | Perform HVAC preventive maintenance<br>Change filter<br>Clean drain pans; proper slope and drainage<br>Use portable water for humidification<br>Keep duct lining dry; move lining outside of duct if possible<br>Fix leaks/clean spills<br>Maintain spotless mechanical room (not a storage area)<br>Avoid back drafting<br>Check/maintain flues from boiler to outside<br>Keep combustion appliances properly tuned<br>Disallow unvented combustion appliances<br>Perform polluting activities during unoccupied hours |
| *Moisture* | |
| mold | Keep building dry |
| *Vehicles* | |
| underground/attached garage | Use exhaust ventilation<br>Maintain garage under negative pressure relative to the building<br>Check air flow patterns frequently<br>monitor CO |

Table 4.4 summarizes management protocols for major sources of pollution in buildings.

## Indoor Contaminant Transport

Contaminants reach worker breathing-zones by traveling from the source to the worker by various pathways. Normally, the contaminants travel with the flow of air.

As mentioned, air moves from areas of high pressure to areas of low pressure. That is why controlling workplace air pressure is an integral part of controlling pollution and enhancing building IAQ performance.

**Table 4.3. Outdoor Sources of Contaminants**

| Category/Common Sources | Mitigation and Control |
|---|---|
| *Ambient Outdoor Air* | |
| air quality in the general area | Filtration or air cleaning of intake air |
| *Vehicular sources* | |
| local vehicular traffic<br>vehicle idling areas<br>loading dock | Locate air intake away from source<br>Require engines shut off at loading dock<br>Pressurize building/zone<br>Add vestibules/sealed doors near source |
| *Commercial/Manufacturing Sources* | |
| laundry or dry cleaning<br>paint shop<br>restaurant<br>photo-processing<br>automotive shop/gas station<br>electronics manufacture/assembly<br>various industrial operations | Locate air intake away from source<br>Pressurize building relative to outdoors<br>Consider air cleaning options for outdoor air intake<br>Use landscaping to block or redirect flow of contaminants |
| *Utilities/Public Works* | |
| utility power plant<br>incinerator<br>water treatment plant | |
| *Agricultural* | |
| pesticide spraying<br>processing or packing plants<br>ponds | |
| *Construction/Demolitions* | |
| | Pressurize building<br>Use walk-off mats |
| *Building Exhaust* | |
| bathrooms exhaust<br>restaurant exhaust<br>air handler relief vent<br>exhaust from major tenant (e.g., dry cleaner) | Separate exhaust or relief from air intake<br>Pressurize building |
| *Water Sources* | |
| pools of water on roof<br>cooling tower mist | Proper roof drainage<br>Separate air intake from source of water<br>Treat and maintain cooling tower water |

*(continued)*

# INDOOR AIR QUALITY AND MOLD CONTROL

**Table 4.3. Outdoor Sources of Contaminants** *(continued)*

| Category/Common Sources | Mitigation and Control |
|---|---|
| *Birds and Rodents* | |
| fecal contaminants | Bird proof intake grills |
| bird nesting | Consider vertical grills |
| *Building Operations and Maintenance* | |
| trash and refuse area | Separate source from air intake |
| chemical/fertilizer/grounds keeping storage | Keep source area clean/lids on tight |
| painting/roofing/sanding | Isolate storage area from occupied areas |
| *Ground Sources* | |
| soil gas | Depressurize soil |
| sewer gas | Seal foundation and penetrations to foundations |
| underground fuel storage tanks | Keep air ducts away from ground sources |

**Table 4.4. Protocols for Managing Major Sources of Pollution in Buildings**

| Type of Protocol | Recommended Solution |
|---|---|
| Remodeling and Renovation | Isolate construction activity from occupants |
| Painting | Establish a protocol for painting and insure that the protocol is followed by both in-house personnel and by contractors |
| | • Use low VOC emission, fast drying paints where feasible |
| | • Paint during unoccupied hours |
| | • Keep lids on paint containers when not in use |
| | • Ventilate the building with significant quantities of outside air during and after painting; insure a complete building flush prior to occupancy |
| | • Use more than normal outside air ventilation for some period after occupancy |
| | • Avoid spraying, when possible |
| Pest Control | Use or require the use of Pest Management by pest control contractors in order to minimize the use of pesticides when managing pests |
| | • Control dirt, moisture, clutter, foodstuff, harborage, and building penetrations to minimize pests |

*(continued)*

Table 4.4. **Protocols for Managing Major Sources of Pollution in Buildings** *(continued)*

| Type of Protocol | Recommended Solution |
|---|---|
|  | • Use baits and traps rather than pesticide sprays where possible<br>• Avoid periodic pesticide application for "prevention" of pests<br>• Use pesticides only where pests are located<br>• Use pesticide specifically formulated for the targeted pest |
|  | Apply pesticides only during unoccupied hours<br>Ventilate the building with significant quantities of outside air during and after applications<br>Insure a complete building flush prior to occupancy<br>Use more than normal outside air ventilation for some period after occupancy<br>Notify occupants prior to occupation<br>If applying outside, keep away from air intake |
| Shipping and Receiving | Establish and enforce a program to prevent vehicle contaminants form entering the building |
|  | • Do not allow idling of vehicles at the loading dock; post signs and enforce the ban<br>• Pressurize the receiving area relative to the outside to insure that contaminants from the loading area do not enter the building; use pressurized vestibules and air locks if necessary<br>• Periodically check the pressure relationships and compliance with the protocol<br>• Notify delivery company supervisors of policy |
| Establish and Enforce a Smoking Policy | Environmental tobacco smoke (ETS) is a major indoor air contaminant. A smoking policy may take one of two forms:<br>• A smoke-free policy which does not allow smoking in any part of the building<br>• A policy that restricts smoking to designated smoking lounges only |

*(continued)*

# INDOOR AIR QUALITY AND MOLD CONTROL

**Table 4.4. Protocols for Managing Major Sources of Pollution in Buildings** *(continued)*

| Type of Protocol | Recommended Solution |
|---|---|
| Smoking Lounge Requirements | A designated smoking lounge must have the following features to be effective in containing ETS: |
| | • The lounge should be fully enclosed |
| | • The lounge should be sealed off from the return air plenum |
| | • The lounge should have exhaust ventilation directly to the outside at 60 cfm per occupant (using maximum occupancy) |
| | • Transfer air from occupied spaces may be used as make up air |
| | • The lounge should be maintained under negative pressure relative to the surrounding occupied spaces |

Air movements should be from occupants, toward a source, and out of the building rather than from the source to the occupants and out the building. Pressure differences will control the direction of air motion and the extent of occupant exposure.

*Driving forces* change pressure relationships and create airflow. Common driving forces are identified in table 4.5.

## Common Airflow Pathways

Contaminants travel along pathways—sometimes over great distances. Pathways may lead from an indoor source to an indoor location or from an outdoor source to an indoor location.

The location experiencing a pollution problem may be close by, in the same or an adjacent area, but it may be a great distance from, and/or on a different floor from a contaminant source.

Knowledge of common pathways helps to track down the source and/or prevent contaminants from reaching building occupants (see table 4.6).

## Ventilation

Ventilation can be used to either exhaust contaminants from a fixed source, or dilute contaminants from all sources within a space. Exhaust and dilution ventilation along with ventilation measurements are discussed later in the text.

Table 4.5.  Major Driving Forces

| Driving Force | Effect |
|---|---|
| Wind | Positive pressure is created on the windward side causing infiltration, and negative pressure on the leeward side causing exfiltration, though wind direction can be varied due to surrounding structures. |
| Stack effect | When the air inside is warmer than outside, it rises, sometimes creating a column of rising—up stairwells, elevator shafts, vertical pipe chases etc. This buoyant force of the air results in positive pressure on the higher floors and negative pressure on the lower floors and a neutral pressure plane somewhere between. |
| HVAC/fans | Fans are designed to push air in a directional flow and create positive pressure in front, and negative pressure behind the fan. |
| Flues and Exhaust | Exhausted air from a building will reduce the building air pressure relative to the outdoors. Air exhausted will be replaced either through infiltration or through planned outdoor air intake vent. |
| Elevators | The pumping action of a moving elevator can push air out of or draw air into the elevator shaft as it moves. |

## MAJOR IAQ CONTAMINANTS

Industrial hygienists spend a large portion of their time working with and mitigating air contaminant problems in the workplace. The list of potential contaminants workers might be exposed to while working is extensive. There are, however, a few major chemical- or material-derived air contaminants (other than those poisonous gases and materials that are automatically top priorities for the industrial hygienist to investigate and mitigate) that are considered extremely hazardous. These too garner the industrial hygienist's immediate attention and remedial action(s). In this text, we focus specifically on asbestos, silica, formaldehyde, and lead, but keep in mind that there are others.

## ASBESTOS EXPOSURE (OSHA, 2002)

Asbestos is the name given to a group of naturally occurring minerals widely used in certain products, such as building materials and vehicle brakes, to resist heat and corrosion. Asbestos includes chrysotile, amosite, crocidolite, tremolite asbestos, anthophyllite asbestos, actinolite asbestos, and any of these materials that have been chemically treated and/or altered. Typically, asbestos appears as a whitish, fibrous material which may release fibers that range in texture from coarse to silky; however, airborne fibers that can cause health damage may be too small to be seen with the naked eye.

**Table 4.6. Common Airflow Pathways for Contaminants**

| Common Pathway | Comment |
|---|---|
| Indoors | |
| Stairwell | The stack effect brings about air flow by drawing air toward these chases on the lower floors and away from these chases on the higher floors, affecting the flow of contaminants |
| Elevator shaft | |
| Vertical electrical or plumbing chases | |
| Receptacles, outlets, openings | Contaminants can easily enter and exit building cavities and thereby move from space to space |
| Duct or plenum | Contaminants are commonly carried by the HVAC system throughout the occupied spaces |
| Duct or plenum leakage | Duct leakage accounts for significant unplanned air flow and energy loss in buildings |
| Flue or exhaust leakage | Leaks from sanitary exhausts or combustion flues can cause serious health problems |
| Room spaces | Air and contaminants move within a room or through doors and corridors to adjoining spaces |
| Outdoors to Indoors | |
| Indoor air intake | Polluted outdoor air or exhaust air can enter the building through the air intake |
| Windows/doors | A negatively pressurized building will draw air and outside pollutants into the building through any available opening |
| Cracks and crevices | |
| Substructures/slab penetrations | Radon and other soil gases and moisture laden air or microbial contaminated air often travel through crawlspaces and other substructures into the building |

An estimated 1.3 million employees in construction and general industry face significant asbestos exposure on the job. Heaviest exposures occur in the construction industry, particularly during the removal of asbestos during renovation or demolition (abatement). Employees are also likely to be exposed during the manufacture of asbestos products (such as textiles, friction products, insulation, and other building materials) and automotive brake and clutch repair work.

The inhalation of asbestos fibers by workers can cause serious diseases of the lungs and other organs that may not appear until years after the exposure has occurred. For instance, asbestosis can cause a buildup of scarlike tissue in the lungs and result in loss of lung function that often progresses to disability and death. As mentioned, asbestos fibers associated with these health risks are too small to be seen with the naked eye, and smokers are at higher risk of developing some asbestos-related diseases. For example, exposure to asbestos can cause asbestosis; mesothelioma (cancer affecting the membranes lining the lungs and abdomen); lung cancer; and cancers of the esophagus, stomach, colon, and rectum.

OSHA has issued the following three standards to assist industrial hygienists with compliance and to protect workers from exposure to asbestos in the workplace:

- 29 CFR 1926.1101 covers construction work, including alteration, repair, renovation, and demolition of structures containing asbestos.
- 29 CFR 1915.1001 covers asbestos exposure during work in shipyards.
- 29 CFR 1910.1001 applies to asbestos exposure in general industry, such as exposure during brake and clutch repair, custodial work, and manufacture of asbestos-containing products.

The standards for the construction and shipyard industries classify the hazards of asbestos work activities and prescribe particular requirements for each classification:

- Class I is the most potentially hazardous class of asbestos jobs and involves the removal of thermal system insulation and sprayed on or troweled-on surfacing asbestos-containing materials or presumed asbestos-containing materials.
- Class II includes the removal of other types of asbestos-containing materials that are not thermal systems insulation, such as resilient flooring and roofing materials containing asbestos.
- Class III focuses on repair and maintenance operations where asbestos-containing materials are disturbed.
- Class IV pertains to custodial activities where employees clean up asbestos-containing waste and debris.

There are equivalent regulations in states with OSHA-approved state plans.

## Permissible Exposure Limits

Employee exposure to asbestos must not exceed 0.1 fibers per cubic centimeter (f/cc) of air, averaged over an 8-hour work shift. Short-term exposure must also be limited to not more than 1 f/cc, averaged over 30 minutes. Rotation of employees to achieve compliance with either PEL is prohibited.

## Exposure Monitoring

In construction and shipyard work, unless the industrial hygienist is able to demonstrate that employee exposures will be below the PELs (a "negative exposure assessment"), it is generally a requirement that monitoring for workers in Class I and II regulated areas be conducted. For workers in other operations where exposures are expected to exceed one of the PELs, periodic monitoring must be conducted. In general industry, for workers who may be exposed above a PEL or above the excursion limit, initial monitoring must be conducted. Subsequent monitoring at reasonable intervals must be conducted, and in no case at intervals greater than 6 months for employees exposed above a PEL.

## Competent Person

In all operations involving asbestos removal (abatement), employers must name a *competent person* qualified and authorized to ensure worker safety and health, as required by Subpart C, "General Safety and Health Provisions for Construction" (29 CFR 1926.20). Under the requirements for safety and health prevention programs, the competent person must frequently inspect jobsites, materials, and equipment. A fully trained and licensed industrial hygienist often fills this role.

In addition, for Class I jobs the competent person must inspect on-site at least once during each work shift and upon employee request. For Class II and III jobs, the competent person must inspect often enough to assess changing conditions and upon employee request.

## Regulated Areas

In general industry and construction, regulated areas must be established where the 8-hour TWA or 30-minute excursions values for airborne asbestos exceed the PELs. Only authorized persons wearing appropriate respirators can enter a regulated area. In regulated areas, eating, smoking, drinking, chewing tobacco or gum, and applying cosmetics are prohibited. Warning signs must be displayed at each regulated area and must be posted at all approaches to regulated areas.

## Methods of Compliance

In both general industry and construction, employers must control exposures using engineering controls, to the extent feasible. Where engineering controls are not feasible to meet the exposure limit, they must be used to reduce employee exposures to the lowest levels attainable and must be supplemented by the use of respiratory protection.

In general industry and construction, the level of exposure determines what type of *respirator* is required; the standards specify the respirator to be used. Keep in mind that respirators must be used during all Class I asbestos jobs. Refer to 29 CFR 1926.103 for further guidance on when respirators must be worn.

Caution *labels* must be placed on all raw materials, mixtures, scrap, waste, debris, and other products containing asbestos fibers.

For any employee exposed to airborne concentrations of asbestos that exceed the PEL, the employer must provide and require the use of *protective clothing* such as coveralls or similar full-body clothing, head coverings, gloves, and foot covering. Wherever the possibility of eye irritation exists, face shields, vented goggles, or other appropriate protective equipment must be provided and worn.

For employees involved in each identified work classification, *training* must be provided. The specific training requirements depend upon the particular class of work being performed. In general industry, training must be provided to all employees exposed above a PEL. Asbestos awareness training must also be provided to employees who perform housekeeping operations covered by the standard. Warning labels must be placed on all asbestos products, containers, and installed construction materials when feasible.

The employer must keep an accurate record of all measurements taken to monitor employee exposure to asbestos. This *recordkeeping* is to include: the date of measurement, operation involving exposure, sampling and analytical methods used, and evidence of their accuracy; number, duration, and results of samples taken; type of respiratory protective devices worn; name, social security number, and the results of all employee exposure measurements. This record must be kept for 30 years.

Regarding *hygiene facilities and practices*, clean change rooms must be furnished by employers for employees who work in areas where exposure is above the TWA and/or excursion limit. Two lockers or storage facilities must be furnished and separated to prevent contamination of the employees' street clothes from protective work clothing and equipment. Showers must be furnished so that employees may shower at the end of the work shift. Employees must enter and exit the regulated area through the decontamination area.

The equipment room must be supplied with impermeable, labeled bags and containers for the containment and disposal of contaminated protective clothing and equipment.

Lunchroom facilities for those employees must have a positive pressure, filtered air supply and be readily accessible to employees. Employees must wash their hands and face prior to eating, drinking, or smoking. The employer must ensure that employees do not enter lunchroom facilities with protective work clothing or equipment unless surface fibers have been removed from the clothing or equipment.

Employees may not smoke in work areas where they are occupationally exposed to asbestos.

## Medical Exams

*In general industry*, exposed employees must have a preplacement physical examination before being assigned to an occupation exposed to airborne concentrations of asbestos at or above the action level or the excursion level. The physical examination must include chest X-ray, medical and work history, and pulmonary function tests. Subsequent exams must be given annually and upon termination of employment, though chest X-rays are required annually only for older workers whose first asbestos exposure occurred more than 10 years ago.

*In construction*, examinations must be made available annually for workers exposed above the action level or excursion limit for 30 or more days per year or who are required to wear negative pressure respirators; chest X-rays are at the discretion of the physician.

## SILICA EXPOSURE

Crystalline silica ($SiO_2$) is a major component of the earth's crust. In pure, natural form, $SiO_2$ crystals are minute, very hard, translucent, and colorless. Most mined minerals contain some $SiO_2$. "Crystalline" refers to the orientation of $SiO_2$ molecules in a fixed pattern as opposed to a nonperiodic, random molecular arrangement defined as amorphous (e.g., diatomaceous earth). Therefore, silica exposure occurs in a wide variety of settings, such as mining, quarrying, and stone cutting operations; ceramics and vitreous enameling; and in use of filters for paints and rubber. The wide use and multiple applications of silica in industrial applications combine to make silica a major occupational health hazard (silicosis), which can lead to death.

*Silicosis* is a disabling, nonreversible and sometimes fatal lung disease caused by overexposure to respirable crystalline silica. More than one million U.S. workers are exposed to crystalline silica, and each year more than 250 die from silicosis (see table 4.7). There is no cure for the disease, but it is 100 percent preventable if employers, workers, and health professionals work together to reduce exposures.

Table 4.7. Risk of Death from Silica in the Workplace, By Occupation and Industry

| Occupation | PMR |
| --- | --- |
| Miscellaneous metal and plastic machine operators | 168.44 |
| Hand molders and shapers, except jewelers | 64.12 |
| Crushing and grinding machine operators | 50.97 |
| Hand molding, casting, and forming occupations | 35.70 |
| Molding and casting machine operators | 30.60 |
| Mining machine operators | 19.61 |
| Mining occupations (not elsewhere classified) | 15.33 |
| Construction trades (not elsewhere classified) | 14.77 |
| Grinding, abrading, buffing, and polishing machine operators | 8.47 |
| Heavy equipment mechanics | 7.72 |
| Miscellaneous material moving equipment operators | 6.92 |
| Millwrights | 6.56 |
| Crane and tower operators | 6.02 |
| Brickmasons and stonemasons | 4.71 |
| Painters, construction and maintenance | 4.50 |
| Furnace, kiln, oven operators, except food | 4.10 |
| Laborers, except construction | 3.79 |
| Operating engineers | 3.56 |
| Welders and cutters | 3.01 |
| Machine operators, not specified | 2.86 |
| Not specified mechanics and repairers | 2.84 |
| Supervisors, production occupations | 2.73 |
| Construction laborers | 2.14 |
| Machinists | 1.79 |
| Janitors and cleaners | 1.78 |

| Industry | PMR |
| --- | --- |
| Metal mining | 69.51 |
| Miscellaneous nonmetallic mineral and stone products | 55.31 |
| Nonmetallic mining and quarrying, except fuel | 49.77 |
| Iron and steel foundries | 31.15 |
| Pottery and related products | 30.73 |
| Structural clay products | 27.82 |
| Coal mining | 9.26 |
| Blast furnaces, steelworks, rolling and finishing mills | 6.49 |
| Miscellaneous fabricated metal products | 5.87 |
| Miscellaneous retail stores | 4.63 |
| Machinery, except electrical (not elsewhere classified) | 3.96 |
| Other primary metal industries | 3.63 |
| Industrial and miscellaneous chemicals | 2.72 |
| Not specified manufacturing industries | 2.67 |
| Construction | 1.82 |

The first column is the occupation or industry title. The second column (PMR) is the observed number of deaths from silicosis per occupation or industry divided by the expected number of deaths. Therefore, a value of one indicates no additional risk. A value of ten would indicate a risk ten times greater than normal risk of silicosis.

*Source: Work-Related Lung Disease Surveillance Report* (2002). National Institute for Occupational Safety and Health, U.S. Department of Health and Human Services, Public Health Service, Centers for Disease Control and Prevention, Table 3-8; DHHA (NIOSH) Publication No. 96-134. Publications Dissemination, EID, National Institute for Occupational Safety and Health, 4676 Columbia Parkway, Cincinnati, OH.

### Guidelines for Control of Occupational Exposure to Silica

In accordance with OSHA's standard for air contaminants (29 CFR 1910.1000), employee exposure to airborne crystalline silica shall not exceed an 8-hour TWA limit (variable) as stated in 29 CFR 1910.1000, table Z-3, or a limit set by a state agency whenever a state-administered Occupational Safety and Health Plan is in effect.

As mandated by OSHA, the first mandatory requirement is that employee exposure be eliminated through the implementation of feasible engineering controls (e.g., dust suppression and ventilation). After all such controls are implemented and they do not control to the permissible exposure, each employer must rotate its employees to the extent possible in order to reduce exposure. Only when all engineering or administrative controls have been implemented, and the level of respirable silica still exceeds permissible exposure limits, may an employer rely on a respirator program pursuant to the mandatory requirements of 29 CFR1910.134. Generally where working conditions or other practices constitute recognized hazards likely to cause death or serious physical harm, they must be corrected.

## FORMALDEHYDE EXPOSURE

Formaldehyde (HCHO) is a colorless, flammable gas with a pungent suffocating odor. Formaldehyde is common to the chemical industry. It is the most important aldehyde produced commercially, and is used in the preparation of urea-formaldehyde and phenol-formaldehyde resins. It is also produced during the combustion of organic materials and is a component of smoke.

The major sources of HCHO in workplace settings are in manufacturing processes (used in the paper, photographic, and clothing industries) and building materials. Building materials may contain phenol, urea, thiourea, or melamine resins which contain HCHO. Degradation of HCHO resins can occur when these materials become damp from exposure to high relative humidity, or if the HCHO materials are saturated with water during flooding or leaking. The release of HCHO occurs when the acid catalysts involved in the resin formulation are reactivated. When temperatures and relative humidity increase, out-gassing increases (DOH, Wash., 2001).

Formaldehyde exposure is most common through gas-phase inhalation. However, it can also occur through liquid-phase skin absorption. Workers can be exposed during direct production, treatment of materials, and production of resins. Health care professionals, pathology and histology technicians, and teachers and students who handle preserved specimens are potentially at high risk.

Studies indicate that formaldehyde is a potential human carcinogen. Airborne concentrations above 0.1 ppm can cause irritation of the eyes, nose, and throat. The severity irritation increases as concentrations increase; at 100 ppm it is immediately dangerous to life and health. Dermal contact causes various skin reactions including sensitization, which might force persons thus sensitized to find other work.

OSHA requires that the employer conduct initial monitoring to identify all employees who are exposed to formaldehyde at or above the action level or STEL and to accurately determine the exposure of each employee so identified. If the exposure level is maintained below the STEL and the action level, employers may discontinue exposure monitoring, until there is a change which could affect exposure levels. The employer must also monitor employee exposure promptly upon receiving reports of formaldehyde-related signs and symptoms.

In regards to exposure control, the best prevention is provided by source control (if possible). The selection of HCHO-free or low-emitting products, for example, exterior grade plywood which uses phenol HCHO resins for indoor use; this is the best starting point.

Secondary controls include filtration, sealants, and fumigation treatments. Filtration can be achieved using selected adsorbents. Sealants involve coating the materials in question with two or three coats of nitro-cellulose varnish, or water based polyurethane. Three coats of these materials can reduce out-gassing by as much as 90%.

Training is required at least annually for all employees exposed to formaldehyde concentrations of 0.1 ppm or greater. The training will increase employees' awareness of specific hazards in their workplace and of the control measures employed. The training also will assist successful medical surveillance and medical removal programs. These provisions will only be effective if employees know what signs or symptoms are related to the health effects of formaldehyde.

## LEAD EXPOSURE

Lead has been poisoning workers for thousands of years. Most occupational overexposures to lead have been found in the construction trades, such as plumbing, welding and painting. In plumbing, soft solder (banned for many uses in the United States), used chiefly for soldering tinplate and copper pipe joints, is an alloy of lead and tin. Although the use of lead-based paint in residential applications has been banned, since lead-based paint inhibits the rusting and corrosion of iron and steel, it is still used on construction projects. Significant lead exposures can also arise from removing paint from surfaces previously coated with lead-based paint. According to OSHA 93-47 (2003), the operations that generate lead dust and fume include the following:

- Flame-torch cutting, welding, the use of heat guns, sanding, scraping and grinding of lead painted surfaces in repair, reconstruction, dismantling, and demolition work.
- Abrasive blasting of structures containing lead-based paints.
- Use of torches and heat guns, and sanding, scraping, and grinding lead-based paint surfaces during remodeling or abating lead-based paint.
- Maintaining process equipment or exhaust duct work.

## Health Effects of Lead

There are several routes of entry in which lead enters the body. When absorbed into the body in certain doses, lead is a toxic substance. Lead can be absorbed into the body by inhalation and ingestion. Except for certain organic lead compounds—such as tetraethyl lead—not covered by OSHA's Lead Standard (29 CFR 1926.62), lead, when scattered in the air as a dust, fume, or mist, can be absorbed into the body by inhalation.

A significant portion of the lead that can be inhaled or ingested gets into the blood stream. Once in the blood stream, lead is circulated throughout the body and stored in various organs and tissues. Some of this lead is quickly filtered out of the body and excreted, but some remains in the blood and other tissues. As exposure to lead continues, the amount stored in the body will increase if more lead is being absorbed that is being excreted. Cumulative exposure to lead, which is typical in construction settings, may result in damage to the blood, nervous system, kidneys, bones, heart, and reproductive system and contributes to high blood pressure. Some of the symptoms of lead poisoning include poor appetite, dizziness, pallor, headache, irritability/anxiety, constipation, sleeplessness, weakness, "lead line" in gums, fine tremors, hyperactivity, "wrist drop" (weakness of extensor muscles), excessive tiredness, numbness, muscle and joint pain or soreness, nausea, and reproductive difficulties.

## Lead Standard Definitions

According to OSHA's Lead Standard, the terms listed below have the following meanings:

Action level: Employee exposure, without regard to the use of respirators, to an airborne concentration of lead of 30 micrograms per cubic meter of air (30 µg/m$^3$), averaged over an eight-hour period.

**Permissible exposure limit (PEL):** The concentration of airborne lead to which an average person may be exposed without harmful effects. OSHA has established a PEL of fifty micrograms per cubic meter of air (50 µg/m$^3$) averaged over an eight-hour period. If an employee is exposed to lead for more than eight hours in any work day, the PEL, a TWA for that day, shall be reduced according to the following formula:

$$\text{Maximum permissible limit (in µg/m}^3\text{)} = 400 \times \text{hours worked in the day.}$$

When respirators are used to supplement engineering and administrative controls to comply with the PEL and all the requirements of the lead standard's respiratory protection rules have been met, employee exposure, for the purpose of determining whether the employer has complied with the PEL, may be considered to be at the level provided by the protection factor of the respirator for those periods the respirator is worn. Those periods may be averaged with exposure levels during periods when respirators are not worn to determine the employee's daily TWA exposure.

**µg/m$^3$:** Micrograms per cubic meter of air. A microgram is one millionth of a gram. There are 454 grams in a pound.

## Worker Lead Protection Program

The employer is responsible for the development and implementation of a worker lead protection program. This program is essential in minimizing worker risk of lead exposure.

The most effective way to protect workers is to minimize exposure through the use of engineering controls and good work practices.

At the minimum, the following elements should be included in the employer's worker protection program for employees exposed to lead:

- Hazard determination, including exposure assessment
- Engineering and work practice controls
- Respiratory protection
- PPE (protective work clothing and equipment)
- Housekeeping
- Hygiene facilities and practices
- Medical surveillance and provisions for medical removal
- Employee information and training
- Signs
- Recordkeeping

## MOLD CONTROL

Molds can be found almost anywhere; they can grow on virtually any organic substance, as long as moisture and oxygen are present. The earliest known writings that appear to discuss mold infestation and remediation (removal, cleaning up) are found in Leviticus 14 of the Old Testament.

Where are molds "typically" found? Name any spot or place; they have been found growing in office buildings, schools, automobiles, in private homes, and other locations where water and organic matter are left unattended. Mold is not a new issue—just one which, until recently, has received little attention by regulators in the United States. That is, there are no state or federal statutes or regulations regarding molds and IAQ.

Molds reproduce by making spores that usually cannot be seen without magnification. Mold spores waft through the indoor and outdoor air continually. When mold spores land on a damp spot indoors, they may begin growing and digesting whatever they are growing on in order to survive. Molds generally destroy the things they grow on (USEPA, 2001).

The key to limiting mold exposure is to prevent the germination and growth of mold. Since mold requires water to grow, it is important to prevent moisture problems in buildings. Moisture problems can have many causes, including uncontrolled humidity. Some moisture problems in workplace buildings have been linked to changes in building construction practices during the 1970s, '80s, and '90s. Some of these changes have resulted in buildings that are tightly sealed, but may lack adequate ventilation, potentially leading to moisture buildup. Building materials, such as drywall, may not allow moisture to escape easily. Moisture problems may include roof leaks, landscaping or gutters that direct water into or under the building, and unvented combustion appliances. Delayed maintenance or insufficient maintenance are also associated with moisture problems in buildings. Moisture problems in temporary structures have frequently been associated with mold problems.

Building maintenance personnel, architects, and builders need to know effective means of avoiding mold growth which might arise from maintenance and construction practices. Locating and cleaning existing growths are also paramount to decreasing the health effects of mold contamination. Using proper cleaning techniques is important because molds are incredibly resilient and adaptable (Davis, 2001).

Molds can elicit a variety of health responses in humans. The extent to which an individual may be affected depends upon his or her state of health, susceptibility to disease, the organisms with which he or she came in contact, and the duration and severity of exposure (Ammann, 2000). Some people experience temporary effects that disappear when they vacate infested areas (Burge, 1997). In others, the effects of exposure may be long-term or permanent (Yang, 2001).

It should be noted that systemic infections caused by molds are not common. Normal, healthy individuals can resist systemic infection from airborne molds.

Those at risk for system fungal infection are severely immunocompromised individuals such as those with HIV or AIDs, individuals who have had organ or bone marrow transplants, and persons undergoing chemotherapy.

In 1994, an outbreak of Stachybotrys chartarum in Cleveland, Ohio was believed by some to have caused pulmonary hemorrhage in infants. Sixteen of the infants died. The Centers for Disease Control (CDC) sponsored a review of the cases and concluded that the scientific evidence provided did not warrant the conclusion that inhaled mold was the cause of the illnesses in the infants. However, the panel also stated that further research was warranted, as the study design for the original research appeared to be flawed (CDC, 1999).

The following mold components are known to elicit a response in humans:

- *Volatile Organic Compounds* (VOCs): "Molds produce a large number of volatile organic compounds. These chemicals are responsible for the musty odors produced by growing molds" (McNeel and Kreutzer, 1996). VOCs also provide the odor in cheeses, and the "off" taste of mold infested foods. Exposure to high levels of VOCs affect the central nervous system, producing such symptoms as headaches, attention deficit, inability to concentrate, and dizziness (Ammann, 2000). According to McNeel, at present the specific contribution of mold VOCs in building-related health problems has not been studied. Also, mold VOCs are likely responsible for only a small fraction of total VOCs indoors (Davis, 2001).
- *Allergens*: All molds, because of the presence of allergen spores, have the potential to cause an allergic reaction in susceptible humans (Rose, 1999). Allergic reactions are believed to be the most common exposure reaction to molds. These reactions can range from mild, transitory response, like runny eyes, runny nose, throat irritation, coughing, and sneezing; to severe, chronic illnesses such as sinusitis and asthma (Ammann, 2000).
- *Mycotoxins*: Some molds are capable of producing mycotoxins, which are natural organic compounds that are capable of initiating a toxic response in vertebrates (McNeel and Kreutzer, 1996). Molds known to potentially produce mycotoxins and which have been isolated in infestations causing adverse health effects include certain species of *Acremonium, Alternaria, Aspergillus, Chaetomium, Caldosporium, Fusarium, Paecilomyces, Penicillium, Stachybotrys,* and *Trichoderma* (Yang, 2001).

While a certain type of mold or mold strain may have the genetic potential for producing mycotoxins; specific environmental conditions are believed to be needed for the mycotoxins to be produced. In other words, although a given mold might have the potential to produce mycotoxins, it will not produce them if the appropriate environmental conditions are not present (USEPA, 2001).

Currently, the specific conditions that cause mycotoxin production are not fully understood. The USEPA recognizes that mycotoxins have a tendency to concentrate in fungal spores and that there is limited information currently available regarding the process involved in fungal spore release. As a result, USEPA is conducting research in an effort to determine "the environmental conditions required for sporulation, emission, aerosolization, dissemination and transport of [Stachybotrys] into the air" (USEPA, 2001).

## Mold Prevention

As mentioned, the key to mold control is moisture control. Solve moisture problems before they become mold problems.

- Fix leaky plumbing and leaks in the building envelope as soon as possible.
- Watch for condensation and wet spots. Fix sources(s) of moisture problem(s) as soon as possible.
- Prevent moisture due to condensation by increasing surface temperature or reducing the moisture level in air (humidity). To increase surface temperature, insulate or increase air circulation. To reduce the moisture level in air, repair leaks, increase ventilation (if outside air is cold and dry), or dehumidify (if outdoor air is warm and humid).
- Keep heating, ventilation, and air conditioning (HVAC) drip pans clean, flowing properly, and unobstructed.
- Vent moisture-generating appliances, such as dryers, to the outside where possible.
- Perform regular building/HVAC inspections and maintenance as scheduled.
- Maintain low indoor humidity, below 60% relative humidity (RH), ideally 30–50%, if possible.
- Clean and dry wet or damp spots within 48 hours.
- Don't let foundations stay wet. Provide drainage and slope the ground away from the foundation.

## Mold Remediation

At the present time, there are no standardized recommendations for mold remediation; however, USEPA is working on guidelines. There are certain aspects of mold cleanup, however, which are agreed upon by many practitioners in the field.

- A common sense approach should be taken when assessing mold growth. For example, it is generally believed that small amounts of growth, like those commonly found on shower walls, pose no immediate health risk to most individuals.
- Persons with respiratory problems, a compromised immune system, or fragile health, should not participate in cleanup operations.
- Cleanup crews should be properly attired. Mold should not be allowed to touch bare skin. Eyes and lungs should be protected from aerosol exposure.
- Adequate ventilation should be provided while, at the same time, containing the infestation in an effort to avoid spreading mold to other areas.
- The source of moisture must be stopped and all areas infested with mold thoroughly cleaned. If thorough cleaning is not possible due to the nature of the material (porous versus semi- and nonporous), all contaminated areas should be removed.

Safety tips that should be followed when remediating moisture and mold problems include:

- Do not touch mold or moldy items with bare hands.
- Do not get mold or mold spores in your eyes.
- Do not breathe in mold or mold spores.
- Consider using PPE when disturbing mold. The minimum PPE is a respirator, gloves, and eye protection.

A variety of mold cleanup methods are available for remediating damage to building materials and furnishings caused by moisture control problems and mold growth. These include wet vacuum, damp wipe, HEPA vacuum, and the removal of damaged materials (and sealing them in plastic bags). The specific method or group of methods used will depend on the type of material affected.

## Checklist for Mold Remediation

*Investigate and evaluate moisture and mold problems*

- ☐ Assess size of moldy area (square feet)
- ☐ Consider the possibility of hidden mold
- ☐ Clean up small mold problems and fix moisture problems before they become large problems

# INDOOR AIR QUALITY AND MOLD CONTROL

- ☐ Select remediation manager for medium or large size mold problem
- ☐ Investigate areas associated with occupant complaints
- ☐ Identify sources(s) or cause of water or moisture problem(s)
- ☐ Note type of water-damaged materials (wallboard, carpet, etc.)
- ☐ Check inside air ducts and air handling unit
- ☐ Throughout process, consult qualified professional if necessary or desired

*Communicate with building occupants at all stages of process, as appropriate*

- ☐ Designate contact person for questions and comments about medium or large scale remediation as needed.

*Plan remediation*

- ☐ Adapt or modify remediation guidelines to fit your situation; use professional judgment
- ☐ Plan to dry wet, nonmoldy materials within 48 hours to prevent mold growth
- ☐ Select clean methods for moldy items
- ☐ Select PPE
- ☐ Select containment equipment; protect building occupants
- ☐ Select remediation personnel who have the experience and training needed to implement the remediation plan and use PPE and containment as appropriate

*Remediate moisture and mold problems*

- ☐ Fix moisture problem, implement repair plan and/or maintenance plan
- ☐ Dry wet, nonmoldy materials within 48 hours to prevent mold growth
- ☐ Clean and dry moldy materials
- ☐ Discard moldy porous items that can't be cleaned

## IAQ EQUATIONS

In the following, we provide a few of the basic, standard equations used in indoor air quality computations.

1. Percentage of Outside Air in the Air Supply

$$\% \text{ Outside Air} = T_{RA} - T_{MA}/T_{RA} - T_{OA} \qquad (4.1)$$

where:

$T_{RA}$ = temperature of return air (dry-bulb)
$T_{MA}$ = temperature of mixed return and outside air (dry-bulb)
$T_{OA}$ = temperature of outdoor air (dry-bulb)

2. Outdoor Air Volume Flow Rate:

$$Q_{OA} = 13{,}000n/C_{indoors} - C_{OA} \qquad (4.2)$$

where:

$Q_{OA}$ = approximate volume flow rate of outdoor air (cfm)

$n$ = number of people working in an office complex (with about 7 people per 1,000 square feet of office space)

$C_{indoors}$ = measured concentration of $CO_2$ in the office air after a log period of occupancy time, such as near lunch or near the end of the day (ppm)

$C_{OA}$ = concentration of $CO_2$ in the outdoor air (ppm)

3. Air Changes per Hour, OA

$$N = \ln(C_i - C_o) - \ln(C_a - C_o)/h \qquad (4.3)$$

where:

$N$ = air exchange, air changes per hour, OA

$C_i$ = concentration of $CO_2$ at start of test

$C_o$ = outdoor concentration, about 330 ppm

$C_a$ = concentration of $CO_2$ at end of test

ln = the natural log

## REFERENCES

Ammann, H., 2000. "Is Indoor Mold Contamination a Threat?" Washington State Department of Health. http://www.doh.wa.gov/ehp/ocha/mold.html (accessed August 9, 2003).

Burge, H. A., 1997. "The Fungi: How They Grow and Their Effects on Human Health." *Heating/Piping/Air Conditioning* (July): 69–75.

Burge, H. A., and M. E. Hoyer, 1998. "Indoor Air Quality." *The Occupational Environment—Its Evaluation and Control.* Ed. S. R. DiNardi. Fairfax, Va.: American Industrial Hygiene Association.

Byrd, R. R., 2003. *IAQ FAG Part 1.* Glendale, CA: Machado Environmental Corporation.

CDC, 1999. *Reports of Members of the CDC External Expert Panel on Acute Idiopathic Pulmonary Hemorrhage in Infants: A Synthesis.* Washington, D.C.: Centers for Disease Control.

Davis, P. J., 2001. *Molds, Toxic Molds, and Indoor Air Quality.* Sacramento: California Research Bureau, California State Library.

DOH, Wash., (2001). *Formaldehyde.* Washington State Department of Health, Office of Environmental Health and Safety. http://www.doh.wa.gov/ehp/ts/IAQ/Formaldehyde.HTM (accessed August 2003).

Hollowell, C. D., et al., 1979a. "Impact of Infiltration and Ventilation on Indoor Air Quality." *ASHRAE J* (July): 49–53.

Hollowell, C. D. et al., 1979b. "Impact of Energy Conservation in Buildings on Health." *Changing Energy Use Futures.* Ed. R. A. Razzolare and C. B. Smith. New York: Pergamon.

McNeel, S., and R. Kreutzer, 1996. "Fungi & Indoor Air Quality." *Health & Environment Digest* 10, no. 2.

OSHA, 2002. *Better Protection Against Asbestos in the Workplace.* Fact Sheet No. OSHA 92-06. Washington, D.C.: U.S. Department of Labor.

OSHA, 1994. *Lead Exposure in Construction.* Fact Sheet No. OSHA 93–47. Washington, D.C.: U.S. Department of Labor. http://www.osha-slc.gov/pls/oshaweb/owadisp.show_document?p_table=FACT_Sheet (accessed August 2003).

Passon, T. J., Jr., et al., 1996. "Sick-Building Syndrome and Building-Related Illnesses." *Medical Laboratory Observer* 28, no. 7: 84–95.

Rose, C. F., 1999. "Antigens." *ACGIH Bioaerosols Assessment and Control.* Cincinnati: American Conference of Governmental Industrial Hygienists, 25.1–25.11.

USEPA, 2001. *Sick Building Syndrome.* Indoor Air Facts No. 4 (revised). http://www.epa.gov/iaq/pubs/sbs.html (accessed August 9, 2003).

Woods, J. E., 1980. *Environmental Implications of Conservation and Solar Space Heating.* Engineering Research Institute, Iowa State University, Ames, BEUL 80-3. Meeting of the New York Academy of Sciences, New York, January 16.

Yang, C. S., 2001. "Toxic Effects of Some Common Indoor Fungi." http://www.envirovillage.com/Newsletters/Enviros/No4_09.htm (accessed August 6, 2003).

## SUGGESTED READING

American Conference of Governmental Industrial Hygienists. *Guidelines for Assessment and Control of Aerosols.* Cincinnati, Ohio: ACGIH, 1997.

American Industrial Hygiene Association. *Field Guide for the Determination of Biological Contaminants in Environmental Samples.* Fairfax, Va.: AIHA, 1996.

American Industrial Hygiene Association. *Practitioners Approach to Indoor Air Quality Investigations.* Fairfax, Va.: AIHA, 1989.

Dockery, D. W. and Spengler, J. D. "Indoor-Outdoor Relationships of Respirable Sulfates and Particles." *Atmos. Environ.* 15 (1981): 335–343.

Spengler, J. D., et al. "Sulfur Dioxide and Nitrogen Dioxide Levels Inside and Outside Homes and the Implications on Health Effects Research." *Environ. Sci. Technolo.* 13 (1979): 1276–1280.

U.S. Environmental Protection Agency. *Building Air Quality: A Guide for Building Owners and Facility Managers.* Washington, D.C.: EPA, 1991.

U.S. Environmental Protection Agency, Indoor Environmental Management Branch. *Children's Health Initiative: Toxic Mold.* http://epa.gov.appedwww/crb/iemb/child.htm (accessed August 9, 2003).

U.S. Environmental Protection Agency. *Mold Remediation in Schools and Commercial Buildings.* www.epa.gov/iaq/molds (accessed July 8, 2005).

# 5

# Noise and Vibration

High noise levels in the workplace are a hazard to employees. High noise levels are a physical stress that may produce psychological effects by annoying, startling, or disrupting the worker's concentration, which can lead to accidents. High levels can also result in damage to workers' hearing, resulting in hearing loss. In this chapter, we discuss the basics of noise, including those elements the industrial hygienist needs to know to ensure that his or her organization's hearing conservation program is in compliance with OSHA. We also discuss the basics of vibration and its control; vibration is closely related to noise.

## OSHA NOISE CONTROL REQUIREMENTS

In 1983, OSHA adopted a Hearing Conservation Amendment to OSHA 29 CFR 1910.95 requiring employers to implement *hearing conservation programs* in any work setting where employees are exposed to an eight-hour TWA of 85 dBA and above (LaBar, 1989). Employers must monitor all employees whose noise exposure is equivalent to or greater than a noise exposure received in 8 hours where the noise level is constantly 85 dBA. The exposure measurement must include all continuous, intermittent, and impulsive noise within an 80 dBA to 130 dBA range and must be taken during a typical work situation. This requirement is performance-oriented because it allows employers to choose the monitoring method that best suits each individual situation (OSHA, 2002).

In addition to concerns over noise levels, the OSHA Standard also addresses the issue of duration of exposure. LaBar (1989) explains the duration aspects of the regulation as follows:

Duration is another key factor in determining the safety of workplace noise. The regulation has a 50 percent 5 dBA logarithmic tradeoff. That is, for every 5-decibel increase in the noise level, the length of exposure must be reduced by 50 percent. For example, at 90 decibels (the sound level of a lawnmower or shop tools), the limit on "safe" exposure is 8 hours. At 95 dBA, the limit on exposure is 4 hours, and so on. For any sound that is 106 dBA and above—this would include such things as a sandblaster, rock concert, or jet engine—exposure without protection should be less than 1 hour, according to OSHA's rule.

The basic requirements of OSHA's Hearing Conservation Standard are explained here:

- *Monitoring noise levels.* Noise levels should be monitored on a regular basis. Whenever a new process is added, an existing process is altered, or new equipment is purchased, special monitoring should be undertaken immediately.
- *Medical surveillance.* The medical surveillance component of the regulation specifies that employees who will be exposed to high noise levels be tested upon being hired and again at least annually.
- *Noise controls.* The regulation requires that steps be taken to control noise at the source. Noise controls are required in situations where the noise level exceeds 90 dBA. Administrative controls are sufficient until noise levels exceed 100 dBA. Beyond 100 dBA engineering controls must be used.
- *Personal protection.* Personal protective devices are specified as the next level of protection when administrative and engineering controls do not reduce noise hazards to acceptable levels. They are to be used in addition to, rather than instead of, administrative and engineering controls.
- *Education and training.* The regulation requires the provision of education and training to do the following: ensure that employees understand (1) how the ear works, (2) how to interpret the results of audiometric tests, (3) how to select personal protective devices that will protect them against the types of noise hazards to which they will be exposed, and (4) how to properly use personal protective devices (LaBar, 1989).

## NOISE AND HEARING LOSS TERMINOLOGY

There are many specialized terms used to express concepts in noise, noise control, and hearing loss prevention. The industrial hygiene practitioner responsible for ensuring compliance with OSHA's hearing conservation program requirements must be familiar with these terms. The NIOSH (2005) definitions below were written in as nontechnical a fashion as possible.

**acoustic trauma**: A single incident which produces an abrupt hearing loss. Welding sparks (to the eardrum), blows to the head, and blast noise are examples of events capable of providing acoustic trauma.

**action level**: The sound level which when reached or exceeded necessitates implementation of activities to reduce the risk of noise-induced hearing loss. OSHA currently uses an 8-hour time weighted average of 85 dBA as the criterion for implementing an effective hearing conservation program.

**attenuate**: To reduce the amplitude of sound pressure (noise).

**attenuation**: *See* Real Ear Attenuation at Threshold (REAT); Real-world.

**audible range**: The frequency range over which the normal ears hear: approximately 20 Hz through 20,000 Hz.

**audiogram**: A chart, graph, or table resulting from an audiometric test showing an individual's hearing threshold levels as a function of frequency.

**audiologist**: A professional, specializing in the study and rehabilitation of hearing, who is certified by the American Speech-Language-Hearing Association or licensed by a state board of examiners.

**background noise**: Noise coming from sources other than the particular noise sources being monitored.

**baseline audiogram**: A valid audiogram against which subsequent audiograms are compared to determine if hearing thresholds have changed. The baseline audiogram is preceded by a quiet period so as to obtain the best estimate of the person's hearing at that time.

**continuous noise**: Noise of a constant level as measured over at least one second using the "slow" setting on a sound-level meter. Note, that a noise which is intermittent, for example, on for over a second and then off for a period would be both variable and continuous.

**controls, administrative**: Efforts, usually by management, to limit workers' noise exposure by modifying workers' schedule or location, or by modifying the operating schedule of noisy machinery.

**controls, engineering**: Any use of engineering methods to reduce or control the sound level of a noise source by modifying or replacing equipment, or making any physical changes at the noise source or along the transmission path (with the exception of hearing protectors).

**criterion sound level**: A sound level of 90 decibels.

**dB (decibel)**: The unit used to express the intensity of sound. The decibel was named after Alexander Graham Bell. The decibel scale is a logarithmic scale in which 0 dBA approximates the threshold of hearing in the mid-frequencies for young adults and in which the threshold of discomfort is between 85 and 95 dBA SPL and the threshold for pain is between 120 and 140 dBA SPL.

**double hearing protection**: A combination of both ear plug and ear muff type hearing protection devices required for employees who have demonstrated Temporary Threshold Shift during audiometric examination and for those who have been advised to wear double protection by a medical doctor in work areas that exceed 104 dBA.

**dosimeter**: When applied to noise, refers to an instrument that measures sound levels over a specified interval, stores the measures, and calculates the sound as a function of sound level and sound duration and describes the results in terms of dose, time weighted average, and (perhaps) other parameters such as peak level, equivalent sound level, sound exposure level, etc.

**equal-energy rule**: The relationship between sound level and sound duration based upon a 3 dB exchange rate. In other words, the sound energy resulting from doubling or halving a noise exposure's duration is equivalent to increasing or decreasing the sound level by 3 dB, respectively.

**exchange rate**: The relationship between intensity and dose. OSHA uses a 5-dB exchange rate. Thus, if the intensity of an exposure increases by 5 dB, the dose doubles. Sometimes, this is also referred to as the doubling rate. The U.S. Navy uses a 4-dB exchange rate; the U.S. Army and Air Force use a 3-dB exchange rate. NIOSH recommends a 3-dB exchange rate. Note that the equal-energy rule is based on a 3-dB exchange rate.

**frequency**: Rate in which pressure oscillations are produced. Measured in hertz (Hz).

**hazardous noise**: Any sound for which any combination of frequency, intensity, or duration is capable of causing permanent hearing loss in a specified population.

**hazardous task inventory**: A concept based on using work tasks as the central organizing principle for collecting descriptive information on a given work hazard. It consists of a list(s) of specific tasks linked to a database containing the prominent characteristics relevant to the hazard(s) of interest which are associated with each task.

**hearing conservation record**: Employee's audiometric record. Includes name, age, job classification, TWA exposure, date of audiogram, and name of audiometric technician. To be retained for duration of employment for OSHA. Kept indefinitely for workers' compensation.

**hearing damage risk criteria**: A standard which defines the percentage of a given population expected to incur a specified hearing loss as a function of exposure to a given noise exposure.

**hearing handicap**: A specified amount of permanent hearing loss usually averaged across several frequencies which negatively impacts employment and/or social activities. Handicap is often related to an impaired ability to communicate. The degree of handicap will also be related to whether the hearing loss is in one or both ears, and whether the better ear has normal or impartial hearing.

**hearing loss**: Hearing loss is often characterized by the area of the auditory system responsible for the loss. For example, when injury or a medical condition affects the outer ear or middle ear (i.e., from the pinna, ear canal, and ear drum to the cavity behind the ear drum—which includes the ossicles) the resulting hearing loss is referred to as a *conductive* loss. When an injury or medical condition affects the inner ear or the auditory nerve that connects the inner ear to the brain (i.e., the cochlea and the VIIIth cranial nerve) the resulting hearing loss is referred to as a *sensorineural* loss. Thus, a welder's spark which damaged the ear drum would cause a conductive hearing loss. Because noise can damage the tiny hair cells located in the cochlea, it causes a sensorineural hearing loss.

**hearing loss prevention program audit**: An assessment performed prior to putting a hearing loss prevention program into place or before changing an existing program. The audit should be a top-down analysis of the strengths and weaknesses of each aspect of the program.

**hearing threshold level (HTL)**: The hearing level, above a reference value, at which a specified sound or tone is heard by an ear in a specified fraction of the trials. Hearing threshold levels have been established so that 0 dB HTL reflects the best hearing of a group of persons.

**hertz (Hz)**: The unit of measurement for audio frequencies. The frequency range for human hearing lies between 20 Hz and approximately 20,000 Hz. The sensitivity of the human ear drops off sharply below about 500 Hz and above 4,000 Hz.

**impulsive noise**: Used to generally characterize impact or impulse noise which is typified by a sound which rapidly rises to a sharp peak and then quickly fades. The sound may or may not have a "ringing" quality (such as a striking hammer on a metal plate or a gunshot in a reverberant room). Impulsive noise may be repetitive, or may be a single event (as with a sonic boom). However, if occurring in very rapid succession (such as with some jack hammers), the noise would not be described as impulsive.

**loudness**: The subjective attribute of a sound by which it would be characterized along a continuum from "soft" to "loud." Although this is a subjective attribute, it depends primarily upon sound pressure level, and to a lesser extent, the frequency characteristics and duration of the sound.

**material hearing impairment**: As defined by OSHA, a material hearing impairment is an average hearing threshold level of 25 dB HTL at the frequencies of 1000, 2000, and 3000 Hz.

**medical pathology**: A disorder or disease. For purposes of this program, a condition or disease affecting the ear, which a physician specialist should treat.

**noise**: Noise is any unwanted sound.

**noise dose**: The noise exposure expressed as a percentage of the allowable daily exposure. For OSHA, a 100% dose would equal an 8-hour exposure to a continuous 90 dBA noise; a 50% dose would equal an 8-hour exposure to an 85 dBA noise or a 4-hour exposure to a 90 dBA noise. If 85 dBA is the maximum permissible level, then an 8-hour exposure to a continuous 85 dBA noise would equal a 100% dose. If a 3 dB exchange rate is used in conjunction with an 85 dBA maximum permissible level, a 50% dose would equal a 2-hour exposure to 88 dBA, or an 8-hour exposure to 82 dBA.

**noise dosimeter:** An instrument that integrates a function of sound pressure over a period of time to directly indicate a noise dose.

**noise hazard area:** Any area where noise levels are equal to or exceed 85 dBA. OSHA requires employers to designate work areas, post warning signs, and warn employees when work practices exceed 90 dBA as a "Noise Hazard Area." Hearing protection must be worn whenever 90 dBA is reached or exceeded.

**noise hazard work practice:** Performing or observing work where 90 dBA is equaled or exceeded. Some work practices will be specified; however, as a "rule of thumb," when shouting must be employed to hold normal conversation with someone who is one foot away, one can assume that a 90 dBA noise level or greater exists and hearing protection is required. Typical examples of work practices where hearing protection is required are jack hammering, heavy grinding, heavy equipment operations, and similar activities.

**noise-induced hearing loss:** A sensorineural hearing loss that is attributed to noise and for which no other etiology can be determined.

**noise level measurement:** Total sound level within an area. Includes workplace measurements indicating the combined sound levels of tool noise (from ventilation systems, cooling compressors, circulation pumps, etc.).

**noise reduction rating (NRR):** The NRR is a single-number rating method which attempts to describe a hearing protector based on how much the overall noise level is reduced by the hearing protector. When estimating A-weighted noise exposures, it is important to remember to first subtract 7 dB from the NRR and then subtract the remainder from the A-weighted noise level. The NRR theoretically provides an estimate of the protection that should be met or exceeded by 98% of the wearers of a given device. In practice, this does not prove to be the case, so a variety of methods for "de-rating" the NRR have been discussed.

**ototoxic:** A term typically associated with the sensorineural hearing loss resulting from therapeutic administration of certain prescription drugs.

**ototraumatic:** A broader term than ototoxic. As used in hearing loss prevention, refers to any agent (e.g., noise, drugs, or industrial chemicals) which has the potential to cause permanent hearing loss subsequent to acute or prolonged exposure.

**presbycusis:** The gradual increase in hearing loss that is attributable to the effects of aging, and not related to medical causes or noise exposure.

**real ear attenuation at threshold (REAT):** A standardized procedure for conducting psychoacoustic tests on human subjects designed to measure sound protection features of hearing protective devices. Typically, these measures are obtained in a calibrated sound field, and represent the difference between subjects' hearing thresholds when wearing a hearing protector versus not wearing the protector.

**real-world**: Estimated sound protection provided by hearing protective devices as worn in "real-world" environments.

**sensorineural hearing loss**: A hearing loss resulting from damage to the inner ear (from any source).

**sociacusis**: A hearing loss related to nonoccupational noise exposure.

**sound intensity (I)**: Sound intensity at a specific location is the average rate at which sound energy is transmitted through a unit area normal to the direction of sound propagation.

**sound level meter (SLM)**: A device which measures sound and provides a readout of the resulting measurement. Some provide only A-weighted measurements, others provide A- and C-weighted measurements, and some can provide weighted, linear, and octave (or narrower) ban measurements. Some SLMs are also capable of providing time-integrated measurements.

**sound power**: Is the total sound energy radiated by a source per unit time. Sound power cannot be measured directly.

**sound pressure level (SPL)**: A measure of the ratio of the pressure of a sound wave relative to a reference sound pressure. Sound pressure level in decibels is typically referenced to 20 mPa. When used alone (e.g., 90 dB APL) a given decibel level implies an unweighted sound pressure level.

**standard threshold shift (STS)**: OSHA uses the term to describe a change in hearing threshold relative to the baseline audiogram of an average of 10 dB or more at 2000, 3000 and 4000 Hz in either ear. Used by OSHA to trigger additional audiometric testing and related follow up.

**significant threshold shift (STS)**: NIOSH uses this term to describe a change of 15 dB or more at any frequency, 5000 through 6000 Hz, from baseline level that is present on an immediate retest in the same ear and at the same frequency. NIOSH recommends a confirmation audiogram within 30 days with the confirmation audiogram preceded by a quiet period of at least 14 hours.

**threshold shift**: Audiometric monitoring programs will encounter two types of changes in hearing sensitivity, that is, threshold shifts: *permanent threshold shift* (PTS) and *temporary threshold shift* (TTS). As the names imply, any change in hearing sensitivity which is persistent is considered a PTS. Persistence may be assumed if the change is observed on a 30-day follow-up exam. Exposure to loud noise may cause a temporary worsening in hearing sensitivity (i.e., a TTS) that may persist for 14 hours (or even longer in cases where the exposure duration exceeded 12 to 16 hours). Hearing health professionals need to recognize that not all threshold shifts represent decreased sensitivity, and not all temporary or permanent threshold shifts are due to noise exposure. When a PTS can be attributable to noise exposure, it may be referred to as a noise-induced permanent threshold shift (NIPTS).

**velocity (c)**: Is the speed at which the regions of sound producing pressure changes move away from the sound source.

**wavelength (λ):** This term refers to the distance required for one complete pressure cycle to be completed (1 wavelength) and is measured in feet or meters.

**weighted measurements:** Two weighting curves are commonly applied to measures of sound levels to account for the way the ear perceives the "loudness" of sounds. *A-weighting* is a measurement scale that approximates the "loudness" of tones relative to a 40 db SPL 1000 Hz reference tone. A-weighting has the added advantage of being correlated with annoyance measures and is most responsive to the mid frequencies, 500 to 4000 Hz. *C-weighting* is a measurement scale that approximates the "loudness" of tones relative to a 90 dB SPL 1000 Hz reference tone. C-weighting has the added advantage of providing a relatively "flat" measurement scale which includes very low frequencies.

## OCCUPATIONAL NOISE EXPOSURE

As mentioned above, *noise* is commonly defined as any unwanted sound. Noise literally surrounds us every day, and is with us just about everywhere we go. However, the noise we are concerned with here is that produced by industrial processes. Excessive amounts of noise in the work environment (and outside it) cause many problems for workers, including increased stress levels, interference with communication, disrupted concentration, and most importantly, varying degrees of hearing loss. Exposure to high noise levels also adversely affects job performance and increases accident rates.

One of the major problems with attempting to protect workers' hearing acuity is the tendency of many workers to ignore the dangers of noise. Because hearing loss, like cancer, is insidious, it's easy to ignore. It sort of sneaks up slowly and is not apparent (in many cases) until after the damage is done. Alarmingly, hearing loss from occupational noise exposure has been well documented since the eighteenth century, yet since the advent of the industrial revolution, the number of exposed workers has greatly increased (Mansdorf, 1993). However, as a direct result of OSHA's requirements, the picture of hearing loss today is not as bleak as it has been in the past. Now that noise exposure must be controlled in all industrial environments, that well written and well managed hearing conservation programs must be put in place, and that employee awareness must be raised to the dangers of exposure to excessive levels of noise, job-related hearing loss is coming under control.

## DETERMINING WORKPLACE NOISE LEVELS

The unit of measurement for sound is the decibel. *Decibels* are the preferred unit for measuring sound, derived from the bel, a unit of measure in electrical communications engineering. The decibel is a dimensionless unit used to express the logarithm of the ratio of a measured quantity to a reference quantity.

In regards to noise control in the workplace, the industrial hygienist's primary concern is first to determine if any "noise-makers" in the facility exceed the OSHA limits for worker exposure—exactly which machines or processes produce noise at unacceptable levels. Making this determination is accomplished by conducting a noise level survey of the plant or facility. Sound measuring instruments are used to make this determination. These include noise dosimeters, sound-level meters, and octave-band analyzers. The uses and limitations of each kind of instrument are discussed below.

The *noise dosimeters* used by OSHA meet the American National Standards Institute (ANSI) Standard S1.25-1978, "Specifications for Personal Noise Dosimeter," which set performance and accuracy tolerances. For OSHA use, the dosimeter must have a 5-dB exchange rate, use a 90-dBA criterion level, be set at slow response, and use either an 80-dBA or 90-dBA threshold gate, or a dosimeter that has both capabilities, whichever is appropriate for evaluation.

When conducting the noise level survey, the industrial hygienist should use an ANSI-approved *sound level meter* (SLM)—a device used most commonly to measure sound pressure. The SLM measures in decibels. One decibel is one-tenth of a bel and is the minimum difference in loudness that is usually perceptible.

The SLM consists of a microphone, an amplifier, and an indicating meter, which responds to noise in the audible frequency range of about 20 to 20,000 Hz. Sound-level meters usually contain "weighting" networks designated "A," "B," or "C." Some meters have only one weighting network; others are equipped with all three. The A-network approximates the equal loudness curves at low sound pressure levels, the B-network is used for medium sound pressure levels, and the C-network is used for high levels.

In conducting a routine workplace sound level survey, using the A-weighted network (referenced dBA) in the assessment of the overall noise hazard has become common practice. The A-weighted network is the preferred choice because it is thought to provide a rating of industrial noises that indicates the injurious effects such noise has on the human ear (gives a frequency response similar to that of the human ear at relatively low sound pressure levels).

With an approved and freshly calibrated (always calibrate test equipment prior to use) sound-level meter in hand, the industrial hygienist is ready to begin the sound level survey. In doing so, the industrial hygienist is primarily interested in answering the following questions:

What is the noise level in each work area?

What equipment or process is generating the noise?

Which employees are exposed to the noise?

How long are they exposed to the noise?

In answering these questions, industrial hygienists record their findings as they move from workstation to workstation, following a logical step-by-step procedure. The first step involves using the sound-level meter set for A-scale slow response mode to measure an entire work area. When making such measurements, restrict the size of the space being measured to less than 1,000 square feet. If the maximum sound level does not exceed 80 dBA, it can be assumed that all workers in this work area are working in an environment with a satisfactory noise level. However, a note of caution: the key words in the preceding statement are "maximum sound level." To assure an accurate measurement, the industrial hygienist must ensure that all noise-makers are actually in operation when measurements are taken. Measuring an entire work area does little good when only a small percentage of the noise-makers are actually in operation.

The next step depends on the readings recorded when the entire work area was measured. For example, if the measurements indicate sound levels greater than 80 dBA, then another set of measurements needs to be taken at each worker's workstation. The purpose here, of course, is to determine two things: which machine or process is making noise above acceptable levels (i.e., >80 dBA), and which workers are exposed to these levels. Remember that the worker who operates the machine or process might not be the only worker exposed to the noise-maker. You need to inquire about other workers who might, from time to time, spend time working in or around the machine or process. Our experience in conducting workstation measurements has shown us noise levels usually fluctuate. If this is the case, you must record the minimum and maximum noise levels. Any noise level that remains above 90 dBA exceeds the legal limit. However, if your measurements indicate that the noise level is never greater than 85 dBA (OSHA's action level), then the noise exposure can be regarded as satisfactory.

If workstation measurements indicate readings that exceed the 85 dBA level, you must perform another step. This step involves determining the length of time of exposure for workers. The easiest, most practical way to make this determination is to have the worker wear a noise dosimeter, which records the noise energy to which the worker is exposed during the work shift.

What happens next?

You must then determine if the worker is exposed to noise levels that exceed the permissible noise exposure levels listed in table 5.1. The key point to remember is that your findings must be based on a TWA. For example, from table 5.1 you will notice that a noise level of 95 dBA is allowed up to 4 hours per day.

*Note:* This parameter assumes that the worker has good hearing acuity with no loss. If the worker has documented hearing loss, exposure to 95 dBA or higher, without proper hearing protection, may be unacceptable under any circumstances.

Several Type 1 sound-level meters (such as the GenRad 1982 and 1983 and the Quest 155) used by OSHA have built-in octave band analysis capability. These *octave-band noise analyzers* can be used to determine the feasibility of controls for individual noise sources for abatement purposes and to evaluate hearing protectors.

Octave-band analyzers segment noise into its component parts. The octave-band filter sets provide filters with the following center frequencies: 31.5; 63; 125; 250; 500; 1,000; 2,000; 4,000; 8,000; and 16,000 Hz.

The special signature of a given noise can be obtained by taking sound-level meter readings at each of these settings (assuming that the noise is fairly constant over time). The results may indicate those octave-bands that contain the majority of the total radiated sound power.

Octave-band noise analyzers can assist industrial hygienists in determining the adequacy of various types of frequency-dependent noise controls. They also can be used to select hearing protectors because they can measure the amount of attenuation offered by the protectors in the octave-bands responsible for most of the sound energy in a given situation.

## ENGINEERING CONTROL FOR INDUSTRIAL NOISE

When the industrial hygienist investigates the possibility of using engineering controls to control noise, the first thing he or she recognizes is that reducing and/or eliminating all noise is virtually impossible. And this should not be the focus in the first place—eliminating or reducing the "hazard" is the goal. While the primary hazard may be the possibility of hearing loss, the distractive effect (or its interference with communication) must also be considered. The distractive effect or excessive noise can certainly be classified as hazardous whenever the distraction might affect the attention of the worker. The obvious implication of noise levels that interfere with communications is emergency response. If ambient noise is at such a high level that workers can't hear fire or other emergency alarms, this is obviously an unacceptable situation.

So what does all this mean? The industrial hygienist must determine the "acceptable" level of noise, then look into applying the appropriate noise control measures. These include making alterations in engineering design (obviously this can only be accomplished in the design phase) or making modifications after installation. Unfortunately, this latter method is the one the industrial hygienist is usually forced to apply—and also the most difficult, depending upon circumstances.

Let's assume that the industrial hygienist is trying to reduce noise levels generated by an installed air compressor to a safe level. Obviously, the first place to start is at the *source*: the air compressor. Several options are available for the industrial hygienist to employ at the source. First, the industrial hygienist would look at the possibility of modifying the air compressor to reduce its noise output. One option might be to install resilient vibration mounting devices. Another might be to change the coupling between the motor and the compressor.

If the options described for use at the source of the noise are not feasible or are only partially effective, the next component the industrial hygienist would look at is the *path* along which the sound energy travels. Increasing the distance between the air compressor and the workers could be a possibility. (Remember, sound levels decrease with distance.) Another option might be to install acoustical treatments on ceilings, floors, and walls. The best option available (in this case) probably is to enclose the air compressor, so that the dangerous noise levels are contained within the enclosure, and the sound leaving the space is attenuated to a lower, safety level. If total enclosure of the air compressor is not practicable, then erecting a barrier or baffle system between the compressor and the open work area might be an option.

The final engineering control component the industrial hygienist might incorporate to reduce the air compressor's noise problem is to consider the *receiver* (the worker/operator). An attempt should be made to isolate the operator by providing a noise reduction or soundproof enclosure or booth for the operator.

## AUDIOMETRIC TESTING (OSHA, 2002)

Audiometric testing monitors an employee's hearing acuity over time. It also provides an opportunity for employers to educate employees about their hearing and the need to protect it.

The employer must establish and maintain an audiometric testing program. The important elements of the program include baseline audiograms, annual audiograms, training, and follow-up procedures. Employers must make audiometric testing available at no cost to all employees who are exposed to an action level of 85 dB or above, measured as an 8-hour TWA.

The audiometric testing program follow-up should indicate whether the employer's hearing conservation program is preventing hearing loss. A licensed or certified audiologist, otolaryngologist, or other physician must be responsible for the program. Both professionals and trained technicians may conduct audiometric testing. The professional in charge of the program does not have to be present when a qualified technician conducts tests. The professional's responsibilities include overseeing the program and the work of the technicians, reviewing problem audiograms, and determining whether referral is necessary.

The employee needs a referral for further testing when test results are questionable or when related medical problems are suspected. If additional testing is necessary or if the employer suspects a medical pathology of the ear that is caused or aggravated by wearing hearing protectors, the employer must refer the employee for a clinical audiological evaluation or otological exam, as appropriate. There are two types of audiograms required in the hearing conservation program: baseline and annual audiograms.

The *baseline audiogram* is the reference audiogram against which future audiograms are compared. Employers must provide baseline audiograms within 6 months of an employee's first exposure at or above an 8-hour TWA of 85 dB. An exception is allowed when the employer uses a mobile test van for audiograms. In these instances, baseline audiograms must be completed within 1 year after an employee's first exposure to workplace noise at or above a TWA of 85 dB. Employees, however, must be fitted with, issued, and required to wear hearing protectors whenever they are exposed to noise levels above a TWA of 85 dB for any period exceeding 6 months after their first exposure until the baseline audiogram is conducted.

Baseline audiograms taken before the hearing conservation program took effect in 1983 are acceptable if the professional supervisor determines that the audiogram is valid. Employees should not be exposed to workplace noise for 14 hours before the baseline test; to accomplish this, the workers are required to wear hearing protectors during this time period.

Employers must provide *annual* audiograms within 1 year of the baseline. It is important to test workers' hearing annually to identify deterioration in their hearing acuity as early as possible. This enables employers to initiate protective follow-up measures before hearing loss progresses. Employers must compare annual audiograms to baseline audiograms to determine whether the audiogram is valid and whether the employee has lost hearing acuity or experienced a standard threshold shift (STS). An STS is an average shift in either ear of 10 dB or more at 2,000, 3,000, and 4,000 Hz.

The employer must fit or refit any employee showing an STS with adequate hearing protectors, show the employee how to use them, and require the employee to wear them. Employers must notify employees within 21 days after the determination that their audiometric test results show an STS. Some employees with an STS may need further testing if the professional determines that their test results are questionable or if they have an ear problem thought to be caused or aggravated by wearing hearing protectors. If the suspected medical problem is not thought to be related to wearing hearing protection, the employer must advise the employee to see a physician. If subsequent audiometric tests show that the STS identified on a previous audiogram is not persistent, employees whose exposure to noise is less than a TWA of 90 dB may stop wearing hearing protectors.

The employer may substitute an annual audiogram for the original baseline audiogram if the professional supervising the audiometric program determines that the employee's

STS is persistent. The employer must retain the original baseline audiogram, however, for the length of the employee's employment. This substitution will ensure that the same shift is not repeatedly identified. The professional also may decide to revise the baseline audiogram if the employee's hearing improves. This will ensure that the baseline reflects actual hearing thresholds to the extent possible. Employers must conduct audiometric tests in a room meeting specific background levels and with calibrated audiometers that meet ANSI specifications of SC-1969.

## HEARING CONSERVATION WRITTEN PROGRAM

As with all other industrial safety requirements, the industrial hygienist must ensure that the specifics of any safety requirement be itemized and spelled out in a well-written program. Not only is it an OSHA requirement that the hearing conservation program be in writing, the industrial hygienist who works without a written program soon discovers implementing any creditable program is virtually impossible.

What information and guidelines should be included in the organization's hearing conservation program? This question is best answered by referring to OSHA's 29 CFR 1910.95 Standard, "Occupational Noise Exposure."

The introduction to the written program should include a purpose statement, one that clearly declares that protection against the effects of noise exposure will be provided when the sound levels exceed those shown in table 5.1 (when measured on the A-scale of a standard sound-level meter at slow response).

In addition to stating the purpose, the written program should contain a statement about the hearing conservation program itself and define terms pertinent to the written program. For example, a statement declaring that the hearing conservation program is designed to comply with OSHA requirements, and that a continuing, effective hearing conservation program will be administered whenever employee noise exposures equal or exceed an 8-hour TWA sound level of 85 decibels measured on the A-scale (slow response) or, equivalently, a dose of 50 percent clearly defines the perimeters of the program. For the purposes of this program, an 8-hour TWA of 85 decibels or a dose of 50 percent will also be referred to as the *action level*. At this point, the written program, along with action level, should list and define other pertinent terms defined earlier.

In your written hearing conservation program, also list or designate responsibilities—that is, who is responsible for managing and enforcing the various components in effecting compliance with the program. So that you can understand what is required, let's take a look at a sample Designation of Responsibilities.

# NOISE AND VIBRATION

Table 5.1. Permissible Noise Exposures*

| Duration per day, hours | Sound level dBA slow response |
| --- | --- |
| 8 | 90 |
| 6 | 92 |
| 4 | 95 |
| 3 | 97 |
| 2 | 100 |
| 1.5 | 102 |
| 1 | 105 |
| 1/2 | 110 |
| 1/4 or less | 115 |

*When the daily noise exposure is composed of two or more periods of noise exposure of different levels, their combined effect should be considered, rather than the individual effect of each. If the sum of the following fractions $C_1/T_1 + C_2/T_2 + C_n/T_n$ exceeds unity, then, the mixed exposure should be considered to exceed the limit value. $C_n$ indicates the total time of exposure at a specified noise level, and $T_n$ indicates the total time of exposure permitted at that level. Exposure to impulsive or impact noise should not exceed 140 dB peak sound pressure level.
*Source:* OSHA, 1995. *Occupational Safety and Health Standards for General Industry.* 29 CFR 1910.95 (Hearing Conservation Standard). Washington, D.C.: U.S. Department of Labor.

## Sample "Designation of Responsibilities"

The Hearing Conservation Program requires good direction, management, supervision and conduct at all levels within the Company. Assigned duties and responsibilities of Company personnel *should not be* delegated to subordinates.

1. Each Company Director will be responsible for implementing and ensuring compliance with Company's Hearing Conservation Program for their Department. Each Director shall establish the responsibilities for managing this program and for designating the employee classification that will perform the following:
   a. Supervise program within departmental work centers.
   b. Report potential noise hazards to Industrial Hygienist for further evaluation.
   c. Provide hearing protection devices as required.
   d. Maintain work center employee training records.
2. Company Industrial Hygienist is the Hearing Conservation Program Manager. The Industrial Hygienist will be responsible for the following:
   a. Writing and modifying, as necessary, Company's Hearing Conservation Program.
   b. Conducting noise level measurements and maintaining a current and accurate noise level measurement summary of all Company workplaces.

c. Providing training on the Hearing Conservation Program as required, including training in how to wear hearing protection devices.
d. Ensuring that otoscopic and audiometric examinations of all Company personnel who come under the Hearing Conservation Program are conducted.
e. Ensuring Hearing Conservation Records are forwarded to Company Human Resources Manager.
f. Providing work center supervisors an ongoing, current approved listing of hearing protection equipment.
3. The Company Human Resource Manager will be responsible for maintaining Employee Hearing Conservation Records, and will facilitate referral of employees requiring further medical examination by a physician.
4. Assigned Supervisors have the following responsibilities under the Company's Hearing Conservation Program:
   a. Reporting to Industrial Hygienist any installation or removal of equipment that might alter the noise level within designated work centers.
   b. Ensuring that hearing protection devices, as prescribed by Industrial Hygienist, are available to employees.
   c. Ensuring hearing protection devices are utilized as required.
   d. Maintaining current work center training records on Hearing Conservation Program training of employees.
   e. Ensuring that all new employees receive a baseline audiogram within six months of hire date.
   f. Ensuring proper hazard labeling practices for all noise hazard areas.
   g. Ensuring that suitable hearing protection is provided to work center visitors.
5. Company personnel will be responsible for:
   a. Familiarizing themselves with the Hearing Conservation Program.
   b. Wearing hearing protection devices as required.

## Monitoring: Sound Level Survey

The hearing conservation program begins with noise monitoring and sound level surveys. Common sense dictates that if a workplace noise hazard is not identified, it will probably be ignored—and no attempt at protecting workers' hearing will be made.

According to OSHA, when information indicates that any employee's exposure equals or exceeds an 8-hour TWA of 85 decibels, the employer must develop and implement a *monitoring program*. The responsibility for noise monitoring is typically assigned to the organization industrial hygienist.

## Additional OSHA Monitoring Procedural Requirements

The noise monitoring protocol that is to be followed includes fashioning a sampling strategy designed to (a) identify employees for inclusion in the hearing conservation program and (b) to enable the proper selection of hearing protectors.

If circumstances (such as high worker mobility, significant variations in sound level, or a significant component of impulse noise) make area monitoring generally inappropriate, the employer is required to use representative personal sampling to comply with the monitoring requirements, unless the employer can show that area sampling produces equivalent results.

All continuous intermittent and impulsive sound levels from 80 decibels to 130 decibels must be integrated into the noise measurements.

Instruments used to measure employee noise exposure must be calibrated to ensure measurement accuracy.

Monitoring must be repeated whenever a change in production, process, equipment, or controls increases noise exposures to the extent that additional employees may be exposed at or above the action level, or the attenuation provided by hearing protectors being used by employees might be rendered inadequate.

The employer is required to notify each employee of the results of the monitoring if they are exposed at or above an 8-hour TWA of 85 decibels.

The employer is required to provide affected employees or their representatives with an opportunity to observe any noise measurements conducted.

## Audiometric Testing

*Audiometric testing* is an important element of the hearing conservation program for two reasons: First, it helps to determine the effectiveness of hearing protection and administrative and/or engineering controls. Second, audiometric surveillance helps to detect hearing loss before it noticeably affects the employee and before the loss becomes legally compensable under workers' compensation. Audiometric examinations are usually done by an outside contractor, but can be done in-house with the proper equipment. Wherever they are done, they require properly calibrated equipment used by a trained and certified audiometric technician.

The importance of audiometric evaluations cannot be overstated. Not only do they satisfy the regulatory requirement, but they also work to tie the whole program together. One thing is certain—if the hearing conservation program is working, employees' audiometric results will not show changes associated with on-the-job noise-induced hearing damage. If suspicious hearing changes are found, the audiometric technician and the audiologist who reviews the record can counsel the employee to wear hearing protection devices more carefully, can assess whether better hearing protection devices are needed, and can use the test results to point out to the employee the need to be more careful in protecting his or her hearing—both on and off the job.

The organizational industrial hygienist needs to ensure that designation of audiometric evaluation procedures is included in the written hearing conservation program. A sample written procedure is included in the following.

### Sample "Designation of Audiometric Evaluation Procedures"

The Hearing Conservation Program requires audiometric evaluation for all Company employees who come under this program.

1. The Industrial Hygienist will ensure that otoscopic and audiometric examinations are conducted.
2. A baseline audiogram will be required of all employees within six months of hire date.
    a. Testing will be conducted after 14 hours of nonexposure to workplace noise.
    b. Testing will be performed by a certified audiometric technician, using a calibrated (annual requirement) audiometer in an environment of less than 50 dB(A) background noise.
3. Follow-up audiograms are required yearly or within 60 days of Temporary or Permanent Threshold Shift.
4. If follow-up audiogram suggests Permanent Threshold Shift rather than Temporary Threshold shift, employee will be referred to physician for evaluation as to whether damage is presbycusis or sensorineural.
5. All cases of occupationally related hearing loss must be recorded on the OSHA 300 form.
6. Industrial Hygienist will review results of Temporary or Permanent Threshold Shifts to determine if cause was work-related or other-than-work-related.
7. Industrial Hygienist will forward the employee Hearing Conservation Record to the Human Resource Manager for inclusion in the employee's medical record.

### Hearing Protection

The *hearing protection* element of the hearing conservation program provides hearing protection devices for employees and training on how to wear them effectively, as long as hazardous noise levels exist in the workplace. Hearing protection comes in various sizes, shapes, and materials, and the cost of this equipment can vary dramatically. Two general types of hearing protection are used widely in industry: the cup muff (commonly called Mickey Mouse Ears) and the plug insert type. Because feasible engineering noise controls have not been developed for many types of industrial equipment, hearing protection devices are the best option for preventing noise-induced hearing loss in these situations.

As with the other elements of the hearing conservation program, the hearing protective device element must be in writing and included in the Hearing Conservation Program. A sample designation of hearing protection devices is presented in the following.

## Sample "Designation of Hearing Protection Devices"

The Hearing Conservation Program requires providing hearing protection devices for employees in designated Noise Hazard Areas.

1. After audiometric examination, the Industrial Hygienist will fit each employee with the proper hearing protection devices. Hearing protection devices of two types will be made available to the employees: Earmuff and/or plug types. If the employee has demonstrated a TTS, the Industrial Hygienist will issue double hearing protection (plugs and muffs worn at the same time) for those employees who work in areas that exceed 104 dBA. A complete listing of each employee's hearing protection requirements will be provided in writing to the work center supervisor.
2. The work center Supervisor is responsible for ensuring that employees wear approved hearing protection devices in designated areas.

## Training

For a hearing conservation program (or any other safety program) to be effective, the participants in the program must be trained. OSHA requires that the employer include this important element in the written Hearing Conservation Program. The training program must be repeated annually for each employee included in the program. The industrial hygienist needs to ensure that the information included in the training program is current. In addition, information must be provided that includes informing the employees of the effects of noise on hearing; the purpose of hearing protectors; the advantages, disadvantages, and attenuation of various types of hearing protectors; and instructions on selection, fitting, use, and care of hearing protectors. The purpose of audiometric testing and an explanation of the test procedures must also be included.

To facilitate compliance with all regulatory standards and the company's safety and health requirements (including the hearing conservation program), organizational management and the industrial hygienist should ensure that emphasis for compliance is made *a condition of employment.* Remember—ensure and document employee participation.

## Safe Work Practices

*Safe work practices* are an important element in the hearing conservation program. Written safe work practices for hearing conservation should focus on relaying noise hazard information to the employee. For instance, if an employee is required to perform some kind of maintenance function in a high noise hazard area, the written procedure for doing the maintenance should include a statement that warns the employee about the noise hazard and lists the personal protective devices that he or she should use. Experience has shown that when such warnings (safe work practices) are placed in preventive maintenance procedures (e.g., noise hazard area, confined space, lockout/tagout required), not only is the program much more efficient, the repeated reminder also helps workers to maintain compliance with regulatory standards.

## Recordkeeping

Under OSHA's 29 CFR 1910.95 (Hearing Conservation Standard), the employer is required to keep and maintain certain records. Along with an accurate record of all employee exposure measurements, the employer is required to retain all employee audiometric test records. Audiometric test records must include:

- name and job classification of the employee;
- date of the audiogram;
- examiner's name;
- date of the last acoustic or exhaustive calibration of the audiometer; and
- employee's most recent noise exposure assessment.

In addition, the employer must maintain accurate records of the measurements of the background sound pressure levels in audiometric test rooms.

The employer is required to retain records of noise exposure measurement for two years. Audiometric test records must be retained for the duration of the affected employee's employment. Employee noise exposure records must be made available to employees whenever they request them. Also, whenever an employee is transferred, the employer is required to transfer the records to the employee's successive employer.

## Administrative and Engineering Controls

Another important element that must be included in any hearing conservation program is *administrative* and *engineering controls*. Administrative controls, simply stated, involve controlling the employee's exposure to noise. If a certain work area has a noise source that exceeds safe exposure levels, the employee is allowed within such a space only up to the time in which he or she has reached their maximum allowed TWA exposure limit. For example, if the noise hazard area consistently produces noise at the 100-dBA level, the employee would only be allowed in such an area up to 2 hours per 8-hour shift. *Note:* A word of caution is advised here—keep in mind that we are referring to an employee who has *no* recorded hearing loss. If an employee has suffered permanent hearing loss, then his or her time exposure at such high noise levels should be significantly reduced. Under no circumstance should the employee with documented hearing loss be exposed to high noise hazards without proper hearing protection.

In hearing conservation, engineering controls play a vital role in providing the level of protection employees need. Again, the preferred hazard control method is to "engineer out" the hazard. Not only should existing equipment be evaluated for possible engineering control applications, new equipment should be evaluated for noise emissions before purchase.

Engineering controls used in controlling hazardous noise levels can be accomplished at the source of the noise through preventive maintenance, speed reduction, vibration isolation, mufflers, enclosures, and substitution of machines. In the air (noise) path, engineering controls such as absorptive material, sound barriers, and increasing the distance between the source and the receiver can be employed. At the receiver, the best engineering control is to enclose and isolate the employee from the noise hazard.

# FREQUENTLY ASKED QUESTIONS ON HEARING LOSS PREVENTION (NIOSH, 2005)

**Q** Don't we lose our hearing as we age?

**A** It's true that most people's hearing worsens as they get older. But for the average person, aging does not cause impaired hearing before at least the age of 60. People who are not exposed to noise and are otherwise healthy keep their hearing for many years. People who are exposed to noise and do not protect their hearing begin to lose their hearing at an early age. For example, by age 25 the average carpenter has "50-year-old" ears. That is, by age 25, the average carpenter has the same hearing as someone who is 50 years old and has worked in a quiet job.

**Q** Can you poke out your eardrums with earplugs?

**A** That is unlikely for two reasons. First, the average ear canal is about 1¼ inches long. The typical ear plug is between ½ to ¾ of an inch long. So even if you inserted the entire earplug, it would still not touch the eardrum. Second, the path from the opening of the ear canal to the eardrum is not straight. In fact, it is quite irregular. This prevents you from poking objects into the eardrum.

**Q** We work in a dusty, dirty place. Should I worry that our ears will get infected by using earplugs?

**A** Using earplugs will not cause an infection. But use common sense. Have clean hands when using earplugs that need to be rolled or formed with your fingers in order for you to insert them. If this is inconvenient, there are plenty of earplugs that are pre-molded or that have stems so that you can insert them without having to touch the part that goes into the ear canal.

**Q** Can you hear warning sounds, such as backup beeps, when wearing hearing protectors?

**A** The fact is that there are fatal injuries because people do not hear warning sounds. However, this is usually because the background noise was too high or because the person had severe hearing loss, not because someone was wearing hearing protectors. Using hearing protectors will bring both the noise and the warning sound down equally. So if the warning sound is audible without the hearing protection, it will usually be audible when wearing the hearing protector. For the unusual situations where this is not the case, the solution may be as simple as using a different type of hearing protector. Also, many warning systems can be adjusted or changed so warning signals are easier to detect.

**Q** Won't hearing protectors interfere with our ability to hear important sounds our machinery and equipment make?

**A** Hearing protectors will lower the noise level of your equipment; it won't eliminate it. However, some hearing protectors will reduce certain frequencies more than others, so wearing them can make noises sound different. In cases where it's important that the sound just be quieter without any other changes, there are hearing protectors that can provide flat attenuation.

There are also noise-activated hearing protectors that allow normal sounds to pass through the ear and only "turn-on" when the noise reaches hazardous levels. There are even protectors that professional concert musicians use that can lower the sound level while retaining sound fidelity.

**Q** Will we be able to hear each other talk when wearing hearing protectors?

**A** Some people find they can wear hearing protectors and still understand speech. Others will have trouble hearing speech while wearing hearing protectors. Being able to hear what other people say depends on many things: distance from the speaker, ability to see the speaker's face, general familiarity with the topic; level of background noise, and whether or not one has an existing hearing impairment. In some cases, wearing hearing protectors can make it easier to understand speech.

In other instances, people may be using hearing protectors to keep out too much sound. You may need a protector that reduces the sound enough to be safe without reducing the sound too much to hear speech at a comfortably loud level. For those people who work in noise and must communicate, it may also be necessary to use communication headsets. Allow your employees to try different protectors. Some will work better than others at helping them to hear speech, and different protectors may work better for different people.

Q  How long does it take to get used to hearing protectors?

A  Think about getting a new pair of shoes. Some shoes take no time to get used to. Others—even though they are the right size—can take a while to get used to. Hearing protectors are no different from other safety equipment in terms of getting used to them. But if hearing protectors are the wrong size, or are worn out, they will not be comfortable. Also, workers may need more than one kind of protector at their job. For example, no one would wear golf shoes to go bowling. If hearing protectors are not suitable for the work being done, they probably won't feel comfortable.

Q  How long can someone be in a loud noise before it is hazardous?

A  The degree of hearing hazard is related to both the level of the noise as well as to the duration of the exposure. But this question is like asking how long can people look at the sun without damaging their eyes. The safest thing to do is to ensure workers always protect their ears by wearing hearing protectors anytime they are around loud noise.

Q  How can I tell if a noise situation is too loud?

A  There are two rules: First, if you have to raise your voice to talk to someone who is an arm's length away, then the noise is likely to be hazardous. Second, if your ears are ringing or sounds seem dull or flat after leaving a noisy place, you probably were exposed to hazardous noise.

Q  How often should your hearing be tested?

A  Anyone regularly exposed to hazardous noise should have an annual hearing test. Also, anyone who notices a change in his or her hearing (or who develops tinnitus) should have his or her ears checked. People who have healthy ears and who are not exposed to hazardous noise should get a hearing test every three years.

Q  Since I already have hearing loss and wear a hearing aid, hearing prevention programs don't apply to me, right?

A  If you have hearing loss, it's important to protect the hearing that you have left. Loud noises can continue to damage your hearing, making it even more difficult to communicate at work and with your family and friends.

## HEARING CONSERVATION PROGRAM ANNUAL EVALUATION

### Hearing Conservation Program Review

After completion of this review, any deficiencies must be corrected as soon as possible. This copy should be maintained in the Company's Hearing Conservation Program for at least two years.

Industrial Hygienist: _____

Date Completed: _____

### Training and Education

Failures or deficiencies in hearing conservation programs (hearing loss prevention programs) can often be traced to inadequacies in the training and education of noise-exposed employees and those who conduct elements of the program.

YES/NO  Has training been conducted at least once a year?

YES/NO  Was the training provided by a qualified instructor?

YES/NO  Was the success of each training program evaluated?

YES/NO  Is the content revised periodically?

YES/NO  Are managers and supervisors directly involved?

YES/NO  Are posters, regulations, handouts, and employee newsletters used as supplements?

YES/NO  Are personal counseling sessions conducted for employees having problems with hearing protection devices or showing hearing threshold shifts?

### Supervisor Involvement

Data indicate that employees who refuse to wear hearing protectors or who fail to show up for hearing tests frequently work for supervisors who are not totally committed to the hearing loss prevention programs.

YES/NO  Have supervisors been provided with the knowledge required to supervise the use and care of hearing protectors by subordinates?

# NOISE AND VIBRATION

YES/NO  Do supervisors wear hearing protectors in appropriate areas?

YES/NO  Have supervisors been counseled when employees resist wearing protectors or fail to show up for hearing tests?

YES/NO  Are disciplinary actions enforced when employees repeatedly refuse to wear hearing protectors?

## Noise Measurement

For noise measurements to be useful, they need to be related to noise exposure risks or the prioritization of noise control efforts, rather than merely filed away. In addition, the results need to be communicated to the appropriate personnel, especially when follow-up actions are required.

YES/NO  Were the essential/critical noise studies performed?

YES/NO  Was the purpose of each noise study clearly stated? Have noise-exposed employees been notified of their exposures and apprised of auditory risks?

YES/NO  Are the results routinely transmitted to supervisors and other key individuals?

YES/NO  Are results entered into health/medical records of noise exposed employees?

YES/NO  Are results entered into shop folders?

YES/NO  If noise maps exist, are they used by the proper staff?

YES/NO  Are noise measurement results considered when contemplating procurement of new equipment? Modifying the facility? Relocating employees?

YES/NO  Have there been changes in areas, equipment, or processes that have altered noise exposure? Have follow-up noise measurements been conducted?

YES/NO  Are appropriate steps taken to include (or exclude) employees in the hearing loss prevention programs whose exposures have changed significantly?

## Engineering and Administrative Controls

Controlling noise by engineering and administrative methods is often the most effective means of reducing or eliminating the hazard. In some cases engineering controls will remove requirements for other components of the program, such as audiometric testing and the use of hearing protectors.

YES/NO  Have noise control needs been prioritized?

YES/NO  Has the cost-effectiveness of various options been addressed?

YES/NO Are employees and supervisors apprised of plans for noise control measures? Are they consulted on various approaches?

YES/NO Do we have to hire outside consultants to execute the plans?

YES/NO Have employees and supervisors been counseled on the operation and maintenance of noise control devices?

YES/NO Are noise control projects monitored to ensure timely completion?

YES/NO Has the full potential for administrative controls been evaluated? Are noisy processes conducted during shifts with fewer employees? Do employees have sound-treated lunch or break areas?

**Monitoring Audiometry and Recordkeeping**

The skills of audiometric technicians, the status of the audiometer, and the quality of audiometric test records are crucial to hearing loss prevention program success. Useful information may be ascertained from the audiometric records as well as from those who actually administer the tests.

YES/NO Has the audiometric technician been adequately trained, certified, and recertified as necessary?

YES/NO Do on-the-job observations of the technicians indicate that they perform a thorough and valid audiometric test, instruct and consult the employee effectively, and keep appropriate records?

YES/NO Are records complete?

YES/NO Are follow-up actions documented?

YES/NO Are hearing threshold levels reasonably consistent from test to test? If not, are the reasons for inconsistencies investigated promptly?

YES/NO Are the annual test results compared to baseline to identify the presence of an OSHA standard threshold shift?

YES/NO Is the annual incidence of standard threshold shift greater than a few percent? If so, are problem areas pinpointed and remedial steps taken?

YES/NO Are audiometric trends (deteriorations) being identified, both in individuals and in groups of employees? (NIOSH recommends no more than 5% of workers showing 15 dB Significant Threshold Shift, same ear, and same frequency.)

# NOISE AND VIBRATION

YES/NO  Do records show that appropriate audiometer calibration procedures have been followed?

YES/NO  Is there documentation showing that the background sound levels in the audiometer room were low enough to permit valid testing?

YES/NO  Are the results of audiometric tests being communicated to supervisors and managers as well as to employees?

YES/NO  Has corrective action been taken if the rate of no-shows for audiometric test appointments is more than about 5%?

YES/NO  Are employees incurring STS notified in writing within at least 21 days? (NIOSH recommends immediate notification if retest shows 15 dB STS, same ear, and same frequency.)

**Referrals**

Referrals to outside sources for consultation or treatment are sometimes in order, but they can be an expensive element of the hearing loss prevention program, and should not be undertaken unnecessarily.

YES/NO  Are referral procedures clearly specified?

YES/NO  Have letters of agreement between the company and consulting physicians for audiologists been executed?

YES/NO  Have mechanisms been established to ensure that employees needing evaluation or treatment actually receive the service (i.e., transportation, scheduling, and reminders)?

YES/NO  Are records properly transmitted to the physician or audiologist, and back to the company?

YES/NO  If medical treatment is recommended, does the employee understand the condition requiring treatment, the recommendation, and methods of obtaining such treatment?

YES/NO  Are employees being referred unnecessarily?

**Hearing Protection Devices**

When noise control measures are infeasible, or until such time as they are installed, hearing protection devices are the only way to prevent hazardous levels of noise form damaging the inner ear. Making sure that these devices are worn effectively requires continuous attention on the part of supervisors and program implementors as well as noise-exposed employees.

YES/NO  Have hearing protectors been made available, or until such time as they are installed, hearing protection devices are the only way to prevent hazardous levels of noise from damaging the inner ear. Making sure that these devices are worn effectively requires continuous attention on the part of supervisors and program implementors as well as noise-exposed employees.

YES/NO  Are employees given the opportunity to select from a variety of appropriate protectors?

YES/NO  Are employees fitted carefully with special attention to comfort?

YES/NO  Are employees thoroughly trained, not only initially but at least once a year?

YES/NO  Are the protectors checked regularly for wear or defects, and replaced immediately if necessary?

YES/NO  If employees use disposable hearing protectors, are replacements readily available?

YES/NO  Do employees understand the appropriate hygiene requirements?

YES/NO  Have any employees developed ear infections or irritations associated with the use of hearing protectors? Are there any employees who are unable to wear these devices because of medical conditions? Have these conditions been treated promptly and successfully?

YES/NO  Have alternative types of hearing protectors been considered when problems with current devices are experienced?

YES/NO  Do employees who incur noise-induced hearing loss receive intensive counseling?

YES/NO  Are those who fit and supervise the wearing of hearing protectors competent to deal with the many problems that can occur?

YES/NO  Do workers complain that protectors interfere with their ability to do their jobs? Do they interfere with spoken instructions or warning signals? Are these complaints followed promptly with counseling, noise control, or other measures?

YES/NO  Are employees encouraged to take their hearing protectors home if they engage in noisy nonoccupational activities?

YES/NO  Are new types of or potentially more effective protectors considered as they become available?

# NOISE AND VIBRATION

YES/NO  Is the effectiveness of the hearing protector program evaluated regularly?

YES/NO  Have at-the-ear protection levels been evaluated to ensure that either over- or under-protection has been adequately balanced according to the anticipated ambient noise levels?

YES/NO  Is each hearing protector user required to demonstrate that he or she understands how to use and care for the protector? The results documented?

**Administrative**

Keeping organized and current on administrative matters will help the program run smoothly.

YES/NO  Have there been any changes in federal or state regulations? Have hearing loss prevention program policies been modified to reflect these changes?

YES/NO  Are copies of company policies and guidelines regarding the hearing loss prevention program available in the offices that support the various program elements? Are those who implement the program elements aware of these policies? Do they comply?

YES/NO  Are necessary materials and supplies being ordered with a minimum of delay?

YES/NO  Are procurement officers overriding the hearing loss prevention program implementor's requests for specific hearing protection or other hearing loss prevention equipment? It so, have corrective steps been taken?

YES/NO  Is the performance of key personnel evaluated periodically? If such performance is found to be less than acceptable, are steps taken to correct the situation?

YES/NO  Regarding safety, has the failure to hear warning shouts or alarms been tied to any accidents or injuries? If so, have remedial steps been taken?

## NOISE UNITS, RELATIONSHIPS, AND EQUATIONS

1. **Sound power** (w) of a source is the total sound energy radiated by the source per unit time. It is expressed in terms of the sound power level ($L_w$) in decibels referenced to $10^{-12}$ watts ($w_0$). The relationship to decibels is shown below:

$$L_w = 10 \log w/w_0$$

where:

$L_w$ = sound power level (decibels)
w = sound power (watts)
$w_0$ = reference power ($10^{-12}$ watts)
log = a logarithm to the base 10

2. Units used to describe **sound pressures** are

$$1 \text{ μbar} = 1 \text{ dyne/cm}^2 = 0.1 \text{ N/cm}^2 = 0.1 \text{ Pa}$$

3. **Sound pressure level** or SPL = 10 log $p^2/p_0$
where:

SPL = sound pressure level (decibels)
p = measured root-mean-square (rms) sound pressure (N/m², μbars). Root-mean-square (rms) value of a changing quantity, such as sound pressure, is the square root of the mean of the squares of the instantaneous values of the quantity.
$p_0$ = reference rms sound pressure 20 μPa, N/m², μbars)

4. **Speed of sound** (c) = c = fλ

5. **Wavelength** (λ) = c/f

6. Calculation of **frequency of octave bands** can be calculated using the following formulae:

**Upper frequency band**: $f_2 = 2f_1$

where:

$f_2$ = upper frequency band
$f_1$ = lower frequency band

**One-half octave band**: $f_2 = \sqrt{2}(f_1)$

where:

$f_2$ = ½ octave band
$f_1$ = lower frequency band

**One-third octave band**: $f_2 = \sqrt[3]{2}(f_1)$

where:

$f_2$ = ⅓ octave band
$f_1$ = lower frequency band

# NOISE AND VIBRATION

7. Formula for **adding noise sources**, when sound power is known:

$$L_w = 10 \log (w_1 + w_2)/(w_0 + w_0)$$

where:

$L_w$ = sound power in watts
$w_1$ = sound power of noise source 1 in watts
$w_2$ = sound power of noise source 2 in watts
$w_0$ = reference sound power (reference $10^{-12}$) watts

8. Formula for **sound pressure additions**, when sound pressure is known:

$$SPL = 10 \log p^2/p_0^2$$

where:

$$p^2/p_0^2 = 10^{SPL/10}$$

and:

SPL = sound pressure level (decibels)
p = measured root-mean-square (rms) sound pressure (N/m², µbars)
$p_0$ = reference rms sound pressure (20 µPa, N/m², µbars)

For three sources, the equation becomes:

$$SPL = 10 \log (10^{SPL_1/10}) + (10^{SPL_2/10}) + (10^{SPL_3/10})$$

When adding any number of sources, whether the sources are identical or not, the equation becomes:

$$SPL = 10 \log (10^{SPL_1/10} \ldots + 10^{SPL_n/10})$$

Determining the sound pressure level from multiple identical sources:

$$SPL_f = SPL_i + 10 \log n$$

where:

$SPL_f$ = total sound pressure level (dB)
$SPL_i$ = individual sound pressure level (dB)
n = number of identical sources

9. The equation for determining **noise levels in a free field** is expressed as:

$$SPL = L_w - 20 \log r - 0.5$$

where:

SPL = sound pressure (reference 0.00002 N/m²)
$L_w$ = sound power (reference $10^{-12}$ watts)
r = distance in feet

10. Calculation for **noise levels with directional characteristics** is expressed as:

$$SPL = L_w - 20 \log r - 0.5 + \log Q$$

where:

SPL = sound pressure (reference 0.00002 N/m²)
$L_w$ = sound power (reference $10^{-12}$ watts)
r = distance in feet
Q = directivity factor
    Q = 2, for one reflecting plane
    Q = 4, for two reflecting planes
    Q = 8, for three reflecting planes

11. Calculating the **noise level at a new distance** from the noise source can be computed as follows:

$$SPL = SPL_1 + 20 \log (d_1)/(d_2)$$

where:

SPL = sound pressure level at new distance ($d_2$)
$SPL_1$ = sound pressure level at $d_1$
$d_n$ = distance from source

12. Calculating **Daily Noise Dose** can be accomplished using the following formula, which combines the effects of different sound pressure levels and allowable exposure times.

$$\text{Daily Noise Dose} = \frac{C_1}{T_1} + \frac{C_2}{T_2} + \frac{C_3}{T_3} \ldots + \frac{C_n}{T_n}$$

where:

$C_i$ = number of hours exposed at given $SPL_i$
$T_i$ = number of hours exposed at given $SPL_i$

13. Formula for calculating **OSHA Permissible Noise Levels**:

$$T_{SPL} = 8/2^{(SPL-90)/5}$$

where:

$T_{SPL}$ = time in hours at given SPL
SPL = sound pressure level (dBA)

14. Formula for converting noise dose measurements to the **equivalent eight-hour TWA**:

$$TWA_{eq} = 90 + 16.61 \log (D)/(100)$$

where:

$TWA_{eq}$ = eight-hour equivalent TWA in dBA
D = noise dosimeter reading in %

## INDUSTRIAL VIBRATION CONTROL

*Vibration* is often closely associated with noise, but is frequently overlooked as a potential occupational health hazard. Vibration is defined as the oscillatory motion of a system around an equilibrium position. The system can be in a solid, liquid, or gaseous state, and the oscillation of the system can be periodic or random, steady state or transient, continuous or intermittent (NIOSH, 1973). Vibrations of the human body (or parts of the human body) are not only annoying, they also affect worker performance, sometimes causing blurred vision and loss of motor control. Excessive vibration can cause trauma when external vibrating forces accelerate some part or all of the body so that amplitudes and restraining capacities by tissues are exceeded.

These two types of vibration are called *whole-body vibration* (affects, e.g., vehicle operators) and *segmental vibration* (occurs, e.g., in foundry operations, mining, stonecutting, and a variety of assembly operations). Vibration originates from mechanical motion, generally occurring at some machine or series of machines. This mechanical vibration can be transmitted directly to the body or body part or it may be transmitted through solid objects to a worker located at some distance away from the actual vibration.

The effect of vibration on the human body is not totally understood; however, we do know that vibration of the chest may create breathing difficulties, and that an inhibition of tendon reflexes is a result of vibration. Excessive vibration can cause reduced ability on the part of the worker to perform complex tasks, and indications of potential damage to other systems of the body also exist.

More is known about the results of segmental vibration (typically transmitted through hand to arm), and a common example is the vibration received when using a pneumatic hammer (jackhammer). One recognized effect of segmental vibration is impaired circulation, a condition known as *Raynaud's Syndrome*, or "dead fingers" or "white fingers." Segmental vibration can also result in the loss of the sense of touch in the affected area. Some evidence suggests that decalcification of the bones in the hand can result from vibration transmitted to that part of the body. Muscle atrophy has also been identified as a result of segmental vibration.

As with noise, the human body can withstand short-term vibration, even though this vibration might be extreme. The dangers of vibration are related to certain frequencies that are resonant with various parts of the body. Vibration outside these frequencies is not nearly as dangerous as vibration that results in resonance.

Control measures for vibration include substituting some other device (one that does not cause vibration) for the mechanical device that causes the vibration. An important corrective measure (often overlooked) that helps in reducing vibration is proper maintenance of tools, or support mechanisms for tools, including coating the tools with materials that attenuate vibrations. Another engineering control often employed to reduce vibration is the application of balancers, isolators, and dampening devices/materials that help to reduce vibration.

## REFERENCES

LaBar, G., 1989. "Sound Policies for Protecting Workers' Hearing," *Occupational Hazards* (July): 46.

Mansdorf, S. Z., 1993 *Complete Manual of Industrial Hygiene.* New York: Prentice Hall.

NIOSH, 1973. *The Industrial Environment: In Evaluation and Control.* Cincinnati, Ohio: National Institute for Occupational Safety and Health.

NIOSH, 2005. *Common Hearing Loss Prevention Terms.* National Institute for Occupational Safety and Health. www.cdc. gov/nioxh/hpterms.html (accessed July 2005).

OSHA, 2002. *Hearing Conservation.* 3074. Washington, D.C.: U.S. Department of Labor.

## SUGGESTED READING

Centers for Disease Control. *Noise and Hearing Loss Prevention.* www.cdc.gov/niosh/topics/noise/faq/faq.html (accessed July 2005).

Gasaway, D. C. *Hearing Conservation: A Practical Manual and Guide.* Englewood Cliffs, NJ: Prentice-Hall, 1985.

Royster, J. D. and L. H. Royster. *Hearing Conservation Programs: Practical Guidelines for Success.* Chelsea, Michigan: Lewis Publishers, 1990.

# 6
# Radiation

The type of radiation most of us are familiar with, used by the nuclear power industry and in nuclear weapons, is *ionizing radiation*. Very few people have difficulty in recognizing the potential destructive power of this type of radiation. However, fewer individuals are aware of another type of radiation, *nonionizing radiation*, which we are exposed to each day. Even fewer people can differentiate between the two types. The industrial hygienist must be familiar with the nature of radiation and understand the detection of radiation, permissible exposure limits, biological effects of radiation, monitoring techniques, and control measures and procedures.

## RADIATION SAFETY PROGRAM ACRONYMS AND TERMS

The following listing of acronyms and definitions that are typically included in radiation safety programs is adapted from the *Radiation Safety Manual* of the U.S. Department of Health and Human Services Public Health Service Centers for Disease Control and Prevention (1999).

### Abbreviations

**ALARA**: As low as reasonably achievable
$^3$**H**: Tritium (hydrogen-3)
**10 CFR 19**: NRC's Title 10, Chapter 1, Code of Federal Regulations, Part 19
**10 CFR 20**: NRC's Title 10, Chapter 1, Code of Federal Regulations, Part 20
$^{14}$**C**: Carbon-14

$^{32}$P: Phosphorous-32

$^{33}$P: Phosphorous-33

$^{35}$S: Sulfur-35

$^{51}$Cr: Chromium-51

$^{60}$Co: Cobalt-60

$^{125}$I: Iodine-125

$^{129}$I: Iodine-129

$^{131}$I: Iodine-131

$^{137}$CS: Iodine-137

**ALI**: annual limit on intake

**AU**: authorized uUser

**CDC**: Centers for Disease Control and Prevention

**Ci**: Curie

**cm2**: square centimeters

**cpm**: counts per minute

**DAC**: derived air concentration

**dpm**: disintegrations per minute

**GM**: Geiger-Muller

**kg**: kilogram

**lfm**: linear feet per minute

**LSC**: liquid scintillation counter

**mCi**: milliCurie

**MeV**: mega electron-volts

**ml**: milliliters

**mrem**: millirem (0.001 rem)

**NaI**: Sodium Iodide

**NRC**: Nuclear Regulatory Commission

**OHC**: Occupational Health Clinic

**OHS**: Office of Health and Safety

**PSA**: Physical Security Activity

**RIA**: radioimmunoassay

**RSC**: Radiation Safety Committee

**RSO**: Radiation safety officer

**TLD**: Thermoluminescent dosimeter

## Terms

**absorbed dose**: The energy imparted by ionizing radiation per unit mass of irradiated material. The units of absorbed dose are the rad and the gray (Gy).

**activity**: The rate of disintegration (transformation) or decay of radioactive material. The units of activity are the curie (Ci) and the Becquerel (Bq).

**alpha particle**: A strongly ionizing particle emitted form the nucleus of an atom during radioactive decay, containing 2 protons and neutrons and having a double positive charge.

**alternate authorized user**: Serves in the absence of the authorized user and can assume any duties as assigned.

**authorized user**: An employee who is approved by the RSO and RSC and is ultimately responsible for the safety of those who use radioisotopes under his or her supervision.

**beta particle**: An ionizing charge particle emitted from the nucleus of an atom during radioactive decay, equal in mass and charge to an electron.

**bioassay**: The determination of kinds, quantities, or concentrations, and, in some cases, the locations of radioactive material in the human body, whether by direct measurement (in vivo counting) or by analysis and evaluation of materials excreted or removed from the human body.

**biological half-life**: The length of time required for one-half of a radioactive substance to be biologically eliminated from the body.

**bremsstrahlung**: Electromagnetic (x-ray) radiation associated with the deceleration of charged particles passing through matter.

**contamination**: The deposition of radioactive material in any place where it is not wanted.

**controlled area**: An area, outside of a restricted area but inside the site boundary, access to which can be limited by the licensee for any reason.

**counts per minute (cpm)**: The number of nuclear transformations from radioactive decay able to be detected by a counting instrument in a one minute time interval.

**curie (Ci)**: A unit of activity equal to 37 billion disintegrations per second.

**declared pregnant woman**: A woman who has voluntarily informed her employer, in writing, of her pregnancy and the estimated date of conception.

**disintegrations per minute (dpm)**: The number of nuclear transformations from radioactive decay in a one-minute time interval.

**dose equivalent**: A quantity of radiation dose expressing all radiation on a common scale for calculating the effective absorbed dose. The units of dose equivalent are the rem and sievert (SV).

**dosimeter**: A device used to determine the external radiation dose a person has received.

**effective half-life**: The length of time required for a radioactive substance in the body to lose one-half of its activity present through a combination of biological elimination and radioactive decay.

**exposure:** The amount of ionization in air from x-rays and gamma rays.

**extremity:** Hand, elbow, arm below the elbow, foot, knee, or leg below the knee.

**gamma rays:** Very penetrating electromagnetic radiations emitted from a nucleus and an atom during radioactive decay.

**half-life:** The length of time required for a radioactive substance to lose one-half of its activity by radioactive decay.

**limits:** Dose limits; the permissible upper bounds of radiation doses.

**permitted worker:** A laboratory worker who does not work with radioactive materials but works in a radiation laboratory.

**photon:** A type of radiation in the form of an electromagnetic wave.

**rad:** A unit of radiation absorbed dose. One rad is equal to 100 ergs per gram.

**radioactive decay:** The spontaneous process of unstable nuclei in an atom disintegrating into stable nuclei, releasing radiation in the process.

**radiation:** Ionizing radiation—alpha particles, beta particles, gamma rays, X-rays, neutrons, high-speed electrons, high-speed protons, and other particles capable of producing ions.

**radiation workers:** Those personnel authorized (as listed on the Authorized User Form) to conduct work with radioactive materials.

**radioisotope:** A radioactive nuclide of a particular element.

**rem:** A unit of dose equivalent. One rem is approximately equal to one rad of beta, gamma, or x-ray radiation, or 1/20 of alpha radiation.

**restricted area:** An area to which access is limited by the licensee for the purpose of protecting individuals against undue risks from exposure to radiation and radioactive materials.

**roentgen:** A unit of radiation exposure. One roentgen is equal to 0.00025 Coulombs of electrical charge per kilogram of air.

**thermoluminescent dosimeter (TLD):** Dosimeter worn by radiation workers to measure their radiation dose. The TLD contains crystalline material which stores a fraction of the absorbed ionizing radiation and releases this energy in the form of light photons when heated.

**total effective dose equivalent (TEDE):** The sum of the deep-dose equivalent (for external exposures) and the committed effective dose equivalent (for internal exposures).

**unrestricted area:** An area to which access is neither limited nor controlled by the licensee.

**X-rays:** A penetrating type of photon radiation emitted from outside the nucleus of a target atom during bombardment of a metal with fast electrons.

## IONIZING RADIATION

*Ionization* is the process by which atoms are made into ions by the removal or addition of one or more electrons; they produce this effect by the high kinetic energies of the quanta (discrete pulses) they emit. Simply, ionizing radiation is any radiation capable of producing ions by interaction with matter. Direct ionizing particles are charged particles (e.g., electrons, protons, alpha particles) having sufficient kinetic energy to produce ionization by collision. Indirect ionizing particles are uncharged particles (e.g., photons, neutrons) that can liberate direct ionizing particles. Ionizing radiation sources can be found in a wide range of occupational settings, including health care facilities, research institutions, nuclear reactors and their support facilities, nuclear weapon production facilities, and various other manufacturing settings, just to name a few.

These ionizing radiation sources can pose a considerable health risk to affected workers if not properly controlled. Ionization of cellular components can lead to functional changes in the tissues of the body. Alpha, beta, neutral particles, x-rays, gamma rays, and cosmic rays are ionizing radiations.

Three mechanisms for external radiation protection include time, distance, and shielding. A shorter time in a radiation field means less dose. From a point source, dose rate is reduced by the square of the distance and expressed by the inverse square law:

$$I_1(d_1)^2 = I_2(d_2)^2$$

where:

$I_1$ = dose rate or radiation intensity at distance $d_1$
$I_2$ = dose rate or radiation intensity at distance $d_2$

Radiation is reduced exponentially by thickness of shielding material.

## Effective Half-Life

The half-life is the length of time required for one-half of a radioactive substance to disintegrate. The formula depicted below is used when the industrial hygienist is interested in determining how much radiation is left in a worker's stomach after a period of time. Effective half-life is a combination of radiological and biological half-lives and is expressed as:

$$T_{eff} = \frac{(T_b)(T_r)}{T_b + T_r}$$

where:

$T_b$ = biological half-life

$T_r$ = radiological half-life

It is important to point out that $T_{eff}$ will always be shorter than either $T_b$ or $T_r$. $T_b$ may be modified by diet and physical activity.

## ALPHA RADIATION

Alpha radiation is used for air ionization—elimination of static electricity (Po-210), clean room applications, and smoke detectors (Am-241). It is also used in air density measurement, moisture meters, nondestructive testing, and oil well logging. Naturally occurring alpha particles are also used for physical and chemical properties, including uranium (coloring of ceramic glaze, shielding) and thorium (high temperature materials).

The characteristics of Alpha radiation are listed below.

- Alpha (a) radiation is a particle composed of two protons and neutrons with source: Ra-226 → Rn 222 → Accelerators.
- Alpha radiation is not able to penetrate skin.
- Alpha-emitting materials can be harmful to humans if the materials are inhaled, swallowed, or absorbed through open wounds.
- A variety of instruments have been designed to measure alpha radiation. Special training in use of these instruments is essential for making accurate measurements.
- A civil defense instrument (CD V-700) cannot detect the presence of radioactive materials that produce alpha radiation unless the radioactive materials also produce beta and/or gamma radiation.
- Because alpha radiation is not penetrating, instruments cannot detect alpha radiation through even a thin layer of water, blood, dust, paper, or other material.
- Alpha radiation travels a very short distance through air.
- Alpha radiation is not able to penetrate turnout gear, clothing, or a cover on a probe. Turnout gear and dry clothing can keep alpha emitters off of the skin.

### Alpha Radiation Detectors

The types of high sensitivity portable equipment used to evaluate alpha radiation in the workplace include:

- Geiger-Mueller counter
- Scintillators
- Solid-state analysis
- Gas proportional devices

## BETA RADIATION

Beta radiation is used for thickness measurements for coating operations; radioluminous signs; tracers for research; and for air ionization (gas chromatograph, nebulizers).

The characteristics of Beta radiation are listed below.

- Beta (ß) is a high energy electron particle with source: Sr-90 → Y-90 → Electron beam machine.
- Beta radiation may travel meters in air and is moderately penetrating.
- Beta radiation can penetrate human skin to the "germinal layer," where new skin cells are produced. If beta-emitting contaminants are allowed to remain on the skin for a prolonged period of time, they may cause skin injury.
- Beta-emitting contaminants may be harmful if deposited internally.
- Most beta emitters can be detected with a survey instrument (such as a DC V-700), provided the metal probe cover is open). Some beta emitters, however, produce very low energy, poorly penetrating radiation that may be difficult or impossible to detect. Examples of these are carbon-14, tritium, and sulfur-35
- Beta radiation cannot be detected with an ionization chamber such as CD V-715.
- Clothing and turnout gear provide some protection against most beta radiation. Turnout gear and dry clothing can keep beta emitters off of the skin.
- Beta radiation presents two potential exposure methods, external and internal. External beta radiation hazards are primarily skin burns. Internal beta radiation hazards are similar to alpha emitters.

### Beta Detection Instrumentation

The types of equipment used to evaluate beta radiation in the workplace include:

- Geiger-Mueller counter
- Gas proportional devices
- Scintillators

- Ion chambers
- Dosimeters

**Shielding for Beta Radiation**

Shielding for beta radiation is best accomplished by using materials with a low atomic number (low z materials) to reduce bremsstrahlung radiation (secondary x-radiation produced when a beta particle is slowed down or stopped by a high-density surface). The thickness is critical to stop maximum energy range, and varies with the type of material used. Typical shielding material includes lead, water, wood, plastics, cement, Plexiglas, and wax.

## GAMMA RADIATION AND X-RAYS

Gamma radiation and X-rays are used for sterilization of food and medical products; radiography of welds, castings, and assemblies; gauging of liquid levels and material density; oil well logging; and material analysis.

The characteristics of gamma radiation and X-rays are listed below.

- Gamma ($\gamma$) is not a particle (electromagnetic wave) composed of high energy electron with source: Tc-99.
- X-ray is composed of photons (generated by electrons leaving an orbit) with source: most radioactive materials, X-ray machines, secondary to ß.
- Gamma radiation and X-rays are electromagnetic radiation like visible light, radio waves, and ultraviolet light. These electromagnetic radiations differ only in the amount of energy they have. Gamma rays and X-rays are the most energetic of these.
- Gamma radiation is able to travel many meters in air and many centimeters in human tissue. It readily penetrates most materials and is sometimes called "penetrating radiation."
- X-rays are like gamma rays. They, too, are penetrating radiation.
- Radioactive materials that emit gamma radiation and X-rays constitute both an external and internal hazard to humans.
- Dense materials are needed for shielding from gamma radiation. Clothing and turnout gear provide little shielding from penetrating radiation but will prevent contamination of the skin by radioactive materials.
- Gamma radiation is detected with survey instruments, including civil defense instruments. Low levels can be measured with a standard Geiger counter, such as the CD V-700. High levels can be measured with an ionization chamber, such as a CD V-715.

- Gamma radiation or X-rays frequently accompany the emission of alpha and beta radiation.
- Instruments designed solely for alpha detection (such as an alpha scintillation counter) will not detect gamma radiation.
- Pocket chamber (pencil) dosimeters, film badges, thermoluminescent, and other types of dosimeters can be used to measure accumulated exposure to gamma radiation.
- The principal health concern associated with gamma radiation is external exposure by penetrating radiation and physically strong source housing. Sensitive organs include the lens of the eye and the gonads; bone marrow is also damaged.

### Gamma Detection Instrumentation

The types of equipment used to evaluate gamma radiation in the workplace include:

- Ion chamber
- Gas proportional
- Geiger Mueller

### Shielding for Gamma and X-rays

Shielding gamma and x-radiation depends on energy level. Protection follows an exponential function of shield thickness. At low energies, absorption can be achieved with millimeters of lead. At high energies, shielding can attenuate gamma radiation.

## RADIOACTIVE DECAY EQUATIONS

Radioactive materials emit alpha particles, beta particles, and photon energy, and lose a proportion of their radioactivity with a characteristic half-life. This is known as radioactive decay. To calculate the amount of radioactivity remaining after a given period of time, use the following basic formulae for decay calculations:

$$\text{Later activity} = (\text{earlier activity})\, e^{-\lambda}\, (\text{elapsed time})$$

$$A = A_i e^{-\lambda t}$$

where:

$$\lambda = LN2/T$$

and:

$\lambda$ = lambda decay constant (probability of an atom decaying in a unit time)
t = time
LN2 = 0.693
T = radioactive half-life (time period in which half of a radioactive isotope decays)
A = new or later radioactivity level
$A_i$ = initial radioactivity level

In determining time required for a radioactive material to decay ($A_o$ to A) use:

$$t = (-LN\ A/A_i)\ (T/LN2)$$

where:

$\lambda$ = lambda decay constant (probability of an atom decaying in a unit time)
t = time
LN2 = 0.693
T = radioactive half-life (time period in which half of a radioactive isotope decays)
A = new or later radioactivity level
$A_i$ = initial radioactivity level

Basic Rule of Thumb: In seven half-lives, reduced to <1%; 10 half-lives <0.1%

In determining the rate of radioactive decay, keep in mind that radioactive disintegration is directly proportional to the number of nuclei present. Thus, the radioactive decay rate is expressed in nuclei disintegrated per unit time.

$$A_i = (0.693/T)\ (N_i)$$

where:

$A_i$ = initial rate of decay
$N_i$ = initial number of radionuclei
T = half life

As mentioned earlier, half-life is defined as the time it takes for a material to lose 50% of its radioactivity. The following equation can be used to determine half-life.

$$A = A_i\ (0.5)^{t/T}$$

where:

A = activity at time t

$A_i$ = initial activity
t = time
T = half-life

## RADIATION DOSE

In the United States, radiation *absorbed dose, dose equivalent*, and *exposure* are often measured and stated in the traditional units called *rad, rem*, or *roentgen* (R). For practical purposes with gamma and x-rays, these units of measure for exposure or dose are considered equal. This exposure can be from an external source irradiating the whole body, an extremity, or other organ or tissue resulting in an *external radiation dose*. Alternately, internally deposited radioactive material may cause an *internal radiation dose* to the whole body or other organ or tissue.

A prefix is often used for smaller measured fractional quantities, such as milli (m) for 1/1,000. For example, 1 rad = 1,000 mrad. Micro (µ) means 1/1,000,000. So, 1,000,000 µrad = 1 rad, or 10 µR = 0.000010 R.

The SI system (System International) for radiation measurement is now the official system of measurement and uses the "gray" (Gy) and "sievert" (Sv) for absorbed dose and equivalent dose respectively. Conversions are as follows:

- 1 Gy = 100 rad
- 1 mGy = 100 mrad
- 1 Sv = 100 rem
- 1 mSv = 100 mrem

Radioactive transformation event (radiation counting systems) can be measured in units of "disintegrations per minute" (dpm) and, because instruments are not 100% efficient, "counts per minute" (cpm). Background radiation levels are typically less than 10 µR per hour, but die to differences in detector size and efficiency, the cpm reading on fixed monitors and various handheld survey meters will vary considerably.

## NONIONIZING RADIATION

*Nonionizing radiation* is described as a series of energy waves composed of oscillating electric and magnetic fields traveling at the speed of light. Nonionizing radiation includes those electromagnetic regions extending from ultraviolet to radio waves, and usually refers to the portion of the spectrum commonly known as the radio frequency range. Nonionizing radiation does not cause ionization. In this text we are concerned with six types of nonionizing radiation that can cause injury: ultraviolet, light, infrared, laser, microwave, and radiofrequency radiation. In humans, ultraviolet radiation causes problems that range from serious sun burns (sometimes ultimately causing skin cancers) to photochemical damage to the eyes; high intensity visible light damages the eyes; infrared radiation leads to skin burns, dehydration, and eye damage; and microwave radiation causes thermal damage to body tissues and internal organs, and leads to cataracts or other eye injury.

In comparison to ionizing radiation, nonionizing radiation is incapable of dislodging orbital electrons, but may leave the atom in an "excited state." All lower energy (frequency) radiation is nonionizing. Nonionizing radiation is expressed as a relationship of frequency, wavelength, and the speed of light. The higher the frequency, the higher the energy.

Nonionizing radiation is found in a wide range of occupational settings and can pose a considerable health risk to potentially exposed workers if not properly controlled. The various types of nonionizing radiation sources are listed below.

*Extremely low frequency radiation* (ELF) at 60 Hz is produced by power lines, electrical wiring, and electrical equipment. Common sources of intense exposure include ELF induction furnaces and high-voltage power lines. Wavelength is in the 50 to 60 Hz range. ACFIH exposure standards are based on understood, verifiable health effects (e.g., magnetophosphenes, induced currents, and potential interference with electronic devices, like pacemakers). ELF electric extremely applied to surfaces of the body induce electric currents and fields inside the body and excite cells.

*Radiofrequency* (RF) and *Microwave* (MW) *radiation* at high enough intensities will damage tissue through heating. MW radiation is absorbed near the skin, while RF radiation may be absorbed throughout the body. Sources of RF and MW radiation include radio emitters and cell phones. Microwave and radiofrequency radiation includes frequencies ranging from 0.1 cm to 300 meters or 1 mHz to 300,000 MHz. Microwaves create heat by causing water molecules to vibrate, get agitated, and heat up. Microwaves are reflected by metal but pass through glass, paper, and plastic. Materials containing water absorb them.

*Infrared radiation* (IR) is emitted by all objects to other objects that have lower surface temperature. Infrared radiation has a wavelength from 700 nanometers to 1 millimeter. The skin and eyes absorb infrared radiation as heat. Workers normally notice excessive exposure through heat sensation and pain. Sources of IR radiation include furnaces, glass blowing, heat lamps, and IR lasers. Infrared light is heat. Exposure standards can be found in the

ACGIH TLV Booklet. To use this information, the wavelength, geometry of source, and length of exposure must be known.

*Visible light radiation* is from the different visible frequencies of the electromagnetic (EM) spectrum, "seen" by our eyes as different colors. Good lighting is conducive to increased employee productivity, and can help prevent incidents related to poor lighting conditions. Excessive visible light radiation can damage the eyes and skin. Visible light wavelength ranges from 400 nanometers to 700 nanometers. Lasers, compact arc lamps, quartz-iodide-tungsten lamps, gas and vapor discharge tubes, and flash lamps are all sources of visible light. Visible light exposure standards are outlined in the ACGIH TLV Booklet. They depend on wavelength and exposure duration.

*Ultraviolet radiation* (UV) has a high photon energy range and is particularly hazardous because there are usually no immediate symptoms of excessive exposure. Sources of UV radiation include the sun, black lights, fluorescent lamps, welding arcs, and UV lasers. The wavelength range of UV extends from 100 nanometers to 400 nanometers. The ozone layer only allows wavelengths greater than 290 nanometers to reach the earth. Exposure standards for UV are wavelength dependent. UV-A: 1 mW/cm$^2$ for $10^3$ seconds measuring UV-A at the source. UV-B and UV-C: wavelength dependent on action spectrum, most active at 200 nm. Sunglasses, clothing, sunblock, and enclosing the source provides the best protection against UV.

*Laser hazards* are primarily eye and skin hazards. LASER is an acronym for Light Amplification by Stimulated Emission of Radiation. The photon of one atom can cause an excited electron of a neighboring atom to drop to the same energy level, thus causing the emission of another identical photon. Lasers typically emit optical (UV, visible, IR) radiations. Common lasers include carbon dioxide IR laser; helium; neon; neodymium YAG; ruby visible lasers; and the Nitrogen UV laser. The American National Standards Institute (ANSI Z136.1) has classified lasers into specific categories. The categories range from I–IV, with Class I being less hazardous than Class IV.

- Class I lasers, such as laser printers, are considered to be incapable of producing damaging radiation levels, and are therefore exempt from most control measures or other forms of surveillance.
- Class II lasers emit radiation in the visible portion of the spectrum, and protection is normally afforded by the normal human aversion response (blink reflex) to bright radiant sources. They—some laser printers, for example—may be hazardous if viewed directly for extended periods of time.

- Class IIIa lasers are those that normally would not produce injury if viewed only momentarily with the unaided eye. They may present a hazard if viewed using collecting optics, like telescopes, microscopes, or binoculars. Example: HeNe lasers above 1 milliwatt but not exceeding 5 milliwatts radiant power.
- Class IIIb lasers can cause severe eye injuries if beams are viewed directly or specular reflections are viewed. A Class III laser is not normally a fire hazard. Example: visible HeNe lasers above 5 milliwatts but not exceeding 500 milliwatts radiant power. Class IIIa and IIIb lasers require "Caution" signs and well-lighted areas to decrease pupil size.
- Class IV lasers are a hazard to the eye from the direct beam and specular reflections and sometimes even from diffuse reflections. Class IV lasers can also start fires and can damage skin. Class IV lasers require "Danger" signs.

## Optical Density

Optical density (OD) is a parameter for specifying the attenuation afforded by a given thickness of any transmitting medium. Since laser beam intensities may be a factor of a thousand or a million above safe exposure levels, percent transmission notation can be unwieldy and is not used. As a result, laser protective eyewear fitters are specified in terms of the logarithmic units of optical density.

Because of the logarithmic factor, a filter attenuating a beam by a factor of 1,000 (or 10(3)) has an OD of 3, and attenuating a beam by 1,000,000 or (10(6)) has an OD of 6. The required optical density is determined by the maximum laser beam intensity to which the individual could be exposed. The optical density of two highly absorbing filters when stacked together is essentially the linear sum of two individual optical densities. The OD for welding goggles may be 14. A pair of specific protective goggles may have an OD of 7.

The formula for calculating optical density is shown below.

$$OD = Log (I_o/I)$$

where:

OD = optical density
$I_o$ or $I_1$ = initial beam intensity
I or $I_2$ = final beam intensity

Finally, as with ionizing radiation (and all other workplace hazards), industrial hygienists must understand the principles of electromagnetic radiation, its uses in the workplace, its hazards and effective control measures. The industrial hygienist will usually find him- or herself responsible for the radiation safety program, if one is needed in the organization.

# RADIATION

## OSHA'S RADIATION SAFETY REQUIREMENTS

OSHA has standards for both ionizing radiation (29 CFR 1910.96) and nonionizing radiation (29 CFR 1910.97). In order to understand the hazards associated with radiation, safety engineers need to understand the basic terms and concepts summarized in the following paragraphs, adapted from 29 CFR 1910.96.

*Radiation* consists of energetic nuclear particles and includes alpha rays, beta rays, gamma rays, X-rays, neutrons, high-speed electrons, and high-speed protons. *Radioactive material* emits corpuscular or electromagnetic emanations as the result of spontaneous nuclear disintegration. A *restricted area* is any area to which access is restricted in an attempt to protect employees from exposure to radiation or radioactive materials. Thus, an *unrestricted area* is any area to which access is not controlled because there is no radioactivity hazard present.

*Dose* is the amount of ionizing radiation absorbed per unit of mass by part of the body or the whole body. A *rad*—a measure of the dose of ionizing radiation absorbed by body tissues—is stated in terms of the amount of energy absorbed per unit of mass of tissue. One rad equals the absorption of 100 ergs per gram of tissue. A *rem* is a measure of the dose of ionizing radiation to body tissue stated in terms of its estimated biological effect relative to a dose of one roentgen (r) to X-rays. *Air dose* means that the dose is measured by an instrument in air at or near the surface of the body in the area that has received the highest dosage. *Personal monitoring devices* are worn or carried by an individual to measure radiation doses received. Widely used devices include film badges, pocket chambers, pocket dosimeters, and film rings.

The *radiation area* is any accessible area in which radiation hazards exist that could deliver doses as follows: (1) within one hour a major portion of the body could receive more than 5 millirem; or (2) within five consecutive days a major portion of the body could receive more than 100 millirem. A *High-radiation area* is any accessible area in which radiation hazards exist that could deliver a dose in excess of 100 millirem within one hour.

OSHA's requirements for *ionizing* radiation (according to 29 CFR 1910.96) include the following:

- The employer must ensure that no individual in a restricted area receives higher levels of radiation than those summarized in table 6.1.
- The employer is responsible for ensuring that no employee under 18 years of age receives, in one calendar year, a dose of ionizing radiation in excess of 10% of the values shown in table 6.2.

Table 6.1. Levels of Radiation

| Part of Body | Dose (Rems/Quarter) |
|---|---|
| Whole body; head and trunk, active, blood-forming organs, lens of eyes, or gonads | 1.25 |
| Hands and forearms; feet and ankles | 8.75 |
| Skin of whole body | 0.5 |

Source: *Code of Federal Regulations* Title 29 Parts 1900–1910, Office of Federal Register, Washington, D.C.: 1985.

- The employer is responsible for the provision and use of radiation, and the use of radiation monitoring devices such as film badges.
- Where a potential for exposure to radioactive materials exists, appropriate warning signs must be posted.

For normal environmental conditions, OSHA requirements for nonionizing radiation (according to 29 CFR 1910.97) include guidelines for electromagnetic energy of frequencies

Table 6.2. Controls for Ionizing Radiation

| Types of Controls | Accomplished by |
|---|---|
| Limit radiation emissions at the source | Limiting the quantity of ionizing material |
| Limited time exposure | Limiting employees' time of exposure; preventing access to locations where radiation sources exist; having written procedures to limit exposures |
| Extend the distance from a source | Increasing employees' distance from a source to reduce exposure |
| Shielding | Reducing radiation levels with shielding made of concrete, lead, steel or water |
| Barriers | Using walls or fences to keep people out who should not be near or around radiation sources |
| Warnings | Clearly marking radiation areas |
| Evacuation | Familiarizing employees with a written evacuation plan to be used in the event of a significant release of radioactive material |
| Security | Using physical monitoring and security procedures |
| Training | Mandating training for employees who work with or around radiation on the hazards of ionizing radiation |

Source: 29 CFR 1910.96, OSHA.

between 10 MHz and 100 GHz: Power density—10 mW/cm$^2$ for periods of 0.1 hour or more; energy density—1 Mw-hr/cm$^2$ (milliwatt hour per square centimeter) during any 0.1-hour period. Note that this guide applies whether the radiation is continuous or intermittent. Appropriate warning signs must also be posted.

## RADIATION EXPOSURE CONTROLS

*Controls*, both engineering and administrative, are an important element in any radiation safety program. The industrial hygienist can employ some controls (depending upon the situation) to protect employees and the public. Again, as we have stated throughout this text, engineering controls are the preferred methodology, when they are appropriate and possible. Tables 6.2 and 6.3 list the kinds of engineering and other control methods that can be employed to protect people from ionizing radiation, as well as the controls for nonionizing radiation. The information contained in these tables comes primarily from publications by the American National Standards Institute (ANSI), New York (readers should refer to a complete listing of ANSI standards) and OSHA 29 CFR 1910.96.

**Table 6.3. Controls for Nonionizing Radiation**

| *Nonionizing Radiation Source* | *Controls* |
|---|---|
| Microwaves | Limit the intensity of microwaves (frequency or wavelength one is exposed to) or limit the duration of exposure; increase the distance from a source and use shielding to limit intensity of exposure; use warning signs about radiation hazard or dangers; provide insulated gloves to minimize shock and burn hazards; properly ground microwave equipment to reduce hazards |
| Ultraviolet radiation | Limit exposure to most harmful wavelengths; use absorbing materials to shield skin and eyes |
| Infrared radiation | Limit duration of exposure and intensity of exposure; avoid looking into infrared sources; use shielding (eyewear that absorbs and reflects the impact of infrared radiation on the eyes) to reduce the intensity of exposure |
| Lasers | Depends on the class of laser (The Food and Drug Administration [FDA] has standards for the classification and safety design features of lasers).<br>  Controls may include: enclose the laser source; control potentially reflective surfaces; use interlocks on doors to locations where lasers are used; install fail-safe pulsing controls to prevent accidental actuation; design remote firing room and controls; use baffles to limit location of beams; provide suitable protective eyewear and clothing |

*Source*: 29 CFR 1910.97, OSHA.

## RADIATION EXPOSURE TRAINING

Rationalizing the need for extensive employee training is not at all difficult when it comes to working with or around ionizing radiation sources and materials. However, to date, not enough emphasis has been placed on training employees on the hazards involved with non-ionizing radiation sources. The safety engineer must ensure that training becomes a key component of the organizational radiation safety program.

## REFERENCES

OSHA, 29 CFR 1910.1096. "Ionizing Radiation," current edition. Washington, D.C.: U.S. Department of Labor.

OSHA, 29 CFR 1910.97. "Nonionizing Radiation," current edition. Washington, D.C.: U.S. Department of Labor.

USHHS (1999), *Radiation Safety Manual.* Atlanta, Ga.: Centers for Disease Control and Prevention.

## SUGGESTED READING

American National Standards Institute. *Safety Levels with Respect to Human Exposure to Radio Frequency Electromagnetic Fields.* C95.1. New York: ANSI, 2000. Meyer, E. *The Chemistry of Hazardous Materials*, 2nd ed. Englewood Cliffs, N.J.: Prentice Hall, 1989.

Michaelson, S. M. and J. C. Lin. *Biological Effects and Health Implications of Radiofrequency Radiation.* New York: Plenum, 1987.

Weber, M. J., ed. *Handbook of Laser Science and Technology.* Vols. 1–5. Boca Raton, Fla.: CRC Press, 1982–1987.

# 7

# Thermal Stress

Appropriately controlling the temperature, humidity, and air distribution in work areas is an important part of providing a safe and healthy workplace. A work environment in which the temperature is not properly controlled can be uncomfortable. Extremes of either heat or cold can be more than uncomfortable—they can be dangerous.

Operations involving high air temperatures, radiant heat sources, high humidity, direct physical contact with hot objects, or strenuous physical activities have a high potential for inducing heat stress in employees engaged in such operations. Such places include: iron and steel foundries, nonferrous foundries, brick-firing and ceramic plants, glass products facilities, rubber products factories, electrical utilities (particularly boiler rooms), bakeries, confectioneries, commercial kitchens, laundries, food canneries, chemical plants, mining sites, smelters, and steam tunnels (OSHA, 2003).

Outdoor activities conducted in hot weather, such as construction, refining, asbestos abatement, and hazardous waste site activities, especially those that require workers to wear semipermeable or impermeable protective clothing, are also likely to cause heat stress among exposed workers.

## CAUSAL FACTORS (OSHA, 2003)

The occurrence of thermal stress to workers in the workplace can be attributed to various causal factors. Age, weight, degree of physical fitness, degree of acclimatization, metabolism, use of alcohol or drugs, and a variety of medical conditions such as hypertension all affect a person's sensitivity to heat. However, even the type of clothing worn must be considered. Prior heat injury predisposes an individual to additional injury.

It is difficult to predict just who will be affected and when, because individual susceptibility varies. In addition, environmental factors include more than the ambient air temperature. Radiant heat, air movement, conduction, and relative humidity all affect an individual's response to heat.

Heat stress and cold stress are major concerns of modern health and thus of concern to industrial hygienists. This chapter provides the information they need to know in order to overcome the hazards associated with extreme temperatures.

## THERMAL COMFORT

Thermal comfort in the workplace is a function of a number of different factors. Temperature, humidity, air distribution, personal preference, and acclimatization are all determinants of comfort in the workplace. However, determining optimum conditions is not a simple process (Alpaugh, 1988).

To fully understand the hazards posed by temperature extremes, industrial hygienists must be familiar with several basic concepts related to thermal energy and comfort. The most important of these are summarized here:

- The American Conference of Governmental Industrial Hygienists (1992) states that workers should not be permitted to work when their deep body temperature exceeds 38°C (100.4°F).
- *Heat* is a measure of energy in terms of quantity.
- A *calorie* is the amount of energy in terms of quantity. A kilocalorie (kcal) is equal to the amount of energy required to raise one liter of water one degree Celsius.
- *Evaporative cooling* takes place when sweat evaporates from the skin. High humidity reduces the rate of evaporation and thus reduces the effectiveness of the body's primary cooling mechanism.
- *Conduction* is the transfer of heat between two bodies that are touching or from one location to another within a body. For example, if an employee touches a workpiece that has just been welded and is still hot, heat will be conducted from the workpiece to the hand. Of course, the result of this heat transfer is a burn.
- *Convection* is the transfer of heat from one location to another by way of a moving medium (a gas or a liquid). Convection ovens use this principle to transfer heat from an electrode by way of gases in the air to whatever is being baked.
- *Metabolic heat* is a by-product of the body's activity (i.e., produced within a body as a result of activity that burns energy). All humans produce metabolic heat. This is why a

room that is comfortable when occupied by just a few people may become uncomfortable when it is crowded. Unless the thermostat is lowered to compensate, the metabolic heat of a crowd will cause the temperature of a room to rise to an uncomfortable level.
- *Environmental heat* is produced by external sources. Gas or electric heating systems produce environmental heat, as do sources of electricity and a number of industrial processes.
- *Radiant heat* is the result of electromagnetic nonionizing energy that is transmitted through space without the movement of matter within that space.

The human body is equipped to maintain an appropriate balance between the metabolic heat it produces and the environmental heat to which it is exposed. Sweating and the subsequent evaporation of the sweat are the body's way of trying to maintain an acceptable temperature balance.

According to Alpaugh (1988), this balance can be expressed as a function of the various factors in equation 7.1.

$$H = M \pm R \pm C - E \qquad (7.1)$$

where:

$H$ = body heat
$M$ = internal heat gain (metabolic)
$R$ = radiant heat gain
$C$ = convection heat gain
$E$ = evaporation (cooling)

The ideal balance when applying the equation is no new heat gain. As long as heat gained from radiation, convection, and metabolic processes do not exceed that lost through the evaporation induced by sweating, the body experiences no stress or hazard. However, when heat gain from any source or sources is more than the body can compensate for by sweating, the result is *heat stress.*

## HEAT DISORDERS AND HEALTH EFFECTS

According to OSHA (2003), heat stress can manifest itself in a number of ways depending on the level of stress. The most common types of heat stress are heat stroke, heat exhaustion, heat cramps, heat rash, transient heat fatigue, and chronic heat fatigue. These various types of heat stress can cause a number of undesirable bodily reactions including prickly heat, inadequate venous return to the heart, inadequate blood flow to vital body parts, circulatory shock, cramps, thirst, and fatigue.

## Heat Stroke

Heat stroke occurs when the body's system of temperature regulation fails and body temperature rises to critical levels. This condition is caused by a combination of highly variable factors, and its occurrence is difficult to predict. Heat stroke is very dangerous and should be dealt with immediately because it can be fatal. The primary signs and symptoms of heat stroke are confusion; irrational behavior; loss of consciousness; convulsions; a lack of sweating (usually); hot, dry skin; and an abnormally high body temperature, e.g., a victim of heat stroke will have a rectal temperature of 104.5°F or higher that will typically continue to climb.

If a worker shows signs of possible heat stroke, professional medical treatment should be obtained immediately. The worker should be placed in a shady area and the outer clothing should be removed. The worker's skin should be wetted and air movement around the worker should be increased to improve evaporative cooling until professional methods of cooling are initiated and the seriousness of the condition can be assessed. Fluids should be replaced as soon as possible. The medical outcome of an episode of heat stroke depends on the victim's physical fitness and the timing and effectiveness of first aid treatment.

## Heat Exhaustion

Heat exhaustion is a type of heat stress that occurs as a result of water and/or salt depletion. When people sweat in response to exertion and environmental heat, they lose more than just water; they also lose salt and electrolytes. *Electrolytes* are minerals that are needed in order for the body to maintain the proper metabolism and in order for cells to produce energy. Loss of electrolytes causes these functions to break down. For this reason it is important for employees to use commercially produced drinks that contain water, salt, sugar, potassium, or electrolytes to replace those lost through sweating.

The signs and symptoms of heat exhaustion are headache, nausea, vertigo, weakness, thirst, and giddiness. Fortunately, this condition responds readily to prompt treatment. Heat exhaustion should not be dismissed lightly, however. One of the more obvious reasons is because the fainting associated with heat exhaustion can be dangerous if the victim was operating machinery or controlling an operation that should not be left unattended.

A victim of heat exhaustion should be moved to a cool but not cold environment and allowed to rest lying down. Fluids should be taken slowly but steadily by mouth until the urine volume indicates that the body's fluid level is once again in balance.

## Heat Cramps

Performing hard physical labor in a hot environment usually causes heat cramps.

This type of heat stress occurs as a result of salt and potassium depletion. Observable symptoms are primarily muscle spasms, typically felt in the arms, legs, and abdomen.

To prevent heat cramps, acclimatize workers to the hot environment gradually over a period of at least a week. Then ensure that fluid replacement is accomplished with a commercially available carbohydrate-electrolyte replacement product that contains the appropriate amount of salt, potassium, and electrolytes.

## Heat Rash

Heat rashes are the most common problem in hot work environments. This is a type of heat stress that manifests itself as small raised bumps or blisters that cover a portion of the body and give off an uncomfortable prickly sensation. It is caused by prolonged exposure to hot and humid conditions in which the body is continuously covered with sweat that can not evaporate because of the high humidity. In most cases, a heat rash will disappear when the affected individual returns to a cool environment.

## Heat Fatigue

Heat fatigue is a type of heat stress that manifests itself primarily because of the victim's lack of acclimatization. Well-conditioned employees who are properly acclimatized will suffer this form of heat stress less frequently and less severely than poorly conditioned employees will. Consequently, preventing heat fatigue involves physical conditioning and acclimatization, because there is no treatment for heat fatigue except to remove the cause of the heat stress before a more serious heat-related condition develops.

## Heat Collapse

In heat collapse (i.e., fainting), the brain does not receive enough oxygen because blood pools in the extremities. As a result, the exposed individual may lose consciousness. This reaction is similar to that of heat exhaustion and does not affect the body's heat balance. However, the onset of heat collapse is rapid and unpredictable. To prevent heat collapse, the worker should gradually become acclimatized to the hot environment.

## INVESTIGATION GUIDELINES

The following guidelines for evaluating heat stress are adapted from *Threshold Limit Values for Chemical Substances and Physical Agents* (ACGIH, 1992–1993).

### Employer and Employee Interviews

Prior to beginning interviews, the inspector should review the OSHA 300 Log and, if possible, the OSHA 301 forms for indications of prior heat stress problems.

Following are some questions for *employer interviews*: What type of action, if any, has the employer taken to prevent heat stress problems? What are the potential sources of heat? What employee complaints have been made?

Following are some questions for *employee interviews*: What heat stress problems have been experienced? What type of action has the employee taken to minimize heat stress? What is the employer's involvement (e.g., Does employee training include information on heat stress)?

### Walkaround Inspection

During the walkaround inspection, the investigator should determine building and operation characteristics; determine whether engineering controls are functioning properly; verify information obtained from the employer and employee interviews; and perform temperature measurements and make other determinations to identify potential sources of heat stress. Investigators may wish to discuss any operations that have the potential to cause heat stress with engineers and other knowledgeable personnel. The walkaround inspection should cover all affected areas. Heat sources, such as furnaces, ovens, and boilers, and relative heat load per employee should be noted.

### Work-Load Assessment

Under conditions of high temperature and heavy workload, the industrial hygienist should determine the work-load category of each job (refer to applicable ACGIH [1992] tables). Work-load category is determined by averaging metabolic rates for the tasks and then ranking them:

- Light work: up to 200 kcal/hour
- Medium work: 200–350 kcal/hour
- Heavy work: 350–500 kcal/hour

# THERMAL STRESS

Where heat conditions in a *cool rest area* are different from those in the work area, the metabolic rate (M) should be calculated using a time-weighted average, as shown in equation 7.2:

$$\text{Average Metabolic Rate} = (M_1)(t_1) + (M_2)(t_2) + \ldots + (M_n)(t_n)/t_1 + t_2 + \ldots + t_n \quad (7.2)$$

where:

   M = metabolic rate
   t = time in minutes

## SAMPLING METHODS

*Body temperature measurements* are done with the various instruments available that estimate deep body temperature by measuring the temperature in the ear canal or on the skin. However, these instruments are not sufficiently reliable to use in compliance evaluations and therefore body temperature is not typically measured in the workplace.

*Environmental heat measurements* should be made at, or as close as possible to, the specific work area where the worker is exposed. When a worker is not continuously exposed in a single hot area but moves between two or more areas having different levels of environmental heat, or when the environmental heat varies substantially at a single hot area, environmental heat exposures should be measured for each area and for each level of environmental heat to which the employee is exposed.

*Wet Bulb Globe Temperature Index* (WBGT) should be calculated using the appropriate formula (see equations 7.3–7.5). The WBGT for continuous all-day or several-hour exposures should be averaged over a 60-minute period. Intermittent exposures should be averaged over a 120-minute period. These averages should be calculated using equation 7.3.

$$\text{Average WBGT} = \frac{(\text{WBGT}_1)(t_1) + (\text{WBGT}_2)(t_2) + \ldots + (\text{WBGT}_n)(t_n)}{t_1 + t_2 + \ldots + t_n} \quad (7.3)$$

For indoor and outdoor conditions with no solar load, WBGT is calculated as

$$\text{WBGT} = 0.7\text{NWB} + 0.3\text{GT} \quad (7.4)$$

For outdoors with a solar load, WBGT is calculated as

$$\text{WBGT} = 0.7\text{NWB} + 0.2\text{GT} + 0.1\text{DB} \quad (7.5)$$

*Measurement* is conducted using portable heat stress meters or monitors. These instruments can calculate both the indoor and outdoor WBGT index according to established

ACGIH Threshold Limit Value equations. With this information and information on the type of work being performed, heat stress meters can determine how long a person can safely work or remain in a particular hot environment.

It should be noted that measurement is often required for those environmental factors (i.e., dry temperature and relative humidity) that most nearly correlate with deep body temperature and other physiological responses to heat. The WBGT is the most often used technique to measure these environmental factors. As mentioned, WBGT values are calculated by using equations 7.4 and 7.5 for indoor and outdoor solar load.

The determination of WBGT requires the use of a black globe thermometer, a natural (static) wet-bulb thermometer, and a dry-bulb thermometer. The measure of environmental factors is performed as follows:

1. The range of the dry-bulb and the natural wet-bulb thermometers should be -5°C to +50°C, within accuracy of ±0.5°C. The dry-bulb thermometer must be shielded from the sun and the other radiant surfaces of the environment without restricting the airflow around the bulb. The wick of the natural wet-bulb thermometer should be kept wet with distilled water for at least one-half hour before the temperature reading is made. It is not enough to immerse the other end of the wick into a reservoir of distilled water and wait until the whole wick becomes wet by capillarity. The wick must be wetted by direct application of water from a syringe one-half hour before each reading. The wick must cover the bulb of the thermometer and an equal length of additional wick must cover the stem above the bulb. The wick should always be clean, and new wicks should be washed before using.
2. A globe thermometer, consisting of a 15-cm (6-inch) diameter hollow copper sphere painted on the outside with a matte black finish, or equivalent, must be used. The bulb or sensor of a thermometer (range -5°C to +100°C with an accuracy of ±0.5°C) must be fixed in the center of the sphere. The globe thermometer should be exposed at least 25 minutes before it is read.
3. A stand should be used to suspend the three thermometers so that they do not restrict free air flow around the bulbs and the wet-bulb and globe thermometer are not shaded.
4. It is permissible to use any other type of temperature sensor that gives a reading similar to that of a mercury thermometer under the same conditions.
5. The thermometers must be placed so that the readings are representative of the employee's work or rest areas, as appropriate.

Once the WBGT has been estimated, industrial hygienists can estimate workers' metabolic heat load using the ACGIH method to determine the appropriate work/rest regimen, clothing, and equipment to use to control the heat exposures of workers in their facilities.

# THERMAL STRESS

Other thermal stress indices include the *Effective Temperature Index* (ET) and the *Heat Stress Index* (HSI). The ET index combines the temperature, the humidity of the air, and air velocity. This index has been used extensively in the field of comfort ventilation and air-conditioning. ET remains a useful measurement technique in mines and other places where humidity is high and radiant heat is low. The HSI was developed by Belding and Hatch in 1965. Although the HSI considers all environmental factors and work rate, it is not completely satisfactory for determining an individual worker's heat stress and it is difficult to use.

## HEAT STRESS SAMPLE MEASUREMENT AND CALCULATION

As mentioned, the Wet Bulb Globe Temperature Index (WBGT) is the most widely used algebraic approximation of an "Effective Temperature" currently in use today. It is an index that can be determined quickly, requiring a minimum of effort and operator skill. As an approximation to an effective temperature, the WBGT takes into account virtually all the commonly accepted mechanisms of heat transfer (radiant, evaporative, etc.). It does not account for the cooling effect of wind speed. Because of its simplicity, WBGT has been adopted by the American Conference of Government Hygienists (ACGIH) as it principal index for use in specifying a heat stress related Threshold Limit Value (TLV).

## Problem 1

What would be the WBGT Index, in °C, for a quarry worker in Connecticut, who must work on a sunny morning when the outdoor dry-bulb temperature is 88°F, the wet-bulb temperature, 72°F, and the globe temperature, 102°F?

Solution:

$$WBGT = 0.7(NWB) + 0.2(GT) + 0.1(DB)$$
$$= 0.7(72) + 0.2(102) + 8.8$$
$$= 50.4 + 20.4 + 8.8$$
$$= 78.8°F \ (26°C)$$

## Problem 2

Later the same afternoon, at the same quarry identified in problem 1, rain clouds have gathered, and rain has started to fall. The quarry manager has covered the work area in the quarry pit with a large tarpaulin to protect his employees. If the wet-bulb temperature under the tarp has increased to 78°F, while the globe temperature has remained unchanged, what will be the new WBGT Index for this slightly different situation? (*Hint*: This is an indoor environment.)

Solution:

$$WBGT = 0.7(NWB) + 0.3(GT)$$

$$= 0.7(78) + 0.3(102)$$

$$= 54.6 + 30.6$$

$$= 85.2°F$$

## THERMAL STRESS CONTROL

Ventilation, air cooling, fans, shielding, and insulation are the five major types of engineering controls used to reduce heat stress in hot work environments.

Heat reduction can also be achieved by using power assists and tools that reduce the physical demands placed on a worker. However, for this approach to be successful, the metabolic effort required for the worker to operate these devises must be less than the effort required without them. It may be possible to reduce the effort necessary to operate power assists.

The worker should be allowed to take frequent rest breaks in a cooler environment.

### Acclimatization

The human body can adapt to heat exposure to some extent. This physiological adaptation is called acclimatization. After a period of acclimatization, the same activity will produce fewer cardiovascular demands. The worker will sweat more efficiently (causing better evaporative cooling), and thus will more easily be able to maintain normal body temperatures.

A properly designed and applied acclimatization program decreases the risk of heat-related illnesses. Such a program basically involves exposing employees to work in a hot environment for progressively longer periods. NIOSH (1986) says that, for workers who have had previous experience in jobs where heat levels are high enough to produce heat stress, the regiment should be 50% exposure on day one, 60% on day two, 80% on day three, and 100% on day four. For new workers who will be similarly exposed, the regimen should be 20% on day one, with a 20% increase in exposure each day thereafter.

## Fluid Replacement

Cool (50°F–60°F) water or any cool liquid (except alcoholic beverages) should be made available to workers to encourage them to drink small amounts frequently, such as one cup every 20 minutes. Ample supplies of liquids should be placed close to the work area. Although some commercial replacement drinks contain salt, this is not necessary for acclimatized individuals because most people add enough salt to their summer diets.

## Engineering Controls

*General ventilation* is used to dilute hot air with cooler air (generally cooler air that is brought in from the outside). This technique clearly works better in cooler climates than in hot ones. A permanently installed ventilation system usually handles large areas or entire buildings. Portable or local exhaust systems may be more effective or practical in smaller areas.

*Air treatment/air cooling* differs from ventilation because it reduces the temperature of the air by removing heat (and sometimes humidity) from the air.

*Air conditioning* is a method of air cooling, but it is expensive to install and operate. An alternative to air conditioning is the use of chillers to circulate cool water through heat exchangers over which air from the ventilation system is then passed; chillers are more efficient in cooler climates or in dry climates where evaporate cooling can be used.

*Local air cooling* can be effective in reducing air temperature in specific areas. Two methods have been used successfully in industrial settings. One type, cool rooms, can be used to enclose a specific workplace or to offer a recovery area near hot jobs. The second type is a portable blower with built-in air chiller. The main advantage of a blower, aside from portability, is minimal set-up time.

Another way to reduce heat stress is to increase the air flow or *convection* using fans in the work area (as long as the air temperature is less than the worker's skin temperature). Changes in air speed can help workers stay cooler by increasing both the convective heat exchange (the exchange between the skin surface and the surrounding air) and the rate of evaporation. Because this method does not actually cool the air, any increases in air speed must impact the worker directly to be effective.

If the dry-bulb temperature is higher than 35°C (95°F), the hot air passing over the skin can actually make the worker hotter. When the temperature is more that 35°C and the air is dry, evaporative cooling may be improved by air movement, although this improvement will be offset by the convective heat. When the temperature exceeds 35°C and the relative humidity is 100%, air movement will make the worker hotter. Increases in air speed have no effect on the body temperature of workers wearing vapor-barrier clothing.

*Heat conduction* methods include insulating the hot surface that generates the heat and changing the surface itself.

Simple *engineering controls*, such as shields, can be used to reduce radiant heat (heat coming from hot surfaces) within the worker's line of sight. Surfaces that exceed 35°C (95°F) are sources of infrared radiation that can add to the worker's heat load. Flat black surfaces absorb heat more than smooth, polished ones. Having cooler surfaces surround the worker assists in cooling because the worker's body radiates heat toward them.

With some sources of radiation, such as heating pipes, it is possible to use both insulation and surface modifications to achieve a substantial reduction in radiation heat. Instead of reducing radiation from the source, shielding can be used to interrupt the path between the source and the worker. Polished surfaces make the best barriers, although special glass or metal mesh surfaces can be used if visibility is a problem.

Shields should be located so that they do not interfere with air flow, unless they are also being used to reduce convective heating. The reflective surface of the shield should be kept clean to maintain its effectiveness.

## Administrative Controls and Work Practices

Training is the key to good work practices. Unless all employees understand the reasons for using new, or changing old, work practices, the chances of such a program succeeding are greatly reduced.

NIOSH (1986) states that a good heat stress training program should include at least the following components:

- knowledge of the hazards of heat stress
- recognition of predisposing factors, danger signs, and symptoms
- awareness of first-aid procedures for, and the potential health effects of, heat stroke
- responsibility of employees to avoid heat stress
- instruction in dangers of using drugs, including therapeutic ones, and alcohol in hot work environments
- use of protective clothing and equipment
- instruction in purpose and coverage of environmental and medical surveillance programs, and the advantages of worker participation in such programs.

# THERMAL STRESS

Hot jobs should be scheduled for the cooler part of the day, and routine maintenance and repair work in hot areas should be scheduled for the cooler seasons of the year.

## Worker Monitoring

Every worker who works in extraordinary conditions that increase the risk of heat stress should be personally monitored. These types of conditions include wearing semipermeable or impermeable clothing when the temperature exceeds 21°C (69.8°F), or working at extreme metabolic loads (greater than 500 kcal/hour).

Personal monitoring can be done by checking the heart rate, recovery heart rate, oral temperature, or extent of body water loss.

- To check heart rate, count the radial pulse for 30 seconds at the beginning of the rest period. If the heart rate exceeds 110 beats per minute, shorten the next work period by one-third and maintain the same rest period.
- The recovery heart rate can be checked by comparing the pulse rate taken at 30 seconds ($P_1$) with the pulse rate taken at 2.5 minutes ($P_3$) after the rest break starts.
- Oral temperature can be checked with a clinical thermometer after work but before the employee drinks water. If the oral temperature taken under the tongue exceeds 37.6°C, shorten the next worker cycle by one-third.
- Body water loss can be measured by weighing the worker on a scale at the beginning and end of each work day. The worker's weight loss should not exceed 1.5% of total body weight in a work day. If a weight loss exceeding this amount is observed, fluid intake should increase.

## Other Administrative Controls

The following administrative controls can be used to reduce heat stress:

- reduce the physical demands of work (e.g., excessive lifting or digging with heavy objects)
- provide recovery areas (e.g., air-conditioned enclosures and rooms)
- use shifts (e.g., early morning, cool part of the day, or night work)
- use intermittent rest periods with water breaks
- use relief workers
- use worker pacing
- assign extra workers and limit worker occupancy, or the number of workers present, especially in confined or enclosed spaces

## PERSONAL PROTECTIVE EQUIPMENT

### Reflective Clothing

Reflective clothing can vary from aprons and jackets to suits that completely enclose the worker from neck to feet, and can stop the skin from absorbing radiant heat. However, since most reflective clothing does not allow air exchange through the garment, the reduction of radiant heat must more than offset the corresponding loss in evaporative cooling. For this reason, reflective clothing should be worn as loosely as possible. In situations where radiant heat is high, auxiliary cooling systems can be used under the reflective clothing.

### Auxiliary Body Cooling

Commercially available *ice vests*, though heavy, may accommodate as many as 72 ice packets, which are usually filled with water. Carbon dioxide (dry ice) can also be used as a coolant. The cooling offered by ice packets lasts only 2 to 4 hours at moderate to heavy heat loads, and frequent replacement is necessary. However, ice vests do not encumber the worker and thus permit maximum mobility. Cooling with ice is also relatively inexpensive.

*Wetted clothing* is another simple and inexpensive personal cooling technique. It is effective when reflective or other impermeable protective clothing is worn. The clothing may be wetted terry cloth coveralls or wetted two-piece, whole-body cotton suits. This approach to auxiliary cooling can be quite effective under conditions of high temperature and low humidity, where evaporation from the wetted garment is not restricted.

*Water-cooled garments* range from a hood, which cools only the head, to vests and "long johns," which offer partial or complete body cooling. Use of this equipment requires a battery-driven circulating pump, liquid-ice coolant, and a container. Although this system has the advantage of allowing wearer mobility, the weight of the components limits the amount of ice that can be carried and thus reduces the effective use time. The heat transfer rate in liquid cooling systems may limit their use to low-activity jobs; even in such jobs, their service time is only about 20 minutes per pound of cooling ice. To keep outside heat from melting the ice, an outer insulating jacket should be an integral part of these systems.

*Circulating air* is the most highly effective, as well as the most complicated, personal cooling system. By directing compressed air around the body from a supplied air system, both evaporative and convective cooling are improved. The greatest advantage occurs when circulating air is used with impermeable garments or double cotton overalls.

One type, used when respiratory protection is also necessary, forces exhaust air from a supplied-air hood ("bubble hood") around the neck and down inside an impermeable suit.

The air then escapes through openings in the suit. Air can also be supplied directly to the suit without using a hood either by a single inlet, by a distribution tree, or by a perforated vest. In addition, a vortex tube can be used to reduce the temperature of circulating air. The cooled air from this tube can be introduced either under the clothing or into a bubble hood. The use of a vortex tube separates the air stream into a hot and cold stream; these tubes also can be used to supply heating in cold climates. Circulating air, however, is noisy and requires a constant source of compressed air supplied through an attached air hose.

One problem with this system is the limited mobility of workers whose suits are attached to an air hose. Another is that of getting air to the work area itself. These systems should therefore be used in work areas where workers are not required to move around much or to climb. Another concern with these systems is that they can lead to dehydration. The cool, dry air feels comfortable and the worker may not realize that it is important to drink liquids frequently.

## Other PPE

The weight of a self-contained breathing apparatus (SCBA) increases stress on a worker, and this stress contributes to overall heat stress. Chemical protective clothing such as totally encapsulating chemical protection suits will also add to the heat stress problem.

## HEAT STRESS COMPLIANCE CHECKLIST

Listed below are sample questions that the industrial hygienist may wish to consider when investigating heat stress in the workplace.

### *Workplace Description*
    YES/NO  Is type of business described?
    YES/NO  Is heat-producing equipment or process used?
    YES/NO  Is there previous history of heat-related problems?

    At "hot" spots:
    _____   Is the heat steady or intermittent?
    _____   Number of employees exposed?
    _____   For how many hours per day?
    YES/NO  Is potable water available?
    YES/NO  Are supervisors trained to detect/evaluate heat stress symptoms?

***Are Exposures Typical for a Workplace at This Locale in This Industry?***

_____ Weather at time of review?
_____ Temperature?
_____ Humidity?
_____ Air Velocity?

YES/NO Is day typical of recent weather conditions?
YES/NO Heat-reducing engineering controls employed?
YES/NO Ventilation in place?
YES/NO Ventilation operating?
YES/NO Air conditioning in place?
YES/NO Air conditioning operating?
YES/NO Fans in place?
YES/NO Fans operating?
YES/NO Shields or insulation between sources and employees?
YES/NO Reflective faces of shields clean?

***Work Practices to Detect, Evaluate, and Prevent or Reduce Heat Stress***

YES/NO Training programs?
_____ Content?
_____ Where given?
YES/NO Liquid replacement program?
YES/NO Acclimatization program?
YES/NO Work/rest schedule?
YES/NO Efficient work schedule (during cooler parts of shift, cleaning and maintenance during shut-downs, etc.)?
YES/NO Cool rest areas (including shelter at outdoor work sites)?
YES/NO Heat monitoring program?
YES/NO Personal protective equipment?
YES/NO Reflective clothing in use?
YES/NO Ice and/or water-cooled garments in use?
YES/NO Wetted undergarments (used with reflective or impermeable clothing) in use?
YES/NO Circulating air systems in use?
YES/NO First Aid Program?
YES/NO Trained personnel?
YES/NO Provision for rapid cool-down?
YES/NO Procedures for getting medical attention?

YES/NO  Transportation to medical facilities readily available for heat stroke victims?

YES/NO  Medical Screening and Surveillance Program?

_____  Content?

_____  Who manages program?

Additional Comments?

*Heat Stress-Related Illness or Accident Follow-up*

Describe events leading up to the episode.

YES/NO  Evaluation/comments by other workers at the scene?

_____  Work at time of episode (heavy, medium, light)?

_____  How long was affected employee working at site prior to episode?

_____  Medical history of affected worker, if known.

YES/NO  Appropriate engineering controls in place?

YES/NO  Appropriate engineering controls in operation?

YES/NO  Appropriate work practices used by affected employee(s)?

YES/NO  Appropriate PPE available?

YES/NO  Appropriate PPE in use?

YES/NO  Medical screening for heat stress and continued surveillance for signs of heat stress given other employees?

Additional comments regarding specific episode(s).

## COLD HAZARDS

Temperature hazards are generally thought of as relating to extremes of heat. This is natural because most workplace temperature hazards do relate to heat. However, temperature extremes at the other end of the spectrum—cold—can also be hazardous. Employees who work outdoors in colder climates and employees who work indoors in such jobs as meatpacking are subjected to cold hazards.

There are four factors that contribute to cold stress: cold temperature, high or cold wind, dampness, and cold water. These factors, alone or in combination, draw heat away from the body (Greaney, 2000). The *windchill factor* increases the level of hazard posed by extremes of cold. Safety engineers need to understand this concept and how to make it part of their deliberations when developing strategies to prevent cold stress injuries. Other cold stress factors include age, disease, and overall physical condition.

OSHA (1998) expresses cold stress through its cold stress equation. That is,

Low Temperature + Wind Speed + Wetness = Injuries & Illness

## Evaluation of Cold Stress

ACGIH recommends protective clothing for temperatures less than 41°F (5°C). To estimate the amount of clothing insulation required (in clo units), the following formula is used:

$$I_{clo} = 11.5 \, (33 - T_{db})/M \qquad (7.6)$$

where:

$I_{clo}$ = clo units (1 clo = 0.156 m² × C/W)
$T_{db}$ = dry bulb temperature (°C)
M = metabolic rate in watts

The major injuries associated with extremes of cold can be classified as being either generalized or localized. A generalized injury from extremes of cold is hypothermia. Localized injuries include frostbite, frostnip, and trenchfoot.

*Hypothermia* results when the body is unable to produce enough heat to replace the heat lost to the environment. It may occur at air temperatures up to 65°F, the body uses its defense mechanisms to help maintain its core temperature.

*Frostbite* is an irreversible condition in which the skin freezes, causing ice crystals to form between cells. The toes, fingers, nose, ears, and cheeks are the most common sites of freezing cold injury.

*Frostnip* is less severe than frostbite. It causes the skin to turn white and typically occurs on the face and other exposed parts of the body. There is not tissue damage; however, if the exposed area is not either covered or removed from exposure to the cold, frostnip can become frostbite.

*Trenchfoot* is caused by continuous exposure to cold water. Also known as "immersion foot" or "chilblains," trenchfoot is a medical condition caused by prolonged exposure of feet to damp and cold. It may occur in wet, cold environments or through actual immersion in water.

## Cold Stress Control Measures

- Minimizing air velocity
- Medical surveillance
- Adjusting length of exposure

- Heating
- Insulating workers
- Supplying warm drinks (e.g., soups)
- Education
- Modifying work schedules
- Buddy system
- Insulating metal parts

## REFERENCES

ACGIH, 1992–1993. *Threshold Limit Values for Chemical Substances and Physical Agents and Biological Exposure Indices.* Cincinnati: American Conference of Governmental Industrial Hygienists.

Alpaugh, E. L., 1988. *Fundamentals of Industrial Hygiene*, 3rd ed. Revised by T. J. Hogan. Chicago: National Safety Council. 259–60.

Greaney, P. P., 2000. Ensuring Employee Safety in Cold-Weather Working Environments. Work Care. www.workcare.com/Archive/News_Art_2000_Dec14.htm (accessed 8/17/05).

NIOSH, 1986. *Criteria for a Recommended Standard—Occupational Exposure to Hot Environments.* DHHS (NIOSH) Publication No. 86–113. Cincinnati, Ohio: National Institute for Occupational Safety and Health.

OSHA, 1998. *The Cold Stress Equation.* OSHA 3156. Washington, D.C.: U.S. Department of Labor.

OSHA, 2003. *Heat Stress.* OSHA Technical Manual IV. Washington, D.C.: U.S. Department of Labor. Accessed at www.osha.gov.

## SUGGESTED READING

Cyr, D. L. and S. B. Johnson. *Battling the Elements Safely.* 2002. At http://www.cdc.gov/nasd/docs/d000901-d001000/d000925/d000925.html.

North, C. "Heat Stress," *Safety & Health* 141, no. 4 (April 1991): 55.

Ramsey, J. D., C. L. Buford, M. Y. Beshir, and R. C. Jensen. "Effects of Workplace Thermal Conditions or Safe Work Behavior." *Journal of Safety Research* 14:105–114, 1983.

Zenz, C. *Occupational Medicine: Principles and Practical Applications.* 2nd ed. St. Louis, Mo.: Mosby Year Book, Inc., 1988.

# 8

# Ventilation

A ventilation system is all very well and good (virtually essential, actually), but an improperly designed ventilation system can make the hazard worse. This point cannot be overemphasized. At the heart of an efficient ventilation system are proper design, proper maintenance, and proper monitoring. The industrial hygienist plays a critical role in ensuring that installed ventilation systems are operating at their optimum level.

Because of the importance of ventilation in the workplace, the industrial hygienist must be well versed in the general concepts of ventilation, the principles of air movement, and monitoring practices. The industrial hygienist must be properly prepared (through training and experience) to evaluate existing systems and design new systems for control of the workplace environment. In this chapter, we present the general principles and concepts of ventilation system design, evaluation, and control. This material provides the basic concepts and principles necessary for ensuring the proper operation of industrial ventilation systems, and serves to refresh the knowledge of the practitioner in the field. Probably the best source of information on ventilation is the ACGIH's *Industrial Ventilation: A Manual of Recommended Practice* (current edition)—a must-have reference for every industrial hygienist.

## VENTILATION TERMS

Because ventilation is one of the most important engineering control techniques used by industrial hygienists, the industrial hygienist must be familiar with and understand the following ventilation terms and definitions (OSHA, 2005).

**acfm**: Actual cubic feet per minute of gas flowing at existing temperature and pressure.
**ACH, AC/H**: Air changes per hour; the number of times air is replaced in an hour.

**air density**: The weight of air in lbs per cubic foot. Dry standard air at T = 68° F (20° C) and BP = 2992 in. Hg (760 mm Hg) has a density of 0.075 lb/cu ft.

**aniometer**: A device that measures the velocity of air. Common types include the swinging vane and the hot-wire anemometer.

**area** ($A$): The cross-sectional area through which air moves. Area may refer to the cross-sectional area of a duct, a window, a door, or any space through which air moves.

**atmospheric pressure**: The pressure exerted in all directions by the atmosphere. At sea level, mean atmospheric pressure is 29.92 in. Hg, 14.7 psi, 407 in. w.g., or 760 mm Hg.

**brake horsepower** (bhp): The actual horsepower required to move air through a ventilation system against a fixed total pressure plus the losses in the fan. bhp = ahp × 1/eff, where eff is fan mechanical efficiency.

**branch**: In a junction of two ducts, the branch is the duct with the lowest volume flow rate. The branch usually enters the main at an angle of less than 90.

**canopy hood (receiving hood)**: A one- or two-sided overhead hood that receives rising hot air or gas.

**capture velocity**: The velocity of air induced by a hood to capture emitted contaminants external to the hood.

**coefficient of entry** ($C_e$): A measure of the efficiency of a hood's ability to convert static pressure to velocity pressure; the ratio of actual flow to ideal flow.

**density correction factor**: A factor applied to correct or convert dry air density of any temperature to velocity pressure; the ratio of actual flow to ideal flow.

**dilution ventilation (general exhaust ventilation)**: A form of exposure control that involves providing enough air in the workplace to dilute the concentration of airborne contaminants to acceptable levels.

**evase**: (Pronounced eh-va-say.) A cone-shaped exhaust stack that recaptures static pressure from velocity pressure.

**fan**: A mechanical device that moves air and creates static pressure.

**fan curve**: A curve relating pressure and volume flow rate of a given fan at a fixed fan speed (rpm).

**fan laws**: Relationships that describe theoretical, mutual performance changes in pressure, flow rate, rpm of the fan, horsepower, density of air, fan size, and power.

**flow rate**: Volume flow rates are described by the conservation of mass formula. Q = VA; where Q = volume, V = velocity, and A = cross-sectional area of air flow.

**friction loss**: The static pressure loss in a system caused by friction between moving air and the duct wall, expressed as in w.g./100 ft, or fractions of VP per 100 ft of duct (mm w.g./m; Kpa/m).

**gauge pressure**: The difference between two absolute pressures, one of which is usually atmospheric pressure.

**head**: Pressure, for example, "The head is 1 in w.g."

**hood**: A device that encloses, captures, or receives emitted contaminants.

**hood entry loss** ($H_e$): The static pressure lost (in inches of water) when air enters a duct through a hood. The majority of the loss usually is associated with a vena contracta formed in the duct.

**hood static pressure** ($SP_h$): The sum of the duct velocity pressure and the hood entry loss; hood static pressure is the static pressure required to accelerate air at rest outside the hood into the duct at velocity.

**hvac system**: Heating, ventilation, and air conditioning system; ventilating systems designed primarily to control temperature, humidity, odors, and air quality.

**indoor air quality** (**IAQ**): The study, examination, and control of air quality related to temperature, humidity, and airborne contaminants.

**industrial ventilation** (**IV**): The equipment or operation associated with the supply or exhaust of air by natural or mechanical means to control occupational hazards in the industrial setting.

**in. w.g.**: Inches of water; a unit of pressure. One inch of water is equal to 0.0735 in. of mercury, or 0.036 psi. Atmospheric pressure at standard conditions is 407 in. w.g.

**laminar flow**: Air flow in which air molecules travel parallel to all other molecules; laminar flow is characterized by the absence of turbulence. Also called streamline flow.

**local exhaust ventilation**: An industrial ventilation system that captures and removes emitted contaminants before dilution into the ambient air of the workplace.

**loss**: Usually refers to the conversion of static pressure to heat in components of the ventilation system, for example, "the hood entry loss."

**manometer**: A device that measures pressure difference; usually a U-shaped glass tube containing water or mercury.

**minimum transport velocity** (**MTV**): The minimum velocity that will transport particles in a duct with little settling; MTV varies with air density, particulate loading, and other factors.

**outdoor air** (**OA**): Outdoor air is the "fresh" air mixed with return air (RA) to dilute contaminants in the supply air.

**pitot tube**: A device used to measure total and static pressures in an airstream.

**plenum**: A low-velocity chamber used to distribute static pressure throughout its interior.

**pressure**: Air moves under the influence of differential pressures. A fan is commonly used to create a difference of pressure in duct systems.

**pressure drop**: The loss of static pressure across a point; for example, "the pressure drop across an orifice is 2.0 in. w.g."

**replacement air**: Air supplied to a space to replace exhausted air. Also called compensating air, make-up air.

**return air**: Air that is returned from the primary space to the fan for recirculation.

**scfm:** Standard cubic feet per minute. A measure of air flow at standard conditions, i.e., dry air at 29.92 in. Hg (760 mm Hg) (gauge), 68°F (20°C).

**sick building syndrome** (SBS): A situation in which building occupants experience acute health and comfort effects that appear to be linked to time spent in a building, where no specific illness or cause can be identified. The complaint may be localized in a particular room or zone, or may be widespread throughout the building.

**slot velocity:** The average velocity of air through a slot. Slot velocity is calculated by dividing the total volume flow rate by the slot area (usually, $V_s$ = 2,000 fpm).

**stack:** A device on the end of a ventilation system that disperses exhaust contaminants for dilution by the atmosphere.

**standard air, standard conditions:** Dry air at 70°F (20°C), 29.92 in. Hg (760 mm Hg), 14.7 psi, 407 inches of water.

**static pressure** (SP): The pressure developed in a duct by a fan; the force in inches of water measured perpendicular to flow at the wall of the duct; the difference in pressure between atmospheric pressure and the absolute pressure inside a duct, cleaner, or other equipment; SP exerts influence in all directions.

**suction pressure:** An archaic term that refers to static pressure on the upstream side of the fan. *See* Static pressure.

**tight building syndrome:** A "modern" problem related to the practice of constructing buildings that do not "leak" to the outside atmosphere. This is now common practice to save energy by preventing heating, ventilation, and air conditioning (HVAC) losses.

**total pressure** (TP): The pressure exerted in a duct, i.e., the sum of the static pressure and the velocity pressure; also called impact pressure, dynamic pressure.

**turbulent flow:** Air flow characterized by transverse velocity components as well as velocity in the primary direction of flow in a duct; mixing velocities.

**velocity** (V): The time rate of movement of air; usually expressed as feet per minute.

**velocity pressure** (VP): The pressure attributed to the velocity of air.

**volume flow rate** (Q): Quantity of air flow in cubic feet per minute (cfm), standard cubic feet per minute (scfm), or actual feet per minutes (acfm).

## CONCEPTS OF VENTILATION

The purpose of industrial ventilation is essentially to (under control) recreate what occurs in natural ventilation. Natural ventilation results from differences in pressure. Air moves from high-pressure areas to low pressure areas. This difference in pressure is the result of thermal conditions. We know that hot air rises, which (for example) allows smoke to escape from the smokestack in an industrial process, rather than disperse into areas where workers operate the process. Hot air rises because air expands as it is heated, becoming lighter. The same principle is in effect when air in the atmosphere becomes heated. The air rises and is replaced by air from a higher-pressure area. Thus, convection currents cause a natural ventilation effect through the resulting winds.

What does all of this have to do with industrial ventilation? Actually, quite a lot. Simply put, industrial ventilation is installed in a workplace to circulate the air within and to provide a supply of fresh air to replace undesirable air.

Could this be accomplished simply by natural workplace ventilation? That is, couldn't we just heat the air in the workplace so that it will rise and escape through natural ports—windows, doors, cracks in walls, or mechanical ventilators in the roof (installed wind-powered turbines, for example)? Yes, we could design a natural system like this, but in such a system, air does not circulate fast enough to remove contaminants before a hazardous level is reached, which defeats our purpose in providing a ventilation system in the first place. Thus, we use fans to provide an artificial, mechanical means of moving the air.

Along with controlling or removing toxic airborne contaminants from the air, installed ventilation systems perform several other functions within the workplace:

1. Ventilation is often used to maintain an adequate oxygen supply in an area. In most workplaces, this is not a problem because natural ventilation usually provides an adequate volume of oxygen; however, in some work environments (deep mining and thermal processes which use copious amounts of oxygen for combustion) the need for oxygen is the major reason for an installed ventilation system.
2. An installed ventilation system can remove odors from a given area. As you might guess, this type of system has applications in such places as athletic locker rooms, rest rooms, and kitchens. In performing this function, the noxious air may be replaced with fresh air, or odors may be masked with a chemical masking agent.
3. One of the primary uses of installed ventilation is one that we are familiar with—providing heating, cooling, and humidity control.
4. A ventilation system can remove undesirable contaminants at their source, before they enter the workplace air (e.g., from a chemical dipping or stripping tank). Obviously, this technique is an effective way to ensure that certain contaminants never enter the breathing zone of the worker—exactly the kind of function safety engineering is intended to accomplish.

Earlier we stated that installed ventilation is able to perform its designed function via the use of a mechanical fan. Actually, a mechanical fan is the heart of any ventilation system, but like the human heart, certain ancillaries are required to make it function as a system. Four major components make up a ventilation system:

1. The fan forces the air to move.
2. An inlet or some type of opening allows air to enter the system.
3. An outlet must be provided for air to leave the system.
4. A conduit or pathway (ducting) not only directs the air in the right direction, but also limits the amount of flow to a predetermined level.

An important concept regarding ventilation systems is the difference between exhaust and supply ventilation. An *exhaust system* removes air and airborne contaminants from the workplace. Such a system may be designed to exhaust an entire work area, or it may be placed at the source to remove the contaminant prior to its release into the workplace air. The second type of ventilation system is the *supply system*, which (as the name implies) adds air to the work area, usually to dilute work area contaminants to lower the concentration of these contaminants. However, a supplied-air system does much more; it also provides movement to air within the space. Standard practice is to equip an area with both exhaust and supply systems, because that allows movement of air from inlet to outlet and is important in replenishing exhausted air with fresh air.

Air movement in a ventilation system is a result of differences in pressure. Note that pressures in a ventilation system are measured in relation to atmospheric pressure. In the workplace, the existing atmospheric pressure is assumed to be the zero point. In the supply system, the pressure created by the system is *in addition to* the atmospheric pressure that exists in the workplace (i.e., a positive pressure). In an exhaust system, the objective is to lower the pressure in the system below the atmospheric pressure (i.e., a negative pressure).

When we speak of increasing and decreasing pressure levels within a ventilation system, what we are really talking about is creating small differences in pressure—small when compared to the atmospheric pressure of the work area. Because of the small-scale differences in pressure, air can be assumed to be incompressible. For this reason, these differences are measured in terms of *inches of water* or *water gauge*, which results in the desired sensitivity of measurement.

Since one pound per square inch of pressure is equal to 27 inches of water, one inch of water is equal to 0.036 pounds pressure, or 0.24% of standard atmospheric pressure. Remember the potential for error introduced by considering air to be incompressible is very small at the pressure that exists within a ventilation system.

The industrial hygienist must be familiar with the three pressures important in ventilation: velocity pressure, static pressure, and the total pressure. To understand these three

# VENTILATION

203

pressures and their function in ventilation systems, you must first be familiar with pressure itself. In fluid mechanics, the energy of a fluid (air) that is flowing is termed *head*. Head is measured in terms of unit weight of the fluid or in foot-pounds/pound of fluid flowing. **Note:** The usual convention is to describe head in terms of feet of fluid that is flowing.

*Pressure*, then, is the force per unit area exerted by the fluid. In the English system of measurement, this force is measured in lbs/ft$^2$. Since we have stated that the fluid in a ventilation system is incompressible, the pressure of the fluid is equal to the head.

*Velocity pressure* (VP) is created as air travels at a given velocity through a ventilation system. VP is only exerted in the direction of airflow and is *always* positive (i.e., above atmospheric pressure). When you think about it, VP has to be positive, and obviously the force or pressure that causes it also must be positive.

Note that the velocity of the air moving within a ventilation system is directly related to the velocity pressure of the system. This relationship can be derived into the standard equation for determining velocity (and clearly demonstrates the relationship between velocity of moving air and the velocity pressure):

$$v = 4005\sqrt{VP} \qquad (8.1)$$

*Static pressure* (SP) is the pressure that is exerted in all directions by the air within the system, which tends to burst or collapse the duct. It is expressed in inches of water gauge ("wg). A simple example may help you grasp the concept of SP. Consider the balloon that is inflated at a given pressure. The pressure within the balloon is exerted equally on all sides of the balloon. No air velocity exits within the balloon itself. The pressure in the balloon is totally the result of SP. Note that static pressure can be both negative and positive with respect to the local atmospheric pressure.

*Total pressure* (TP) is defined as the algebraic sum of the static and VP or

$$TP = SP + VP \qquad (8.2)$$

The TP of a ventilation system can be either positive or negative (i.e., above or below atmospheric pressure). Generally, the total pressure is positive for a supply system, and negative for an exhaust system.

For the industrial hygienist to evaluate the performance of any installed ventilation system, he or she must make measurements of pressures in the ventilation system. Measurements are normally made using instruments such as a manometer or a Pitot tube.

The *manometer* is often used to measure the SP in the ventilation system. The manometer is a simple, U-shaped tube, open at both ends, and usually constructed of clear glass or plastic so that the fluid level within can be observed. To facilitate measurement, a graduated scale is usually present on the surface of the manometer. The manometer is filled with a liquid (water, oil, or mercury). When pressure is exerted on the liquid within the manometer, the pressure causes the level of liquid to change as it relates to the atmospheric pressure external to the ventilation system. The pressure measured, therefore, is relative to atmospheric pressure as the zero point.

When manometer measurements are used to obtain positive pressure readings in a ventilation system, the leg of the manometer that opens to the atmosphere will contain the higher level of fluid. When a negative pressure is being read, the leg of the tube open to the atmosphere will be lower, thus indicating the difference between the atmospheric pressure and the pressure within the system.

The *Pitot tube* is another device used to measure static pressure in ventilation systems. The Pitot tube is constructed of two concentric tubes. The inner tube forms the impact portion, while the outer tube is closed at the end and has static pressure holes normal to the surface of the tube. When the inner and outer tubes are connected to opposite legs of a single manometer, the VP is obtained directly. If the engineer wishes to measure SP separately, two manometers can be used. Positive and negative pressure measurements are indicated on the manometer as above.

## Local Exhaust Ventilation

*Local exhaust ventilation* (the most predominant method of controlling workplace air) is used to control air contaminants by trapping and removing them near the source. In contrast to dilution ventilation (which lets the contamination spread throughout the workplace, later to be diluted by exhausting quantities of air from the workspace), local exhaust ventilation surrounds the point of emission with an enclosure, and attempts to capture and remove the emissions before they are released into the workers' breathing zone. The contaminated air is usually drawn through a system of ducting to a collector, where it is cleaned and delivered to the outside through the discharge end of the exhauster. A typical local exhaust system consists of a hood, ducting, an air-cleaning device, fan, and a stack. A local exhaust system is usually the proper method of contaminant control if:

- the contaminant in the workplace atmosphere constitutes a health, fire, or explosion hazard.
- national or local codes require local exhaust ventilation at a particular process.
- maintenance of production machinery would otherwise be difficult.
- housekeeping or employee comfort will be improved.
- emission sources are large, few, fixed and/or widely dispersed.

- emission rates vary widely by time.
- emission sources are near the worker-breathing zone.

The industrial hygienist must remember that determining beforehand precisely the effectiveness of a particular system is often difficult. Thus, measuring exposures and evaluating how much control has been achieved after a system is installed is essential. A good system may collect 80 to 90+ percent, but a poor system may capture only 50 percent or less. Without total enclosure of the contaminant sources (where capture is obviously very much greater), the industrial hygienist must be aware of the limitations and must be familiar with handling problems like these.

Once the system is installed, and has demonstrated that it is suitable for the task at hand, the system must be well maintained. Careful maintenance is a must. In dealing with ventilation problems, the industrial hygienist soon finds out that his or her worst headache in maintaining the system is poor—or no—maintenance.

A phenomenon that many practitioners in the industrial hygiene field forget (or never knew in the first place) is that ventilation, when properly designed, installed, and maintained, can go a long way to ensure a healthy working environment. However, ventilation does have limitations. For example, the effects of blowing air from a supply system and removing air through an exhaust system are different. To better understand the difference and its significance, let's look at an example of air supplied through a standard exhaust duct.

When air is exhausted through an opening, it is gathered equally from all directions around the opening. This includes the area behind the opening itself. Thus, the cross-sectional area of airflow approximates a spherical form, rather than the conical form that is typical when air is blown out of a supply system. To correct this problem, a flange is usually placed around the exhaust opening, which reduces the air contour, from the large spherical contour to that of a hemisphere. As a result, this increases the velocity of air at a given distance from the opening. This basic principle is used in designing exhaust hoods. Remember that the closer the exhaust hood is to the source, and the less uncontaminated air it gathers, the more efficient the hood's percentage of capture will be. Simply put, it is easier for a ventilation system to blow air than it is for one to exhaust it. Keep this in mind whenever you are dealing with ventilation systems and/or problems.

## General and Dilution Ventilation

Along with local exhaust ventilation are two other major categories of ventilation systems: general and dilution ventilation. Each of these systems has a specific purpose, and finding all three types of systems present in a given workplace location is not uncommon.

*General ventilation* systems (sometimes referred to as heat control ventilation systems) are used to control indoor atmospheric conditions associated with hot industrial environments (such as those found in foundries, laundries, or bakeries) for the purpose of preventing acute discomfort or injury. General ventilation also functions to control the comfort level of the worker in just about any indoor working environment. Along with the removal of air that has become process-heated beyond a desired temperature level, a general ventilation system supplies air to the work area to condition (by heating or cooling) the air, or to make up for the air that has been exhausted by dilution ventilation in a local exhaust ventilation system.

A *dilution ventilation system* dilutes contaminated air with uncontaminated air, to reduce the concentration below a given level (usually the TLV of the contaminant) to control potential airborne health hazards, fire and explosive conditions, odors, and nuisance type contaminants. This is accomplished by removing or supplying air, to cause the air in the workplace to move, and as a result, mix the contaminated with incoming uncontaminated air.

This mixing operation is essential. To mix the air there must be, of course, air movement. Air movement can be accomplished by natural draft caused by prevailing winds moving through open doors and windows of the work area.

Thermal draft can also move air. Whether the thermal draft is the result of natural causes or is generated from process heat, the heated air rises, carrying any contaminant present upward with it. Vents in the roof allow this air to escape into the atmosphere. Makeup air is supplied to the work area through doors and windows.

A mechanical air moving device provides the most reliable source for air movement in a dilution ventilation system. Such a system is rather simple. It requires a source of exhaust for contaminated air, a source of air supply to replace the air mixture that has been removed with uncontaminated air, and a duct system to supply or remove air throughout the workplace. Dilution ventilation systems often are equipped with filtering systems to clean and temper the incoming air.

## VENTILATION INVESTIGATION AND ANALYSIS

Industrial hygienists must pay particular attention to installed ventilation systems and/or the lack of installed ventilation systems. In addition to the possibility of installed ventilation systems not operating correctly and efficiently as per design specifications, ventilation systems may be deficient in confined spaces; facilities failing to provide adequate maintenance

# VENTILATION

of ventilation equipment; facilities operated to maximize energy conservation; windowless areas; and areas with high occupant densities. Any ventilation deficiency must be verified by measurement.

OSHA (2005) points out that there are five basic types of ventilation systems:

1. dilution and removal by general exhaust
2. local exhaust
3. makeup air (or replacement)
4. HVAC (primarily for comfort)
5. recirculation systems

Ventilation systems generally involve a combination of these types of systems. For example, a large local exhaust system may also serve as a dilution system, and the HVAC system may serve as a makeup air system.

## Health Effects

Inadequate or improper ventilation is the cause of about half of all indoor air quality (IAQ) problems in nonindustrial workplaces. In this section, we address ventilation in buildings and industrial facilities.

*Indoor air contaminants* include but are not limited to particulates, pollen, microbial agents, and organic toxins. These can be transported by the ventilation system or originate in various parts of the ventilation system—wet filters, wet insulation, wet undercoil pans, cooling towers, or evaporative humidifiers.

People exposed to these agents may develop signs and symptoms related to "humidifier fever," "humidifier lung," "air conditioner lung." In some cases, indoor air quality contaminants cause clinically identifiable conditions such as occupational asthmas, reversible airway disease, and hypersensitivity pneumonitis.

*Volatile organic and reactive chemicals* (for example, formaldehyde) often contribute to indoor air contamination. The facility's ventilation system may transport reactive chemicals from a source area to other parts of the building. Tobacco smoke contains a number of organic and reactive chemicals and is often carried this way. In some instances the containment source may be the outside air. Outside air for ventilation or makeup air for exhaust systems may bring contaminants into the workplace (e.g., vehicle exhaust or fugitive emissions from a neighboring smelter).

## Standards, Codes, and OSHA Regulations

Foremost among OSHA and industry consensus standards are those recommended by the Air Movement and Control Association (AMCA), the American Society of Heating, Refrigerating, and Air-Conditioning Engineers (ASHRAE), the American National Standards Institute (ANSI), the Sheet Metal and Air Conditioning Contractors National Association (SMACNA), the National Fire Protection Association (NFPA), and the American Conference of Governmental Industrial Hygienists (ACGIH). AMCA is a trade association that has developed standards and testing procedures for fans. ASHRAE is a society of heating and air conditioning engineers that has produced, through consensus, a number of standards related to indoor air quality, filter performance and testing, and HVAC systems. ANSI has produced several important standards on ventilation, including ventilation for paint spray booths, grinding exhaust hoods, and open-surface tank exhausts. Four ANSI standards were adopted by OSHA in 1971 and are codified in 29 CFR 1910.94; these standards continue to be important as guides to design. ANSI has recently published a new standard for laboratory ventilation (ANSI Z9.5). SMACNA is an association representing sheet metal contractors and suppliers. It sets standards for ducts and duct installation. NFPA has produced a number of recommendations, which become requirements when adopted by local fire agencies (e.g., NFPA 45 lists a number of ventilation requirements for laboratory fume hood use). The ACGIH has published widely used guidelines for industrial ventilation.

Ventilation criteria or standards are included in OSHA regulatory codes for job- or task-specific worker protection. In addition, many OSHA health standards include ventilation requirements. The four standards in 29 CRF 1910.94 deal with local exhaust systems, and OSHA's construction standards (29 CFR 1926) contain ventilation standards for welding. OSHA's compliance policy regarding violation of ventilation standards is set forth in the *Field Inspection Reference Manual.*

## INVESTIGATIVE GUIDELINES

Workplace investigations of ventilation systems may be initiated by worker complaints of possible overexposures to air contaminants, possible risk of fire or explosion from flammable gas or vapor levels at or near the lower explosive level (LEL), or indoor air quality complaints. The investigation will then involve an examination of the ventilation system's physical and operating characteristics.

## Faulty Ventilation Conditions and Causes

Common faulty ventilation conditions and their probable causes are listed in table 8.1. Specific points to consider during any investigation of a ventilation system include emission sources, air behavior, and employee involvement. Points that should be included in a review of operational efficacy are shown in table 8.2.

Table 8.1. Common Ventilation Conditions and Causes

| Condition | Possible cause(s) |
|---|---|
| Worker complaints, improper use of system, nonuse of system, alteration of system by employees. | The hood interferes with work. The hood provides poor control of contaminants. |
| Excessive employee exposures although flow volumes and capture velocities are at design levels. | Employee work practices need improvement. The ventilation system interferes with work or worker productivity and leads workers to bypass the system. Employee training is not adequate. Design of system is poor. |
| Constant plugging of duct. | Plugged ducts occur when transport velocity is inadequate or when vapor condenses in the duct, wets particles, and causes a build-up of materials. These problems are caused by poor design, open access door close to the fan, fan problems, or other problems. |
| Reduced capture velocities or excessive fugitive emissions. | The cause of these conditions is usually reduced flow rate, unless the process itself has changed. Reduced flow rate occurs in the following situations:<br>• plugged or dented ducts<br>• slipping fan belts<br>• open access doors<br>• holes in ducts, elbows<br>• closed blast gate to branch, or opened branch gates to other branches, or corroded and stuck blast gates<br>• fan turning in reverse direction (This can occur when lead wires are reversed and cause the motor and fan to turn backwards. Centrifugal fans turning backwards may deliver up to only 50% of rated capacity.)<br>• worn out fan blades<br>• additional branches or hoods added to system since initial installation<br>• clogged air cleaner |

Source: OSHA, 2005.

### Table 8.2. Problem Characterization

*Emission Source*

Where are all emission sources or potential emissions sources located?
Which emission sources actually contribute to exposure?
What is the relative contribution of each source to exposure?
Characterization of each contributor:
- chemical composition
- temperature
- rate of emission
- direction of emission
- initial emission velocity
- pattern of emission (continuous or intermittent)
- time intervals of emission
- mass of emitted material

*Air Behavior*

Air temperature
Air movement (direction, velocity)
Mixing potential
Supply and return flow conditions, to include pressure differences between space and surrounding areas
Sources of tempered and untempered make-up air
Air changes per hour
Influence of existing HVAC systems
Effects of wind speed and direction
Effects of weather and season

*Employee*

Worker interaction with emission source
Worker exposure levels
Worker location
Worker education, training, cooperation

*Source*: OSHA, 2005.

## Research and Documentation

The characteristics of the ventilation system that must be documented during an investigation include equipment operability, physical measurements of the system, and use practices.

First, before taking velocity or pressure measurements, note and record the operating status of the equipment. For example, are filters loaded or clean? Are variable-flow devices like dampers, variable-frequency drives, or inlet vanes in use? Are make-up units operating? Are system blueprints available?

***Duct diameters*** are measured to calculate duct areas. Inside duct diameter is the most important measurement, but an outside measurement is often sufficient for a sheet metal

# VENTILATION

duct. To measure the duct, the tape should be thrown around the duct to obtain the duct circumference, and the number should be divided by 3.142 to obtain the diameter of the duct.

***Hood and duct dimensions*** can be estimated from plans, drawings, and specifications. Measurements can be made with measuring tape. If a duct is constructed of 2.5- or 4-foot sections, the sections can be counted (elbows and tees should be included in the length).

***Hood-face velocities*** outside the hood or at the hood face can be estimated with velometers, smoke tubes, and swinging-vane anemometers, all of which are portable, reliable, and require no batteries.

The minimum velocity that can be read by an anemometer is 50 feet per minute (fpm). The meter should always be read in the upright position, and only the tubing supplied with the equipment should be used.

Anemometers often cannot be used if the duct contains dust or mist because air must actually pass through the instrument for it to work. The instrument requires periodic cleaning and calibration at least once per year. Hot-wire anemometers should not be used in airstreams containing aerosols.

To measure hood-face velocity, (1) mark off imaginary areas, (2) measure the velocity at the center of each area, and (3) average all measured velocities.

Smoke is useful for measuring face velocity because it is visible. Nothing convinces management and employees more quickly that the ventilation is not functioning properly than to show smoke drifting away from the hood, escaping the hood, or traveling into the worker's breathing zone. Smoke can be used to provide a rough estimate of face velocity: (1) squeeze off a quick blast of smoke, (2) time the smoke plume's travel over a two-foot distance, and (3) calculate the velocity in feet per minute. For example, if it takes two seconds for the smoke to travel two feet, the velocity is 60 fpm.

$$\text{Velocity} = \text{Distance/Time} \; or \; V = D/T$$

***Hood static pressure*** (SPH) should be measured about 4 to 6 duct diameters downstream in a straight section of the hood take-off duct. The measurement can be made with a pitot tube or by a static pressure tap into the duct sheet metal.

Pressure gauges come in a number of varieties, the simplest being the U-tube manometer.

Inclined manometers offer greater accuracy and greater sensitivity at low pressures than U-tube manometers. However, manometers rarely can be used for velocities less than 800 fpm (i.e. velocity pressures less than 0.05" w.g.). Aneroid-type manometers use a calibrated bellows to measure pressures. They are easy to read and portable but require regular calibration and maintenance.

***Duct velocity measurements*** may be made directly (with velometers and anemometers) or indirectly (with manometers and pitot tubes) using duct velocity pressure.

Air flow in industrial ventilation ducts is almost always turbulent, with a small, nonmoving boundary layer at the surface of the duct.

Because velocity varies with distance from the edge of the duct, a single measurement may not be sufficient. However, if the measurement is taken in a straight length of round duct, 4 to 6 diameters downstream and 2 to 3 diameters upstream from obstructions or directional changes, then the average velocity can be estimated at 90% of the centerline velocity. (The average velocity pressure is about 81% of centerline velocity pressure.)

A more accurate method is the traverse method, which involves taking six to ten measurements on each of two or three passes across the duct, 90° or 60° opposed. Measurements are made in the center of concentric circles of equal area.

Density corrections (e.g., temperature) for instrument use should be made in accordance with the manufacturer's instrument instruction manual and calculation/correction formulas.

***Air cleaner and fan condition*** measurements can be made with a pitot tube and manometer.

## Good Practices

***Hood placement*** must be done close to the emission source to be effective. The maximum distance from the emission source should not exceed 1.5 duct diameters.

Ensure the appropriate relationship of capture velocity ($V_c$) to duct velocity ($V_d$) for a simple plain or narrow flanged hood. For example, if an emission source is one duct diameter in front of the hood and the duct velocity ($V_d$) = 3,000 fpm, then the expected capture velocity ($V_c$) is 300 fpm. At two duct diameters from the hood opening, capture velocity decreases by a factor of 10, to 30 fpm.

A rule of thumb that can be used with simple capture hoods is that if the duct diameter (D) is 6 inches, then the maximum distance of the emission source from the hood should not exceed 9 inches. Similarly, the minimum capture velocity should not be less than 50 fpm. Simply, for simple capture hoods, maximum capture distance should not be more than 1.5 times the duct diameter.

***System effect loss***, which occurs at the fan, can be avoided if the necessary ductwork is in place.

Use of the *six-and-three rule* ensures better design by providing for a minimum loss at six diameters of straight duct at the fan inlet and a minimum loss at three diameters of straight duct at the fan outlet.

**VENTILATION**

**Table 8.3. Good Practices for Reviewing Plans and Specifications**

Investigate the background and objectives of the product.
Understand the scope of the project. What is to be included and why?
Look for conciseness and precision. Mark ambiguous phrases, "legalese," and repetition.
Do the specifications spell out exactly what is wanted? What is expected?
Do plans and specifications adhere to appropriate codes, standards, requirements, policies, and do they recommend good practice as established by the industry?
Will the designer be able to design, or the contractor to build, the system from the plans and specifications?
Will the project meet OSHA requirements if it is built as proposed?

*Source*: OSHA, 2005.

System effect loss is significant if any elbows are connected to the fan at inlet or outlet. For each 2.5 diameters of straight duct between the fan inlet and any elbow, cubic feet per minute (CFM) loss will be 20%.

***Stack height*** should be 10 feet higher than any roof line or air intake located within 50 feet of the stack. For example, a stack placed 30 feet away from an air intake should be at least 10 feet higher than the center of the intake.

***Ventilation system drawings and specifications*** usually follow standard forms and symbols, e.g., as described in the Uniform Construction Index (UCI).

Plan sections include electrical, plumbing, structural, or mechanical drawings. The drawings come in several views: plan (top), elevation (side and front), isometric, or section.

Elevations (side and front views) give the most detail. An isometric drawing is one that illustrates the system in three dimensions. A sectional drawing provides duct or component detail by showing a cross-section of the component.

Drawings are usually to scale. (Check dimensions and lengths with a ruler or a scale to be sure that this is the case. For example, 1/8 inch on the sheet may represent one foot on the ground.) Good practices to follow when reviewing plans and specifications are listed in table 8.3.

## PREVENTION AND CONTROL

A well-designed system and a continuing preventive maintenance program are key elements in the prevention and control of ventilation system problems.

### A Good Maintenance Program

1. Put it on paper. *Establish a safe place to file* drawings, specifications, fan curves, operating instructions, and other papers generated during design, construction, and testing.

2. *Establish a program of periodic inspection.* The types and frequencies of inspections depend on the operation of the system and other factors.
    - *Daily:* Visual inspection of hoods, ductwork, access and clean-out doors, blast gate positions, and hood static pressure drop across air cleaner, and verbal contact with users ("How is the system performing today?"). A quick way to check for settled material in a duct is to take a broomstick and tap the underside of all horizontal ducts. If the tapping produces a "clean" sheet metal sound, the duct is clear. If the tapping produces heavy, thudding sounds and no sheet metal vibration, liquids or settled dust may be in the duct.
    - *Weekly:* Visual inspection of air cleaner capacity, fan housing, pulley belts.
    - *Monthly:* Visual inspection of air cleaner components.
3. *Establish a preventive maintenance program.* Certain elements of any ventilation system should be checked on a regular schedule and replaced if found to be defective.
4. *Provide worker training.* Workers need to be trained in the purpose and functions of the ventilation system. For example, they need to know how to work safely and how best to utilize the ventilation system. Exhaust hoods do little good if the welder does not know that the hood must be positioned close to the work.
5. *Keep written records.* Maintain written documentation not only of original installations but also of all modifications as well as problems and their resolution.

## Dealing with Microorganisms

If you suspect microbial agents, check for stagnant water in the ventilation system. The presence of mold or slime is a possible sign of trouble. Table 8.4 lists preventive measures for controlling microbial problems in ventilation systems.

**Table 8.4. Preventive Measures for Reducing Microbial Problems in Buildings**

Prevent buildup of moisture in occupied spaces (relative humidity of 60% or less).
Prevent moisture collection in HVAC components.
Remove stagnant water and slime from mechanical equipment.
Use steam for humidifying.
Avoid use of water sprays in HVAC systems.
Use filters with a 50%—70% collection efficiency rating.
Find and discard microbe-damaged furnishings and equipment.
Provide regular preventive maintenance.

## Volatile Organic or Reactive Chemicals

If an organic or reactive chemical (e.g., formaldehyde) is believed to be the primary agent in an IAQ problem, potential controls to consider include additional dilution ventilation, removal or isolation of the offending material, and the transfer of sensitized employees.

## Tobacco Smoke in Air

OSHA has published a proposed rule for IAQ (including tobacco smoke in the workplace), and this rulemaking is likely to be completed in the near future. Smoking policies should include provisions for dedicated smoking areas. Dedicated smoking areas should be configured so that migration of smoke into nonsmoking areas will not occur. Such areas should (1) have floor-to-ceiling walls of tight construction, (2) be under negative pressure relative to adjacent areas, and (3) be exhausted outside the building and not recirculated.

## FACTS, CONCEPTS, AND PRACTICE PROBLEMS

Ventilation is an important form of emission/exposure control. It also provides for health, comfort, and well-being. All human occupancies require ventilation.

### Facts and Concepts to Know

Properties of air          MW = 29; weight density = 0.075 lbs/cu ft at STP

STP (ventilation)          standard temperature and pressure
T = 70°F, BP = 29.92 in. Hg, dry air; weight density = 0.075 lbs/cu ft at STP

Density Correction Factor (d), derived from the Ideal Gas equations:

$$d = \frac{T_{TSP}\,(\text{absolute})}{T_{actual}\,(\text{absolute})} \times \frac{BP_{actual}}{BP_{STP}}$$

where:

BP = absolute barometric pressure
°R = degrees Rankine (°R = °F + 460; absolute temperature)
K = Kelvin (K = °C + 273; absolute temperature)

## Example 1

The temperature is 90°F and the barometric pressure is 27.50 in. Hg. What is the density correction factor, d?

Solution:

$$d = \frac{460 + 70}{460 + 90°} \times \frac{B.P.}{29.92 \text{ in. Hg}}$$

$$d = \frac{530}{550} \times \frac{27.50 \text{ in. Hg}}{29.92 \text{ in. Hg}} = 0.886$$

Local exhaust ventilation systems are made up of five components: hood, ductwork, air cleaner, fan, and stack.

## Pressure

Air moves under the influence of pressure differentials. A fan is commonly used to create the pressure difference. At sea level the standard static (barometric) pressure (SP) is 14.7 psia = 29.92 in. Hg = 407 inch w.g. If a fan is capable of creating one inch of negative static pressure (e.g., 1" w.g., or "one inch water gauge"), then the absolute static pressure in the duct will be reduced to 406 inches w.g.

Manometers     Used to measure pressure differences.
Pitot tube         A device used to measure TP and SP
                   "S" = "side" = SP
                   "TP" = "tip" = "top" = TP
                   Attach both legs to the manometer to measure VP

Pressures are related as follows:         TP = SP + VP

|  | TP | SP | VP |
|---|---|---|---|
| Upstream | – | – | + |
| Downstream | + | + | + |

## Example 2

Determine the velocity pressure, VP.

# VENTILATION

$$TP = -0.35" \text{ w.g. } SP = -0.5 \text{ w.g.}$$

Solution:

$$TP = SP + VP \text{ and}$$

$$VP = TP - SP = -0.35 - (-0.50) = 0.15" \text{ w.g.}$$

## Role of velocity pressure

The velocity pressure is related to the velocity of air in the duct. The relationship is given by:

$$V = 4005 \sqrt{VP/d}$$

where:

- VP = velocity pressure, inch w.g. (measured with a Pitot tube, for example)
- V = velocity, feet per minute (fpm)
- d = density correction factor

### Example 3

The velocity pressure of an airstream in a lab fume hood duct is VP = 0.33 inches w.g. What is the velocity? (d = 1)

Solution:

$$V = 4005 \, (VP/d)^{1/2} = 4005 \, (0.33/1)^{1/2} = 2300 \text{ fpm}$$

## Static Pressure

Static pressure is the potential energy of the ventilation system. It is converted to kinetic energy (VP) and other (less useful) forms of energy (heat, vibration, and noise). These are the "losses" of the system.

Volume flow rate can be described by:

$$Q = VA$$

where:

- Q = volume flow rate, cubic feet per minute (cfm)
- A = cross-sectional area of duct, square feet (sf)
- V = velocity, fpm

## Example 4

The cross-sectional area of a duct is A = 0.7854 sq ft. The velocity of air flowing in the duct is V = 2250 ft. per minute. What is the flow rate, Q?

Solution:

$$Q = V \times A$$

$$= 2250 \text{ fpm} \times 0.7854 \text{ sq ft} = 1770 \text{ scfm}$$

## Example 5

The static pressure is measured in a 10-inch square duct at SP = –1.15 inch w.g. The average total pressure is TP = –0.85 inch w.g. Find the velocity and volume flow rate of the air flowing in the duct. (STP; d = 1)

Solution:

$$\text{Area, A} = 10" \times 10"/144" \text{ per sq ft} = 0.6944 \text{ sq ft}$$

$$VP = TP - SP = -0.85" - (-1.15") = 0.30" \text{ w.g.}$$

$$\text{Velocity, V} = 4005 \,(VP/d)^{1/2} = 4005 \,(0.30)^{1/2} = 2194$$

$$\text{Volume flow rate, Q} = VA = (2194)(0.6944) = 1524 \text{ scfm}$$

## Losses

As the air moves through a duct, losses are created (i.e., static pressure is converted to heat, vibration, noise.) Losses include hood entry, friction, elbow, branch entry, system effect, air cleaner, and others. The loss is usually directly related to velocity pressure:

$$SP_{Loss} = K \times VP \times d$$

where:

$SP_{Loss}$ = loss of static pressure, inches w.g.
K = loss factor, unitless
VP = average velocity pressure in duct; d = density correction factor

## Hoods

The hood captures, contains, or receives contaminants generated at an emission source. The hood converts duct static pressure to velocity pressure and hood entry losses (e.g., slot and duct entry losses).

# VENTILATION

$$He = K \times VP \times d = |SP_h| \, VP$$

where:

$H_e$ = hood entry loss
$K$ = loss factor, unitless
$VP$ = velocity pressure in duct, inches w.g.
$|SP_h|$ = absolute static pressure about 5 duct diameters down the duct from the hood, inches w.g
$d$ = density correction factor

## Coefficient of Entry (Ce)

A hood's ability to convert static pressure to velocity pressure is given by

$$Ce = \frac{Q_{actual}}{Q_{ideal}} = \sqrt{VP/SP_h} = \sqrt{1/(1+K_h)}$$

### Example 6

What is the hood static pressure $SP_h$ when the duct velocity pressure is $VP = 0.33$ inches w.g., and the hood entry loss is $He = 0.44$ inches w.g.? What is Ce?

Solution:

$$|SP_h| = VP + He = 0.33 + 0.44 = 0.77 \text{ inch w.g.}$$

$$Ce = \frac{Q_{actual}}{Q_{ideal}} = \left[\frac{VP}{|SP_h|}\right]^{0.5} = \left[\frac{0.33}{0.77}\right]^{0.5} = 0.65$$

## Hood Entry Loss

Hood entry losses normally occur at the hood slots and at the entrance to the duct, due to the vena contracta formed. The most narrowed portion of the vena contracta is usually found about one-half duct diameter inside the duct or plenum.

The hood static pressure, SPh, is the sum total of the acceleration and all losses from the hood face to the point of measurement in the duct.

Head loss is

$$He = K \times VP_d \times d$$

where K is the hood entry loss factor.

Hood entry loss factors (K) have been estimated over the years and have been reported in the literature for many types of hoods, including lab fume hoods.

## *Example 7*

What is the entry loss He for a laboratory fume hood when the average velocity pressure in the duct is $VP_d = 0.30"$ w.g. (assume K = 2.0, d = 1)?

Solution:

$$He = K \times VP_d \times d = 2.0 \times 0.30 \times 1 = 0.60" \text{ w.g.}$$

## Three hood types

Three types of hoods are the enclosing, capture (active, external), and receiving (passive, often canopy type hood).

## Determining Q (flow) for capture hoods

Area approach: Basically, air approaches from all directions toward the source of negative pressure. Imagine a three-dimensional sphere around the end of a small, plain duct hood. See the figure below, a point-source hood. Air molecules don't know if they are in the front, to the side, or to the back of the opening. All they know is that they are experiencing a big push to get over to that spot of negative pressure. The velocity of air moving toward the opening is equal at all points on the surface of the sphere. The surface area of a sphere is given by

$$A = 4\pi X^2$$

Knowing the area and the desired capture velocity at X, we can estimate the volume flow rate from Q = VA.

## *Example 8*

Air enters an ideal 4" plain duct hood. What is the required volume flow rate Q for capture 6" in front of the hood if we need $V_c = 100$ fpm? (Area method)

Solution:

$$Q = V_c \times A \qquad \text{(where area} = A = 4\pi X^2)$$

$$= 100 \times 4\pi (0.5')^2 \qquad \text{(where 6"} = 0.5)$$

$$= 315 \text{ cfm}$$

# VENTILATION

## Fans

The types of fans include the centrifugal (forward curved, backward inclined, radial) and axial.

Fan characteristic curves plot volume flow rate Q against static pressure, horsepower, noise, and efficiency.

Fans are specified by pressure and flow rate—the "system operating point" (SOP). The pressure is that found across the fan—at the inlet and outlet of the fan in the ductwork.

### Fan Total Pressure

The fan total pressure (FTP) represents all energy requirements for moving air through the ventilation system. FTP is calculated by adding the absolute values of the average total pressures found at the fan. If the sign convention is followed, then a formula for FTP is

$$FTP = TP(outlet) - TP(inlet)$$

Substituting for TP = SP + VP gives

$$FTP = SP_{out} + VP_{out} - SP_{in} - VP_{in}$$

If VP (out) equals VP (in), in other words, if the average inlet and outlet velocities are equal, then the VP terms in the above equation cancel, leaving

$$FTP = SP_{out} - SP_{in}$$

The FTP is often referred to as "the fan total static pressure drop."

### Example 9

The inlet and outlet conditions at a fan are $SP_{out} = 0.10"$, $SP_{in} = -0.75"$.

$$VP_{in} = VP_{out} = 0.25" \text{ w.g.}$$

What is FTP?

Solution:

$$FTP = SP_{out} - SP_{in} = 0.10" \text{ w.g.} - (-0.75") \text{ w.g.} = 0.85" \text{ w.g.}$$

## REFERENCES

OSHA, 2005. *Ventilation Investigation.* OSHA Technical Manual III, chapter 3. Washington, D.C.: U.S. Department of Labor.

## SUGGESTED READING

NIOSH, 1987. *Guidance for Indoor Air Quality Investigations.* Cincinnati, Ohio: National Institute for Occupational Safety and Health.

Spellman, F. R., and N. E. Whiting. *Safety Engineering: Principles and Practices,* 2nd ed. Lanham, Md.: Government Institutes, 2005.

USEPA, 1991. *Building Air Quality.* At www.epa.gov/iaq/pubs.

# 9

# Personal Protective Equipment

The primary objective of any health and safety program is worker protection. It is the responsibility of management to carry out this objective. Part of this responsibility includes protecting workers from exposure to hazardous materials and hazardous situations that arise in the workplace. It is best for management to try to eliminate these hazardous exposures through changes in workplace design or engineering controls. When hazardous workplace exposures cannot be controlled by these measures, personal protective equipment (PPE) becomes necessary. When looking at hazardous workplace exposures, keep in mind that government regulations consider PPE the last alternative in worker protection because it does not eliminate the hazards. PPE only provides a barrier between the worker and the hazard. If PPE must be used as a control alternative, a positive attitude and strong commitment by management is required.

—S. Z. Mansdorf

Mansdorf makes a number of important statements concerning personal protective equipment (PPE) that are worth reviewing.

*It is best for management to try to eliminate these hazardous exposures through changes in workplace design or engineering controls.* Sound familiar? We have consistently made this same point throughout this text. A hazard, any hazard, if possible, should be "engineered out" of the system or process. Determining when and how to engineer out a hazard is one of the industrial hygienist's primary functions. However, the industrial hygienist can much more effectively accomplish this if he or she is included in the earliest stages of the design process. Remember, it does little good (and is often very expensive) to attempt to engineer out any hazard once the hazard is in place.

*When hazardous workplace exposures cannot be controlled by these measures, personal protective equipment (PPE) becomes necessary.* While the goal of the industrial hygienist is certainly to engineer out all workplace hazards, we realize that this goal is virtually impossible to achieve in every case. Even in this age of robotics, computers, and other automated equipment and processes, the man-machine-process interface still exists. When people are included in the work equation, the opportunity for their exposure to hazards is very real—as injury statistics make clear.

*Consider PPE the last alternative in worker protection because it does not eliminate the hazards.* This statement is extremely important for two reasons: First, the industrial hygienist's primary goal is (as we have said before) to engineer out the problem. If this is not possible, the second alternative is to implement administrative controls. When neither is possible, PPE becomes the final choice. The key words here are "the final choice." Secondly, PPE is sometimes incorrectly perceived—by the supervisor and/or the worker—as the first line of defense against all hazards. This, of course, is incorrect and dangerous. The worker must be made to understand (by means of enforced company rules, policies, and training, and a large dose of common sense) that PPE affords only minimal protection against most hazards. **It does not eliminate the hazard.**

*PPE only provides a barrier between the worker and the hazard.* Experience shows us that when some workers put on their PPE, they also don a "superhero" mentality. Often, when workers use eye, hand, foot, head, hearing, protective clothing or respiratory protection, they adopt an "I can't be touched—I am safe" attitude. Because they feel safe, they act as if they are magically protected, invincible, and well out of harm's way. Nothing could be further from the truth. Let's look at a classic example.

A work crew was assigned to clear trees, shrubs, and undergrowth from a densely wooded area to provide clear access for valve checkers, who routinely (on a semiannual basis) inspect the operation of mechanically operated values on an underground wastewater interceptor line in the area. Because the pipeline transited a rural forested area, this clearing assignment was both routine and necessary. Many of the workers used chain saws in this clearing operation. All of them had been trained on the proper operation and safety considerations involved in using chain saws, and each worker had been issued the appropriate PPE: gloves, safety shoes, safety glasses, and hardhats with wire mesh face shields and attached ear muffs.

During the clearing operation one of the workers inadvertently cut his left leg quite severely on the inner calf with the chain saw he was using. The victim was transported to the nearest medical facility and received extensive treatment for the deep and ragged wound (remember, chain saws do not cut human flesh cleanly—they gouge out chunks).

# PERSONAL PROTECTIVE EQUIPMENT

During the accident follow-up investigation phase, we asked the victim to explain how he got injured. The answer he gave us did not surprise us, but his honesty did. He stated that he had all his PPE on. "Sure, it was uncomfortable to wear," but he had worn it anyway. And while he was cutting away, he really didn't consider the hazards involved with operating the 20" chain saw. "I knew I was well-protected with my PPE and all, so I just let the ol' saw rip away." And, of course, that is just what happened, the saw ripped away—right into his leg. "Just felt like I was fully protected," he said, shaking his head in disbelief at his own stupidity.

Sound ridiculous? Such incidents happen many times every day. Workers tend to forget that PPE is only a barrier between themselves and the hazard, one that works to dissipate force and keep hazardous materials from contacting vulnerable parts of the body. The hazard is still there, behind the barrier that PPE provides. Workers forget how easily most barriers can be circumvented, torn, or ripped away. Unless the hazard is engineered out, it is always there. All the PPE in the world cannot fully protect a worker who is not vigilant.

## OSHA'S PPE STANDARD

In the past, many OSHA standards have included PPE requirements, ranging from very general to very specific requirements. It may surprise you to know, however, that not until recently (1993–94) did OSHA incorporate a stand-alone primary PPE Standard into its 29 CFR 1910/1926 Guidelines. This relatively new *Personal Protective Equipment Standard* is covered for general industry under 1910.132-138, but you can also find PPE requirements elsewhere in the General Industry Standards. For example, 29 CFR 1910.156, OSHA's Fire Brigades Standard has requirements for firefighting gear. In addition, 29 CFR 1926.95-106 covers the construction industry. OSHA's general PPE requirements mandate that employers conduct a hazard assessment of their workplaces to determine what hazards are present that require the use of PPE, provide workers with appropriate PPE, and require them to use and maintain it in sanitary and reliable condition.

As currently written, the PPE standard focuses on head, feet, eye, hand, body (clothing), respiratory, and hearing protection.

Common PPE classifications and examples include:

- Head protection (hard hats, welding helmets)
- Eye protection (safety glasses, goggles)
- Face protection (face shields)
- Respiratory protection (respirators)
- Arm protection (protective sleeves)

- Hearing protection (ear plugs, muffs)
- Hand protection (gloves)
- Finger protection (cots)
- Torso protection (aprons)
- Leg protection (chaps)
- Knee protection (kneeling pads)
- Ankle protection (boots)
- Foot protection (boots, metatarsal shields)
- Toe protection (safety shoes)
- Body protection (coveralls, chemical suits)

Note that respiratory and hearing protection requirements have been covered extensively under their own standards for quite some time. Respiratory protection is covered under 1910.134 and hearing protection under 1910.95. Noise and hearing protection have already been discussed in chapter 5 of this text. OSHA's respiratory protection requirements are discussed later in this chapter.

Using PPE is often essential, but, as mentioned, it generally is the last line of defense after engineering controls, work practices, and administrative controls. Recall that engineering controls involve physically changing a machine or work environment. Administrative controls involve changing how or when employees do their jobs, such as scheduling work and rotating employees to reduce exposures. Work practices involve training workers how to perform tasks in ways that reduce their exposure to workplace hazards.

## Employer/Employee Requirements

OSHA mandates several requirements for both the employer and the employee under its PPE Standard, including:

- Employers are required to provide employees with personal protective equipment that is sanitary and in good working condition.
- The employer is responsible for examining all PPE used on the job to ensure that it is of a safe (and approved) design and in proper condition.
- The employer must ensure that employees use PPE.
- The employer must provide a means for obtaining additional and replacement equipment; defective and damaged PPE is not to be used.
- The employer must ensure that PPE is inspected on a regular basis.

- The employee must ensure that he or she uses PPE when required.
- Where employees provide their own PPE, the employer must ensure that it is adequate, including properly maintained and sanitized.

*Note:* While the employer must ensure the employee wears PPE when required, both the employer and employee should factor in three considerations: (1) the PPE used must not degrade performance unduly; (2) it must be reliable; and (3) it must be suitable for the hazard involved.

## Hazard Assessment

How does an industrial hygienist determine when and where an employer should provide PPE, and when the employee should use it? This can be determined in three ways:

1. From the manufacturer's guidance (when it comes to equipment and processes produced by a manufacturer, the manufacturer is considered the "expert" on the equipment or process, and is normally best-suited to determine the hazards associated with the equipment and/or processes they manufacture).
2. If the process or equipment the employee is working on or with involves chemicals, the Material Safety Data Sheets (MSDS) for the chemicals involved list the required PPE to be used.
3. OSHA mandates that the employer perform a hazard assessment of the workplace.

The purpose of the *hazard assessment* is to determine if workplace hazards are present or likely to be present that necessitate the use of PPE. If a facility presents such hazards, the employer is required to (1) select, and have each affected employee use the types of PPE that will protect the affected employee from the hazards identified in the hazard assessment; (2) communicate selection decisions to each affected employee; and (3) select PPE that properly fits each affected employee.

The employer is required to verify that the workplace hazard assessment has been conducted through a written certification that identifies (1) the workplace evaluated, (2) the person certifying that the evaluation has been performed, (3) the date of the hazard assessment, and (4) the document itself as a certification of hazard assessment.

*Note:* The industrial hygienist must maintain up-to-date copies of the PPE Hazard Assessment forms. During a recent OSHA audit, the auditor wanted to see copies of the assessments conducted at our work centers.

## PPE Training Requirement

OSHA requires the employer to provide training to each employee required to use PPE. Employees must be trained to:

- Use PPE properly
- Be aware of when PPE is necessary
- Know what kind of PPE is necessary
- Understand the limitations of PPE in protecting employees from injury
- Don, adjust, wear, and doff PPE
- Maintain PPE Properly

*Note:* During an OSHA audit of your facility, the auditor may want to look at a copy of your facility's PPE training program. Almost certainly, the auditor will want to review your training records for PPE training. Remember this: You can conduct all the training in the world, and have it performed by well-known experts in the field—but if you did not document the training, in OSHA's eyes (and the courts), it never occurred. You ***must*** have proof of training conducted.

During an OSHA audit, OSHA requires each employee to demonstrate understanding of his or her training on PPE. This is usually best accomplished by demonstration (e.g., wearing and operating SCBA) and through written or oral examination.

If the employer has reason to believe that any affected employee who has already been trained on PPE does not have the understanding and skill required, the employer must retrain such an employee. In this retraining requirement, remember that everything in life is dynamic (constantly changing), including the workplace and work assignments. OSHA understands this dynamic trend, and thus requires the employer to retrain employees when introducing new processes, equipment, or any new element in a job task that might render previous training obsolete. Changes also occur in PPE itself. Maybe a new type or model of PPE is introduced and used in the workplace. If this is the case, the employer must ensure that employees using such PPE are fully trained on the new PPE.

## HEAD PROTECTION

OSHA requires employers to ensure that employees are protected from head injury whenever they work in areas where there is a possible danger of head injury from impact, from falling or flying objects, or from electrical shock and burns.

According to 29 CFR 1926.100(b), helmets for the protection of workers against impact and penetration of falling and flying objects must meet the specifications contained in the American National Standards Institute (ANSI) publication, Z89.1-1969, *Safety Requirements for Industrial Head Protection.*

According to 29 CRF 1926.100(c), helmets for the head protection of employees exposed to high voltage electrical shock and burns must meet the specification contained in ANSI, Z89.2-1971.

## HAND PROTECTION

Under general requirements listed in 29 CFR 1910.138(a), *Hand Protection*, OSHA mandates that employers must select and require employees to use appropriate hand protection when employees' hands are exposed to hazards such as skin absorption of harmful substances, severe cuts or lacerations, severe abrasions, punctures, chemical or thermal burns, and harmful temperature extremes.

In the selection of protective gloves, 29 CFR 1910.100(b) mandates employers to base the selection of the appropriate hand protection on an evaluation of the performance characteristics of the hand protection relative to the task(s) to be performed, conditions present, duration of use, and the hazards and potential hazards identified.

## EYE AND FACE PROTECTION

Eye and face protection requirements are outlined in 29 CFR 1910.133. ANSI, in its publication *Practice for Occupational and Educational Eye and Face Protection* (Z87.1989), specifies the use and construction of protective eyewear.

Eye and face protection includes safety glasses, chemical goggles, and face shields. Appropriate selection is based on the type of hazard.

*Face shields* or chemical splash goggles are appropriate whenever a worker may be subject to splashing from chemicals. They are excellent for use whenever workers are dipping parts into open-surface tanks containing plating baths, cleaning solutions, organic chemicals, or corrosive chemicals. They are also appropriate whenever a worker may be subject to flying particles, such as from using a portable or pedestal grinder. Face shields are designed only to prevent direct splash exposures to the face and not to provide complete eye protection; they serve as "secondary" eye protection only.

*Safety glasses* are necessary when working with hazardous materials or when operating machinery, air guns, or when there is reasonable probability of injury that can be prevented by use of such equipment. Safety glasses should be affixed with side-shields.

## FOOT PROTECTION

OSHA, in its 29 CFR 1910.136, *Occupational Foot Protection*, states that the employer must ensure that each affected employee wears protective footwear when working in areas where there is a danger of foot injuries due to falling or rolling objects, or objects piercing the sole, and where employees' feet are exposed to electrical hazards.

Protective footwear is required to be durable, comfortable, and designed for expected exposures. Unacceptable footwear in most work settings includes any footwear that permits direct contact between the foot and a foreign agent (e.g., open-toe or open-heel shoes). Canvas athletic shoes are not recommended for those who handle or work around corrosive chemical substances. Steel-toed safety shoes are required whenever there is the potential for items to be dropped onto the foot.

## FULL BODY PROTECTION

According to OSHA (2005), the purpose of chemical protective clothing and equipment is to shield or isolate individuals from the chemical, physical, and biological hazards that may be encountered during hazardous materials operations. During chemical operations, it is not always apparent when exposure occurs. Many chemicals pose invisible hazards without warning properties.

The guidelines provided in this section describe the various types of clothing that are appropriate for use in various chemical operations, and provide recommendations in their selection and use.

It is important that protective clothing users realize that no single combination of protective equipment and clothing is capable of protecting against all hazards. Thus protective clothing should be used in conjunction with other protective methods, for example, engineering or administrative controls to limit chemical contact with personnel. The use of protective clothing can itself create significant wearer hazards, such as heat stress, physical and psychological stress, in addition to impaired vision, mobility, and communication. In general, the greater the level of chemical protective clothing, the greater the associated risks. For any given situation, equipment and clothing should be selected that provide an adequate level of protection. Overprotection as well as underprotection can be hazardous and should be avoided.

## DESCRIPTION OF PROTECTIVE CLOTHING

### Protective Clothing Applications

Protective clothing must be worn whenever the wearer faces potential hazards arising from chemical exposure. Some examples include:

# PERSONAL PROTECTIVE EQUIPMENT

- Emergency response
- Chemical manufacturing and process industries
- Hazardous waste site cleanup and disposal
- Asbestos removal and other particulate operations
- Agricultural application of pesticides

Within each application, there are several operations which require chemical protective clothing. For example, in emergency response, the following activities dictate chemical protective clothing use:

- *Site survey*: the initial investigation of a hazardous materials incident. These situations are usually characterized by a large degree of uncertainty and mandate the highest levels of protection.
- *Rescue*: entering a hazardous materials area for the purpose of removing an exposure victim. Special considerations must be given to how the selected protective clothing may affect the ability of the wearer to carry out rescue and may contribute to the contamination of the victim.
- *Spill mitigation*: entering a hazardous materials area to prevent a potential spill or to reduce the hazards from an existing spill (e.g., applying a chlorine repair kit on railroad tank car). Protective clothing must accommodate the required tasks without sacrificing adequate protection.
- *Emergency monitoring*: outfitting personnel in protective clothing for the primary purpose of observing a hazardous materials incident without entry into the spill site. This may be applied to monitoring contract activity for spill cleanup.
- *Decontamination*: applying decontamination procedures to personnel or equipment leaving the site. In general a lower level of protective clothing is used by personnel involved in decontamination.

## The Clothing Ensemble

The approach in selecting personal protective clothing must encompass an "ensemble" of clothing and equipment items which are easily integrated to provide both an appropriate level of protection and still allow one to carry out activities involving chemicals. In many cases, simple protective clothing by itself may be sufficient to prevent chemical exposure, such as wearing gloves in combination with a splash apron and face shield (or safety goggles).

The following is a checklist of components that may form the chemical protective ensemble:

- Protective clothing (suit, coveralls, hoods, gloves, boots)
- Respiratory equipment (SCBA, combination SCBA/SAR, air purifying respirators)
- Cooling system (ice vest, air circulation, water circulation)
- Communication device
- Head protection
- Eye protection
- Ear protection
- Inner garment
- Outer protection (overgloves, overboots, flashcover)

Factors that affect the selection of ensemble components include:

- How each item accommodates the integration of other ensemble components. Some ensemble components may be incompatible due to how they are worn (e.g., some SCBAs may not fit within a particular chemical protective suit or allow acceptable mobility when worn).
- The ease on interfacing ensemble components without sacrificing required performance (e.g., a poorly fitting overglove that greatly reduces wearer dexterity).
- Limiting the number of equipment items to reduce donning time and complexity (e.g., some communications devices are built into SCBAs, which as a unit are not NIOSH certified. This is important to take into account because only NIOSH-approved respirators should be used).

## Levels of Protection

PPE is categorized into levels, A through D. Level A is the most complete and comprehensive protection and level D is minimal protection.

The PPE levels listed below can be used as the starting point for ensemble creation; however, each ensemble must be tailored to the specific situation in order to provide the most appropriate level of protection. For example, if an emergency response activity involves a highly contaminated area or if the potential contamination is high, it may be advisable to wear a disposable covering, such as Tyvek coveralls or PVC splash suits, over the protective ensemble.

*Level A* protection includes:
- Vapor protective suit
- Two-way radio

# PERSONAL PROTECTIVE EQUIPMENT

- Pressure demand, self-contained breathing apparatus (SCBA)
- Hard hat
- Two pairs of gloves
- Chemical resistant steel toe shank and disposable booties

Use level A protection when:
- The chemical concentration is known to be above a safe level.
- During a confined space entry.
- In the presence of extremely hazardous substances (e.g., cyanide).
- In the presence of skin destructive substances.

*Note:* Level A protective clothing must resist permeation by the chemical or mixture present. Ensemble items must allow integration without loss of performance.

*Level B* protection includes:
- Two-piece liquid splash-protective suit with hood or disposable suit
- Pressure demand, full-facepiece SCBA
- Hard hat
- Two pairs of gloves
- Chemical resistant steel toe shank and disposable booties

Use level B protection when:
- Immediately dangerous to life and health (IDLH) conditions exist.
- Concentrations are above the protection factors provided by a full mask, air-purifying respirator.
- Oxygen levels are less than 19.5%.
- Contaminant skin contact is unlikely to the head and neck.
- An unidentified vapor is suspected.

*Note:* Protective clothing must resist penetration by the chemicals or mixtures present. Ensemble items must allow integration without loss of performance.

*Level C* protection includes:
- Full-face air-purifying respirator
- Hard hat
- Two pairs of gloves

- Chemical resistant steel toe with shank and disposable booties
- Two-piece suit or disposable suit

Use level C protection when:
- The air concentration is known.
- Assigned protective factors offer control with an air-purifying respirator.
- There is no threat of IDLH conditions.
- There is no skin hazard.
- There is no unidentified vapor present.

*Note:* Protective clothing items must resist penetration by the chemical or mixtures present. Chemical airborne concentration must be less than IDLH levels. The atmosphere must contain at least 19.5% oxygen.

*Level D* protection includes:
- Safety glasses
- Hard hat
- One pair of gloves
- Safety shoes
- Coveralls

Use level C protection when:
- There is no measurable concentration.
- No exposure to splash or inhalation will occur.

*Note:* This level should not be worn in the Hot Zone, and is not acceptable for chemical emergency response. The atmosphere must contain at least 19.5% oxygen.

The type of equipment used and the overall level of protection should be evaluated periodically as the amount of information about the chemical situation or process increases, and when workers are required to perform different tasks. Personnel should upgrade or downgrade their level of protection only with concurrence of the site supervisor, safety officer, or plant industrial hygienist.

The recommendations listed above serve only as guidelines. It is important for you to realize that selecting items by how they are designed or configured alone is not sufficient to ensure adequate protection. In other words, just having the right components to form an ensemble is not enough. The USEPA levels of protection do not define what performance the selected clothing or equipment must offer.

## Clothing Selection Factors

*Chemical hazards*: Chemicals present a variety of hazards such as toxicity, corrosiveness, flammability, reactivity, and oxygen deficiency. Depending on the chemicals present, any combination of hazards may exist.

*Physical environment*: Chemical exposure can happen anywhere: in industrial settings, on the highways, or in residential areas. It may occur either indoors or outdoors; the environment may be extremely hot, cold, or moderate; the exposure site may be relatively uncluttered, or rugged, presenting a number of physical hazards; chemical handling activities may involve entering confined spaces, heavy lifting, climbing a ladder, or crawling on the ground. The choice of ensemble components must account for these conditions.

*Duration of exposure*: The protective qualities of ensemble components may be limited to certain exposure levels (e.g., material chemical resistance, air supply). The decision for ensemble use time must be made assuming the worst case exposure with applicable safety margins.

*Protective clothing or equipment available*: Hopefully, an array of different clothing or equipment is available to workers to meet all intended applications. Reliance on one particular clothing or equipment item may severely limit a facility's ability to handle a broad range of chemical exposures. In its acquisition of equipment and clothing, the safety department or other responsible authority should attempt to provide a high degree of flexibility while choosing protective clothing and equipment that is easily integrated and provides protection against each conceivable hazard.

## Classification of Protective Clothing

Personal protective clothing includes the following:

- Fully encapsulating suits
- Nonencapsulating suits
- Gloves, boots, and hoods
- Firefighter's protective clothing
- Proximity, approach clothing
- Blast or fragmentation suits
- Radiation-protective suits

Firefighter turnout clothing, proximity gear, blast suits, and radiation suits by themselves are not acceptable for providing adequate protection from hazardous chemicals.

## Material Chemical Resistance

Ideally, the chosen material(s) must resist permeation, degradation, and penetration by the respective chemicals.

*Permeation* is the process by which a chemical dissolves in or moves through a material on a molecular basis. In most cases, there will be no visible evidence of chemicals permeating a material. Permeation *breakthrough time* is the most common result used to assess material chemical compatibility. The rate of permeation is a function of several factors such as chemical concentration, material thickness, humidity, temperature, and pressure. Most material testing is done with 100% chemical over an estimated exposure period. The time it takes the chemical to permeate through the material is the breakthrough time. An acceptable material is one where the breakthrough time exceeds the expected period of garment use. However, temperature and pressure effects (e.g., small increases in ambient temperature) may accelerate permeation and reduce the effectiveness of this safety factor.

*Degradation* involves physical changes in a material as the result of a chemical exposure, use, or ambient conditions (e.g., sunlight). The most common observations of material degradation are discoloration, swelling, loss of physical strength, or deterioration.

*Penetration* is the movement of chemicals through zippers, seams, or imperfections in a protective clothing material.

It is important to note that no material protects against all chemicals and combinations of chemicals, and that no currently available material is an effective barrier to any prolonged chemical exposure.

## DECONTAMINATION PROCEDURES

*Decontamination* is the process of removing or neutralizing contaminants that have accumulated on personnel and equipment. This process is critical to health and safety at hazardous material response sites. Decontamination protects end users from hazardous substances that may contaminate and eventually permeate the protective clothing, respiratory equipment, tools, vehicles, and other equipment used in the vicinity of the chemical hazard; it protects all plant or site personnel by minimizing the transfer of harmful material into clean areas; it helps prevent mixing of incompatible chemicals;

# PERSONAL PROTECTIVE EQUIPMENT

and it protects the community by preventing uncontrolled transportation of contaminants from the site.

Decontamination can be accomplished to allow for reuse of chemical protective clothing, or *gross decontamination* allows the user to safely doff and discard the chemical protective clothing.

## Prevention of Contamination

The first step in decontamination is to establish standard operating procedures that minimize contact with chemicals and thus the potential for contamination. For example:

- Stress work practices that minimize contact with hazardous substances (e.g., do not walk through areas of obvious contamination; do not directly touch potentially hazardous substances).
- Use remote sampling, handling, and container-opening techniques (e.g., drum grapples, pneumatic impact wrenches).
- Protect monitoring and sampling instruments by bagging. Make openings in the bags for sample ports and sensors that must contact site materials.
- Wear disposable outer garments and use disposable equipment where appropriate.
- Cover equipment and tools with a strippable coating that can be removed during decontamination.
- Encase the source of contaminants (e.g., with plastic sheeting or overpacks).
- Ensure all closures and ensemble component interfaces are completely secured, and that no open pockets that could serve to collect contaminant are present.

## Contaminant Permeation

*Surface contaminants* may be easy to detect and remove. *Permeated contaminants* are contaminants that are difficult or impossible to detect or remove. If contaminants that have permeated a material are not removed by decontamination, they may continue to permeate the material where they can cause an unexpected exposure.

Four major factors affect the extent of permeation:

1. *Contact time.* The longer a contaminant is in contact with an object, the greater the probability and extent of permeation. For this reason, minimizing contact time is one of the most important objectives of a decontamination program.

2. *Concentration.* Molecules flow from areas of high concentration to areas of low concentration. As concentrations of chemicals increase, the potential for permeation of personal protective clothing increases.
3. *Temperature.* An increase in temperature generally increases the permeation rate of contaminants.
4. *Physical state of chemicals.* As a rule, gases, vapors, and low-viscosity liquids tend to permeate more readily than high-viscosity liquids or solids.

## Decontamination Methods

Decontamination methods either (1) physically remove contaminants, (2) inactivate contaminants by chemical detoxification or disinfection/sterilization, or (3) remove contaminants by a combination of both physical and chemical means.

## INSPECTION, STORAGE, AND MAINTENANCE OF PROTECTIVE CLOTHING

The end user in donning protective clothing and equipment must take all necessary steps to ensure that the protective ensemble will perform as expected. During an emergency is not the right time to discover discrepancies in the protective clothing. Teach the end user to care for his or her clothing and other protective equipment in the same manner as parachutists care for parachutes. Following a standard program for inspection, proper storage, and maintenance along with realizing protective clothing/equipment limitations is the best way to avoid chemical exposure during emergency response.

## Inspection

An effective chemical protective clothing inspection program should feature five different inspections:

1. Inspection and operational testing of equipment received as new from the factory or distributor.
2. Inspection of equipment as it is selected for a particular chemical operation.
3. Inspection of equipment after use or training and prior to and after maintenance.
4. Periodic inspection of stored equipment.
5. Periodic inspection when a question arises concerning the appropriateness of selected equipment, or when problems with similar equipment are discovered.

## Storage

Clothing must be stored properly to prevent damage or malfunction from exposure to dust, moisture, sunlight, damaging chemicals, extreme temperatures, and impact. Procedures are needed for both initial receipt of equipment and after use or exposure of that equipment. Many manufacturers specify recommended procedures for storing their products. These should be followed to avoid equipment failure resulting from improper storage.

## Maintenance

Manufacturers frequently confine the sale of certain protective suit parts to the individuals or groups who are specially trained, equipped, or authorized by the manufacturer to purchase them. Explicit procedures should be adopted to ensure that the appropriate level of maintenance is performed only by those individuals who have this specialized training and equipment. In no case should you attempt to repair equipment without checking with the person in your facility that is responsible for chemical protective clothing maintenance.

## PROTECTIVE CLOTHING TRAINING

Training in the use of protective clothing is beneficial because:

- It allows the user to become familiar with the equipment in a nonhazardous, nonemergency condition.
- The user gains confidence in his or her equipment.
- It makes the user aware of the limitations and capabilities of the equipment.
- Worker efficiency in performing various tasks increases.
- The likelihood of accidents during chemical operations is reduced.

## RISKS OF WEARING PROTECTIVE CLOTHING

Wearing full body chemical protective clothing puts the wearer at considerable risk of developing heat stress. This can result in health effects ranging from transient heat fatigue to serious illness or death. Heat stress is caused by a number of interacting factors, including:

- Environmental conditions
- Type of protective ensemble worn
- Work activity required
- Individual characteristics of the responder

When selecting chemical protective clothing and equipment, each item's benefit should be carefully evaluated for its potential for increasing the risk of heat stress. For example, if a lighter, less insulating suit can be worn without a sacrifice in protection, then it should be. Because the incidence of heat stress depends on a variety of factors, all workers wearing full body chemical protective ensembles should be monitored.

In regards to *heart rate*, count the radial pulse during a 30-second period as early as possible in any rest period. If the heart rate exceeds 110 beats per minute at the beginning of the rest period, the next work cycle should be shortened by one-third.

In regards to *oral temperature*, do not permit an end user to wear protective clothing and engage in work when his or her oral temperature exceeds 100.6°F (38.1°C).

*Body water loss* is another important risk factor. Measure the end user's weight on a scale accurate to ±0.25 pounds prior to any response activity. Compare this weight with his or her normal body weight to determine if enough fluids have been consumed to prevent dehydration. Weights should be taken while the end user wears similar clothing, or ideally, in the nude. The body water loss should not exceed 1.5% of total body weight.

## SAMPLE PPE TRAINING GUIDE

This sample PPE training guide has been successfully used for more than six years. While other training guides on the subject may be more inclusive, this guide helps the industrial hygienist formulate his or her facility's PPE training requirement.

### Introduction

This training guide is designed to be used by Company work center supervisors to provide required OSHA PPE training on 29 CFR 1910.132–138. OSHA's new PPE Standard is a "Performance Standard," meaning that employers and employees must meet the minimum requirements in the standard and/or other requirements, as determined by performance and experience, and specified by organizational safety officials.

### Requirements of PPE Standard

1. Employer designated Safety Officials must conduct both a work center and employee job classification hazard assessment to determine who must use PPE and where it must be used.
2. Employers must provide approved PPE to employees who are required to use it in the normal performance of their duties.

# PERSONAL PROTECTIVE EQUIPMENT

3. Employers must train employees on where, how, and when to use PPE. Employers must also train employees on the limitations of PPE.
4. Employees are required to make continued assessments of actual PPE usage to ensure that PPE actually works as designed. Most of the above requirements are to be performed by the Company Industrial Hygienist.

The Industrial Hygienist will train work center Supervisors.

Training of work center employees will be provided by work center Supervisors.

## Employee Information

PPE is designed to protect you from health and safety hazards that cannot be removed from your work environment. PPE is specifically designed to protect many parts of your body, including your eyes, face, head, hands, feet, and hearing.

PPE such as respirators are designed to protect your pulmonary function. Respirators are covered under the Company's Respiratory Protection Program and are not required to be presented in this training session under the new PPE Standard.

## Eye and Face Protection

1. Eye and/or face protection must be worn any time that it is required by a MSDS.
2. Eye and/or face protection must be worn any time that Company Safe Work Practices require it.
3. Both eye and face protection are now required any time you work with:
   - chemicals,
   - hazardous gases,
   - flying particles,
   - molten metals, and
   - whenever deemed necessary and appropriate by the Supervisor.
4. Welder's eye and face protection must be worn when welding. Ensure proper UV rated protective glass is used for welding as follows:
   - For welding operations using less than 60 amps, shade 7 is required.
   - For welding operations using 60–160 amps, shade 8 is required.
   - For welding operations using 160–250 amps, shade 10 is required.
   - For welding operations using more than 250 amps, shade 11 is required.
   - For torch soldering use shade 2.
   - For torch brazing use shade 3.

5. When wearing safety glasses, coverage from front and sides is required.

    If employee prescription glasses meet ANSI standards for safety glasses, the employer must provide employee with protective side shields, and the employee must use them.

    If the employee's prescription glasses do not meet ANSI safety glass requirements, the employer is not required to provide prescription glasses but must provide oversize safety goggles. The oversize safety goggles must fit and seal properly over prescription glasses.

6. If the employee wears contact lenses, he or she may face additional hazards from chemicals or dust.

    Dust caught under the lens can cause painful abrasions. Chemicals can react with contacts to cause permanent injury.

    Under no circumstances are contact lenses to be considered protective devices. Eye protection must be worn in addition to or instead of contact lenses.

7. Face and eye protection devices must be distinctly marked to facilitate identification of manufacturer.

## Head Protection

1. Employees are required to wear protective helmets (hardhats) when working in areas where there is a potential for injury to the head from falling objects.
2. Employees are required to wear protective helmets (hardhats) when working in or around construction projects.
3. Employees are required to wear protective helmets (hardhats) when working in or around areas where flying debris could cause potential head injuries.
4. Employees are required to wear protective helmets (hardhats) when working near exposed electrical conductors, which could contact the head.
5. Employees are required to wear protective helmets (hardhats) when organizational Safety Officials and/or Supervisors determine the need.
6. The hardhat suspension must be designed to absorb some impact. It must be adjusted to fit the wearer and to keep the shell a minimum distance of one-and-one-fourth inches above the wearer's head.
7. Company employees are required to wear Class B hardhats. Class B hardhats are made from insulating material designed to protect you from impact and from electric shock by voltages of up to 20,000 volts.
8. Employees are to periodically check their hardhat suspension. Look for loose or torn cradle straps, loose rivets, broken sewing lines or other defects.

9. Hardhats are to be dated when issued and must be replaced every two to five years, or after major impact.

## Hand Protection

1. Employees must be issued hand protection when they are exposed to hazards such as those from skin absorption of harmful substances, severe cuts or lacerations, severe abrasions, punctures, chemical burns, thermal burns, and harmful temperature extremes.
2. Extreme caution is to be exercised whenever employees are wearing gloves while working on moving machinery.
3. The Supervisor is to provide the employee with the correct type of hand protection required for the job. Check MSDS if unsure.
4. Whatever gloves are selected and provided by the Supervisor, make sure they fit. The Supervisor is responsible for ensuring that gloves selected are the most appropriate gloves for a particular application, for determining how long they can be worn, and whether they can be reused.
5. Employees must be instructed on how to inspect gloves.

## Foot Protection

1. Employees must wear approved safety shoes when the possibility of foot injury could occur from heavy or sharp objects that fall on the feet.
2. Employees must wear approved safety shoes when something could roll over their feet.
3. Employees must wear approved safety shoes when something could pierce the sole of the shoe.
4. Employees must wear approved safety shoes whenever directed by Supervisor or designated Safety Officials.
5. Employees who work around exposed electrical wires or connections need to wear metal-free nonconductive shoes or boots.
6. Employees, who are required to work in a static-free environment (e.g., when working with computers or other electronic equipment) should wear a conductive shoe designed to drain static charges into a mat or the floor.
7. Employees who might have to work in or around chemical spills are required to be supplied with rubber or synthetic footwear (type based on MSDS).

## Cleaning and Maintenance

All PPE must be properly cleaned and maintained.

Cleaning is particularly important for eye and face protection, where dirty or fogged lenses could impair vision.

PPE is to be inspected, cleaned, and maintained at regular intervals so that the PPE provides the requisite protection.

## Summary

It is the employer's responsibility to teach employees about the PPE they require. However, it is the employee's responsibility to wear it. No one can use it for the employee except the employee.

## Sample PPE Quiz

NAME: _____

SIGNATURE: _____

DATE: _____

WORK CENTER: _____

| | | |
|---|---|---|
| TRUE  FALSE | 1. | For PPE to protect you, it must be used, and used properly. |
| TRUE  FALSE | 2. | Wearing PPE, even if improperly, is better than not wearing it at all. |
| TRUE  FALSE | 3. | Safety glasses or goggles are required any time there is a danger of something striking the eye. |
| TRUE  FALSE | 4. | A hard hat's most important part is its suspension, which should keep the shell at least one and one-fourth inches above your head. |
| TRUE  FALSE | 5. | Hard hats and bump caps are interchangeable. |
| TRUE  FALSE | 6. | Leather shoes or boots offer good protection against spills of caustic chemicals. |
| TRUE  FALSE | 7. | If you work with chemicals, regularly inspect your rubber boots and gloves and repair them if they become worn. |

TRUE    FALSE    8. Selection of the proper PPE for handling chemicals should be based on MSDS recommendation.

TRUE    FALSE    9. Any type of glasses (i.e., sunglasses, prescription glasses) can serve as regular safety glasses.

TRUE    FALSE    10. Contact lenses are useful in protecting the eyes from the harmful rays produced by welding.

## PPE: THE BOTTOM LINE

When properly and regularly used, PPE effectively provides the needed barrier for workers against the hazard. But PPE is useless unless the worker wears it and uses it properly.

Not all PPE is found on the jobsite, though. Here's an example, not from industry, but one common in homes across the country.

Alan Baker's life was saved, ten years ago, by a bicycle helmet. Alan is a serious bicyclist whose holidays always involve extensive bicycle touring. He's a skilled, watchful, and experienced bicyclist, but skill alone doesn't keep a rider safe. On a short local ride, while crossing a set of train tracks that did not cross the road on the square, the front tire of Alan's bike dropped into the gap between track and pavement, seized up, and sent the rear wheel high into the air. Alan was flung onto the pavement on his head.

His $30.00 ANSI-approved bicycle helmet was a total loss, the shell and styro crushed. Alan had scrapes and bruises, and road-burn on his chin. However, he did not have a fractured skull. He walked away from that accident.

PPE can't help you unless you use it though. At age 15, Nathaniel Hunt was struck by an unobservant driver, who ran a stop sign and struck him while he was crossing a side street on his bicycle. The head injuries he sustained kept him in a coma for four days and hospitalized for two weeks. Without the protection of a helmet, the physical damage to his brain set him back in school several years. Nathaniel knew he had heard the information the teachers were providing, knew he had read the material—but for a long time, he just couldn't make the mental connections. Nathaniel is okay, now, but he will always carry the scars from that wreck.

Most parents of today's young children are old enough to remember learning to ride their bikes and never seeing a bicycle helmet. They were not in popular use. Those of us who learned to ride twenty-five to thirty years ago (or longer) were, in part, safer on the roads than our children are—there weren't so many roads, or so many cars, or so many people. As traffic and population grew, so did the risk for more serious bicycling accidents.

In the United States, one child dies and another fifty are injured every day in bicycle accidents. Across the country, statistics show that only a small percentage of children actually wear helmets, though the proper use of bicycle helmets can reduce head injuries by as much as 85%. Fifteen states have laws requiring the use of bicycle helmets, and in those states, fatalities and head injuries from bicycle accidents are significantly reduced (Ryan, 1998).

If parents demonstrate they think the proper headgear is important by wearing helmets when they ride, their children almost always do so, too. Only about 30% of children whose parents don't wear helmets wear helmets themselves. An ANSI- or Snell-approved helmet can be purchased for under $20.00. When you consider that, for children, 33% of emergency room visits, 66% of hospitalizations, and 75% of deaths are for head injuries, using helmets for bicycling and for other sports that offer the risk of head injury (e.g., in-line skating and skate boarding) seems like a no-brainer (Ryan, 1998).

## RESPIRATORY PROTECTION

> The basic purpose of any respirator is, simply, to protect the respiratory system from inhalation of hazardous atmospheres. Respirators provide protection either by removing contaminants from the air before it is inhaled or by supplying an independent source of respirable air. The principal classifications of respirator types are based on these categories.
> —NIOSH *Guide to Industrial Respiratory Protection*

> Written procedures shall be prepared covering safe use of respirators in dangerous atmospheres that might be encountered in normal operations or in emergencies. Personnel shall be familiar with these procedures and the available respirators.
> —OSHA 29 CFR 1910.134(c)

Wearing respiratory protective devices to reduce exposure to airborne contaminants is widespread in industry. An estimated 5 million workers wear respirators, either occasionally or routinely. Although it is preferred industrial hygiene practice to use engineering controls to reduce contaminant emissions at their source, there are operations where this type of control is not technologically or economically feasible or is otherwise inappropriate.

Respirators are devices that can allow workers to safely breathe without inhaling particles or toxic gases. Two basic types are (1) *air-purifying*, which filter dangerous substances from the air; and (2) *air-supplying*, which deliver a supply of safe breathing air from a tank (SCBA), or group of tanks (cascade system), or an uncontaminated area nearby via hose or airline to a mask.

Since respirators are not as consistently reliable as engineering and work practice controls, and may create additional problems, they are not the preferred method of reducing exposures below the occupational exposure levels. Accordingly, their use as a primary control is restricted to certain circumstances. In those circumstances where engineering and work practice controls cannot be used to reduce airborne contaminants below their occupational exposure levels (e.g., certain maintenance and repair operations, emergencies, or during periods when engineering controls are being installed), the use of respirators could be justified to reduce worker exposure. In other cases, where work practices and engineering controls alone cannot reduce exposure levels to below the occupational exposure level, the use of respirators would be essential for supplemental protection.

If the industrial hygienist determines that respiratory protection is required, then it is incumbent upon him or her to implement a written respiratory protection program that is in compliance with OSHA's Respiratory Protection Standard (29 CFR 1910.134).

Respirators can only provide adequate protection if they are properly selected for the task; are fitted to the wearer and are consistently donned and worn properly; and are properly maintained so that they continue to provide the protection required for the work situation. These variations can only be controlled if a comprehensive respiratory protection program is developed and implemented in each workplace where respirators are used. When respirator use is augmented by an appropriate respiratory protection program, it can prevent fatalities and illnesses form both acute and chronic exposures to hazardous substances.

We have continuously stressed the vital need to attempt first to engineer-out any hazard. However, when engineering and other methods of control are not feasible, proper selection and use of respiratory protection can be used to protect against airborne hazards.

Unlike past practices—where respiratory protection entailed nothing more than providing respirators to workers who could be exposed to airborne hazards and then expecting workers to use the respirator to protect themselves—today, supplying respirators without the proper training, paperwork, and testing is illegal. Employers are sometimes unaware of that, and supply respirators to their employees without having a comprehensive respiratory protection program. This is a serious mistake—by issuing respirators, they have implied that a hazard actually exists. In a lawsuit, they would become fodder for the lawyers.

OSHA mandates that an effective program must be put in place. This respiratory protection program must not only follow OSHA's guidelines, but must also be well planned and properly managed. A well planned, well-written respiratory protection program must include:

- Procedures for selecting respirators for use in the workplace.
- Medical evaluations of employees required to use respirators.
- Fit-testing procedures for tight-fitting respirators.
- Use of respirators in routine and reasonably foreseeable emergency situations.

- Procedures and schedules for cleaning, disinfecting, storing, inspecting, repairing, and otherwise maintaining respirators.
- Procedures to ensure adequate air quality, air quantity and flow of breathing air for atmosphere-supplying respirators.
- Training of employees in the respiratory hazards to which they are potentially exposed.
- Training of employees in the proper use of respirators, including putting on and removing them, any limitations on their use, and maintenance procedures.
- Procedures for regularly evaluating the effectiveness of the program.

In this section, we discuss these elements and explain what they require by providing a sample written respiratory protection program. Although each individual written program must be site specific and germane to existing conditions, this sample information will aid the industrial hygienist to implement a respiratory protection program that complies with OSHA requirements.

*Note:* For permit-required confined space entry operations, respiratory protection is a key piece of safety equipment, one always required for entry into an Immediately Dangerous to Life or Health (IDLH) space, and one that must be readily available for emergency use and rescue if conditions change in a non-IDLH space. Remember, however, that *only air-supplying respirators should be used in confined spaces where there is not enough oxygen.*

Selecting the proper respirator for the job, the hazard, and the worker is very important, as is thorough training in the use and limitations of respirators. Compliance with OSHA's Respiratory Standard begins with developing written procedures covering all applicable aspects of respiratory protection. Because this requirement is important, we present the following sample written respiratory protection program that includes OSHA's required elements. Again, while this sample program is designed for a fictitious organization named "Company," in reality it has been successfully used for several years (along with the respirator program evaluation checklist presented later) and has proven its worth and effectiveness through worksite testing and OSHA evaluation.

## SAMPLE RESPIRATORY PROTECTION WRITTEN PROGRAM

I. Introduction
   The Occupational Safety and Health Act (OSH Act) requires that every employer provide a safe and healthful work environment. This includes ensuring workers are pro-

tected from unacceptable levels of airborne hazards. While most air is safe to breathe, certain work operations and locations have characteristic problems of air contamination. Control measures are required to reduce airborne hazard concentrations to safe levels. When controls are not feasible, or while they are being implemented, workers must wear approved respiratory protection.

The Company has adopted this "Respiratory Protection Program" to comply with OSHA regulations (as set forth in 29 CFR 1910.134) and to do all that is possible to protect those employees who are filling a job classification that requires respirator use in the performance of their duties. All Departments and workcenters are included and must adhere to the requirements set forth in this program. Company's "Respiratory Protection Program" is an organized approach for assuring employees a safe work place by providing specific requirements in these areas:

A. Designation of individual departmental responsibilities.

B. Definition of various terms used in the "Respiratory Protection Program."

C. Designation of types of respirators and their applications.

D. Designation of procedures for respirator selection and distribution.

E. Designation of procedures to be used for inspection and maintenance of respirators.

F. Designation of procedures for employee respirator fit-testing.

G. Designation of a procedure for medical surveillance.

H. Designation of a training program for personnel participating in Company's "Respiratory Protection Program."

I. Documentation procedure for personnel participating in Company's "Respiratory Protection Program."

II. Responsibilities

A. Department Directors will be responsible for the following:

1. Implement and ensure compliance of departmental personnel with Company's "Respiratory Protection Program."

2. Specify the job classifications that use respirators, and ensure this job requirement is included in job descriptions for these classifications.

B. Company's Safety Division has the following responsibilities under Company's "Respiratory Protection Program."

1. Develop and modify as necessary Company's written "Respiratory Protection Program."

2. Check and review quarterly all work center programs, including the work center respirator inspection record.

3. Compile and maintain a master respirator inventory list for Company.
4. Implement an ongoing respirator-training program.
5. Conduct initial and annual employee fit-testing.
6. Provide *initial* and *annual* spirometric evaluation to ensure that employees are capable of wearing a respirator under their given work conditions.
7. Provide technical assistance in determining the need for respirators and in the selection of appropriate types of respirators.
8. Forward training, fit-test, initial/annual spirometric evaluation, and medical doctor's evaluation for suitability to wear a respirator to Human Resources Manager for inclusion into employee's personnel record.
9. Inspect quarterly the accuracy and proper maintenance of records specified in this program.
10. Conduct air quality tests annually on internal combustion engine-driven airline respirator compressors to ensure proper air quality.

C. Company Supervisory personnel are responsible for the following:
1. Ensure that respirators are available to employees as needed.
2. Ensure that employees wear appropriate respirators as required.
3. Ensure inspection of cartridge type respirators on a monthly basis, and self-contained breathing apparatus (SCBA) and Airline Hose Mask systems on a weekly and monthly basis. Insure records of respirator inspections are maintained.
4. Ensure employees are fit-tested and receive initial/annual spirometric evaluation prior to using a respirator.

D. The Employee is responsible for the following:
1. Use supplied respirators in accordance with instructions and training.
2. Clean, disinfect, inspect, and store assigned respirator(s) properly.
3. Perform self–fit-test prior to each use, and ensure that manageable physical obstructions such as facial hair (mustaches only) do not interfere with respirator fit.
4. Report respirator malfunctions to supervision and conduct "After Use Inspection" of SCBA type respirator.
5. Report any poor health conditions that may preclude safe respirator usage.

E. Company Human Resources Manager is responsible for the following:
1. Schedule required initial medical examination and spirometric evaluation for all new employees who fill job classifications requiring the use of respirators.
2. Maintain records of employee medical, spirometric, and fit-test results.

III. Definition of Terms

Company's "Respiratory Protection Program" defines various terms as follows:

**aerosol**: A suspension of solid particles or liquid droplets in a gaseous medium.

**asbestos**: A broad mineralogical term applied to numerous fibrous silicates composed of silicon, oxygen, hydrogen, and metallic ions like sodium, magnesium, calcium, and iron. At least six forms of asbestos occur naturally. Types of asbestos that are currently regulated—Actinolite, Amosite, Anthophylite, Chrysotile, Crocidolite and Tremolite.

**banana oil**: A liquid which has a strong smell of bananas, used to check for general sealing of a respirator during fit-testing.

**blasting abrasive**: A chemical contaminant composed of silica, silicates, carbonates, lead, cadmium, or zinc and classified as a dust.

**breathing resistance**: The resistance that can build up in a chemical respirator cartridge that has become clogged by particulates.

**chemical hazard**: Any chemical that has the capacity to produce injury or illness when taken into the body.

**cleaning respirators**: Cleaning respirators involves washing with mild detergent and rinsing with potable water.

**dust**: A dispersion of tiny solid airborne particles produced by grinding or crushing operations.

**fit-testing**: An evaluation of the ability of a respiratory device to interface with the wearer in such a manner as to prevent the workplace atmosphere from entering the worker's respiratory system.

**forced expiratory volume** (FEV1): That volume of air which can be forcibly expelled during the first second of expiration.

**forced vital capacity** (FVC): The maximal volume of air which can be exhaled forcefully after a maximal inhalation.

**fume**: Solid particles generated by condensation from the gaseous state.

**gas**: A substance which is in the gaseous state at ordinary temperature and pressure.

**IDLH**: Immediately dangerous to life and health. Any condition that poses an immediate threat to life, or which is likely to result in acute or immediately severe health effects.

**irritant smoke** (stannic oxychloride): A chemical used to check for general sealing of a respirator during a fit-test.

**mist**: A dispersion of liquid particulates.

**oxygen deficiency**: Any level below the PEL of 19.5%.

**particulates**: Dusts, mists, and fumes.

**permissible exposure limit** (PEL): The maximum time-weighted average concentration of a substance in air that a person can be exposed to during an 8-hour shift.

### Sample PEL/IDLH Chart

| Chemical Name | PEL (8 hr. average) | IDLH |
|---|---|---|
| Ammonia | 50 ppm | 300 ppm |
| Carbon Dioxide | 5,000 ppm | 50,000 ppm |
| Carbon Monoxide | 50 ppm | 1,200 ppm |
| Sodium Hydroxide | Must use SCBA to enter | |
| Sulfur Dioxide | 5 ppm | 100 ppm |
| Chlorine | 1 ppm | 10 ppm |
| Hydrogen Chloride | 5 ppm | 100 ppm |
| Hydrogen sulfide | 10 ppm | 100 ppm |
| Propane | 1,000 ppm | 2,100 ppm |
| Oxygen | 19.5% (Min) | — |
| Flammable | 10% LEL | |

**respirator**: A face mask which filters out harmful gases and particles from air, enabling a person to breathe and work safely.

**respiratory hazard**: Any hazard that enters the human body by inhalation.

**saccharin**: A chemical sometimes used to check for general sealing of a respirator during fit-testing.

**smoke**: Particles that result from incomplete combustion.

**spirometric evaluation**: A test used to measure pulmonary function. A measurement of FVC and FEV1 of 70% or greater is satisfactory. A measurement of less than 70% may require further pulmonary function evaluation by a medical doctor.

**vapor**: The gaseous state of a substance which is liquid or solid at ordinary temperature and pressure.

IV. Types of Respirators
  A. Chemical Cartridge Respirators
    1. Description: Chemical cartridge respirators may be considered low-capacity gas masks. They consist of a facepiece, which fits over the nose and mouth of the

wearer. Attached directly to the facepiece is a small replaceable filter-chemical cartridge.

    2. Application: Usually this type of respiratory protection equipment is used where there is exposure to solvent vapors or dust and particulate matter, as with sandblasting, spray coating, or degreasing. They may not be worn in IDLH atmospheres.

B. Airline respirators (helmets, hoods, and masks), Cascade-fed or Compressor-fed.

    1. Description: These devices provide air to the wearer through a small-diameter, high-pressure hose line from a source of uncontaminated air. The source is usually derived from a compressed air line with a valve in the hose to reduce the pressure. A filter must be included in the hose line (between the compressed air line and the respirator) to remove oil and water mists, oil vapors, and any particulate matter that may be present in the compressed air. Internally lubricated compressors require that precautions be taken against overheating, since the heated oil will break down and form carbon monoxide. Where the air supply for airline respirators is taken from the compressed air line, a carbon monoxide alarm must be installed in the air supply system. Completion of prior-to-operation preventive maintenance check on the carbon monoxide alarm system is critical.

    2. Application: Airline respirators used in industrial application for confined space entry (IDLH atmosphere) must be equipped with an emergency escape bottle.

C. Self-Contained Breathing Apparatus (SCBA)

    1. Description: This type of respirator provides Grade D breathing air (not pure oxygen), either from compressed air or breathing air cylinders, or by chemical action in the canister attached to the apparatus. It enables the wearer to be independent of any outside source of air. This equipment may be operable for periods between one-half to two hours. The operation of the self-contained breathing apparatus is fairly complex, and it is therefore necessary that the wearer have special training before being permitted to use it in an emergency situation.

    2. Application: Because the oxygen-producing mechanism is self-contained in the apparatus, it is the only type of equipment that provides complete protection and at the same time permits the wearer to travel for considerable distances from a source of respirable air. SCBAs (with the exception of hot work activities) can be used in many industrial applications.

V. Respirator Selection and Distribution Procedures

Work center supervisors select respirators. Selection is based on matching the proper color-coded cartridge with the type of protection desired. Selection is also dependent upon the quality of fit and the nature of the work being done. Cartridge type respirators are issued to the individuals who are required to use them. Each individually assigned respirator is identified in a way that does not interfere with its performance. Questions about the selection process are to be referred to the Safety Division.

[*Note:* In the following sections, several references are made to various inspection records. You should design site-specific standard record forms and inspection records for use with your respiratory protection program.]

VI. Respirator Inspection, Maintenance, Cleaning and Storage

To retain their original effectiveness, respirators should be periodically inspected, maintained, cleaned, and properly stored.

A. Inspection

1. Respirators should be inspected before and after each use, after cleaning, and whenever cartridges or cylinders are changed. Appropriate entries should be made in a respirator "Inspection After Each Use" record.

2. If a 1/2-face air-purifying respirator is taken out of use, indicate it on the inspection records. The respirator must be inspected thoroughly before it is put back in use.

3. Work center supervisors shall ensure all cartridge type respirators are inspected once per month, and make appropriate entries in a "Supervisor's Monthly Respirator Inspection Checklist." The work center supervisor or designated person shall inspect all SCBAs and airline respirators weekly and monthly, and make appropriate entries in a "SCBA/Air Line Respirator Weekly and Monthly Inspection and Maintenance Checklist" record. These records are to be kept by each work center for a period of three years.

4. Safety Division personnel will inspect these records quarterly.

B. Maintenance

Respirators that do not pass inspection must be replaced or repaired prior to use. Respirator repairs are limited to the changing of canisters, cartridges, cylinders, filters, head straps, and those items as recommended by the manufacturer. No attempt should be made to replace components, or make adjustments, modifications, or repairs beyond the manufacturer's recommendations.

C. Cleaning

Individually assigned cartridge respirators are cleaned as frequently as necessary by the assignee to ensure proper protection is provided. SCBA respirators are cleaned after each use. The following procedure is used for cleaning respirators:

1. Filters, cartridges, or canisters are removed before washing the respirator, and discarded and replaced as necessary.

2. Cartridge-type and SCBA respirator facepieces are washed in a detergent solution, rinsed in clean potable water, and allowed to dry in a clean area. A clean brush is used to scrub the respirator to remove adhering dirt.

D. Storage

After inspection, cleaning, and necessary repairs, respirators are stored to protect against dust, sunlight, heat, extreme heat, extreme cold, excessive moisture, or damaging chemicals. Respirators are to be stored in plastic bags or the original case. Individuals assigned respirators are to store their respirator in assigned personal locker. General use SCBAs are to be stored in designated cabinets, racks, or lockers with other protective equipment. Respirators are not to be stored in toolboxes or in the open. Individual cartridges or masks with cartridges are to be sealed in plastic bags to preserve their effectiveness.

VII. Respirator Fit-Testing

The "Respiratory Protection Program" provides standards for respirator fit-testing. The goal of respirator fit-testing is (1) to provide the employee with a face seal on a respirator that exhibits the most protective and comfortable fit and (2) to instruct the employee on the proper use of respirators and their limitations. There are three levels of fit-testing: Initial, Annual, and Pre-Use Self-Testing.

A. The Initial and Annual fit-tests are rigorous procedures used to determine whether the employee can safely wear a respirator.

The Initial and Annual tests are conducted by the Safety Division. Both tests utilize the Cartridge and SCBA-type respirator to check each employee's suitability for wearing either type. Fit-testing requires special equipment and test chemicals such as banana oil, irritant smoke, or saccharin. In general, any change to the face or mouth may alter respirator fit, and may require the use of a specially fitted respirator; Company's Safety Division will make this determination. Upon completion of Initial fit-testing, the safety division forwards the original of the employee's Fit-Test Record to the Human Resources Manager for inclusion in the employee's file. A copy will be forwarded to the affected work center supervisor.

*Note:* Any change to the face or mouth that may alter respirator fit must be brought to the immediate attention of the work center supervisor. Dental changes—loss of teeth, new dentures, braces, and so forth—may affect respirator fit and may require a new fitting with a different type mask.

B. Pre-Use Self-Testing is a routine requirement for all employees who wear respirators.

Each time the respirator is used, it must be checked for positive and negative seal. The Safety Division will train supervisors on this procedure. Supervisors are responsible for training employees in their individual work centers.

1. Positive Pressure Check Procedure (cartridge style respirator): After the respirator has been put in place and straps adjusted for firm but comfortable tension, the exhalation valve is blocked by the wearer's palm. He or she takes a deep breath and gently exhales a little air. Hold the breath for ten (10) seconds. If the mask fits properly, it will feel as if it wants to pop away from the face, but no leakage will occur.
2. Negative Pressure Check Procedure (cartridge style respirator): While still wearing the respirator, cover both filter cartridges with the palms, and inhale slightly to partially collapse the mask. Hold this negative pressure for 10 seconds. If no air leaks into the mask, it can be assumed the mask is fitting properly.

*Note:* Self-test fit-testing can be conducted, for both positive and negative pressure checks, on the SCBA type respirator by crimping the hoses with fingers, and vice blocking airways with palm of hands.

If either test shows leakage, the following procedure should be followed:
1. Ensure mask is clean. A dirty or deteriorated mask will not seal properly, nor will one that has been stored in a distorted position. Proper cleaning and storage procedures must be used.
2. Adjust the head straps to have snug, uniform tension on the mask. If only extreme tension on the straps will seal the respirator, report this to the Supervisor. Note that a mask with uncomfortably tight straps rapidly becomes obnoxious to the wearer.

*Note:* 1910.134 (g)(1)(A) states: Personnel with facial hair that comes between the sealing surface of the facepiece and the face, or that interferes with valve function shall not be permitted to wear tight-fitting respirators. Thus, respirator wearers with beards or side burns that interfere with the face-seal are prohibited from wearing tight-fitting respirators on the job.

VIII. Medical Surveillance

OSHA states that no one should be assigned a task requiring use of respirators unless they are found medically fit to wear a respirator by competent medical authorities. Company's "Respiratory Protection Program" will include a medical surveillance procedure that includes:

A. Pre-Employment Physical/Spirometric Evaluation/Five Year Follow-Up Physical Exam

All new and regular employees who fill job classifications that require respirator use in the performance of their duties are required to pass an initial medical examination to determine fitness to wear respiratory protection on the job. Annual spirometric evaluations will be conducted to ensure that employees covered under this program meet the OSHA requirements for fitness to wear respirators. On a continuous five-year basis, all Company employees covered under this program will be reexamined by competent medical authorities to ensure their continued fitness to wear respiratory protection on the job.

Each Department Director will specify which job classifications require the employee to use respirators. A medical doctor will conduct pre-employment and five-year follow-up medical evaluation. The Safety Division will conduct spirometric evaluation. The Safety Division will forward the employee's spirometry results to the Human Resources Manager for inclusion in the employee's personnel file.

B. Annual Spirometric Evaluation

Annual spirometric evaluations will be conducted by the Safety Division on all employees filling job classifications requiring the use of respirators in the performance of their duties. Spirometry testing will be used to measure Forced Vital Capacity (FVC) and Forced Expiratory Volume-1 second (FEV1). If FVC is less than 75% and/or FEV1 is less than 70%, the employee will not be allowed to wear a respirator unless a written waiver is obtained from a medical doctor. The supervisor determines whether the employee can be exempted from work functions that require wearing a respirator.

*Note:* Company will make reasonable accommodations to allow employees to retain their current positions with specified medical restrictions on respirator use.

The Safety Division will route annual results of spirometric testing to human resources manager for inclusion in each employee's personnel file, and will notify appropriate supervisors of any employee who fails the test.

IX. Training

No worker may wear a respirator before spirometric evaluation, medical evaluation, fit-testing, and training have all been completed and documented.

A. The Safety Division holds the responsibility for providing employee respirator training.

B. Supervisors are the day-to-day monitors of the program, and have the responsibility to perform refresher training and to ensure self fit-testing is accomplished by their employees as needed.

Available dates for Safety Division administered training sessions will be published on a routine basis. Supervisors are responsible for scheduling their new employees for the next available session. Training on respiratory protection is also conducted at New Employee Safety Orientation sessions.

This respiratory protection program is subject to changes and improvements as new regulations and technologies emerge. The Safety Division will train supervisors and employees as applicable on any new information.

X. Documentation Procedures

Documentation of safety training is very important. OSHA insists that certain records be maintained on all employees. All safety-training records should be considered legal records; the likelihood of having to use safety-training records in a court of law is real.

A. The following information will be maintained by the Safety Division:
   1. Date and location of initial employee training
   2. Inventory records of all Company respirators
B. The following information will be processed by the Human Resources Manager for inclusion in the employee's personnel file.
   1. Results of annual employee fit-testing
   2. Results of new employee medical evaluation and annual spirometric testing (to remain on file for five years)
C. Supervisors will maintain:
   1. A file of respirator inspection records
   2. Respirator inventory records

*Note:* The maintenance and accuracy of all records specified in this will be inspected quarterly by the Safety Division.

XI. Procedure for Safe Use of SCBA/Supplied Air Respirators

To be in compliance with 1910.134(e)(3), Company is providing these written procedures covering the safe use of respirators (SCBA & Supplied Air Respirators only).

SCBAs and/or supplied air (with emergency escape bottles) are to be used in all situations that involve chemical handling, confined space entry during normal operations, and in emergencies.

*Note:* Air purifying/chemical cartridge respirators are to be used only for coatings and sand blasting operations, and **NEVER** for confined space entry or any other activity where oxygen deficiency or atmospheric contaminants are present.

A. Safe Use Procedure in Dangerous Atmospheres

This written procedure is prepared for safe respirator use in IDLH atmospheres that may occur in normal operations or emergencies. All Company personnel covered under this program are to be familiar with these procedures and respirators.

1. Inspect all respirator equipment prior to use to ensure that it is complete and in good repair.
2. Ensure respirator facepiece is correct size for your face; perform a self–fit-test.
3. Ensure that available air is adequate for the expected time to be used.

*Note:* No Company employee should use an SCBA that is not 100% full.

4. Test all alarms on the respirator to ensure that they work.
5. At least two fully trained and certified standby/rescue persons, equipped with proper rescue equipment (including an SCBA) will be present in the nearest safe area for emergency rescue of those wearing respirators in an IDLH atmosphere.
6. Communications (visual, voice, signal line, telephone, radio, or other suitable type) will be maintained among all persons present (those in the IDLH atmosphere and the standby person or persons). The respirator wearers are to be equipped with safety harness and safety lines to permit their removal from the IDLH atmosphere if they are overcome.
7. The atmospheres in a confined space may be IDLH because of toxic air contaminants or lack of oxygen. Before any Company employee enters a confined space, tests must be performed to determine the presence and concentration of any flammable vapor or gas, or any toxic airborne particulate, vapor, or gas, and to determine the oxygen concentration (follow all procedures as outlined in Company's Confined Space Program).
8. No one is to enter if a flammable substance exceeds the lower explosive limit (LEL). No one should enter without wearing the proper type of respirator if any air contaminant exceeds the established permissible exposure limit (PEL), or if there is an oxygen deficiency. Ensure that the confined space is force-ventilated to keep the flammable substance at a safe level.

*Note:* Even if the contaminant concentration is below the established breathing time-weighted average (TWA) limit and there is enough oxygen, the safest procedure is to ventilate the entire space continuously, and to monitor the contaminant and oxygen concentrations continuously if people are to work in the confined space without respirators.

9. If the atmosphere in a confined space is IDLH owing to a high concentration of an air contaminant or oxygen deficiency, those who must enter the space to perform work must wear a pressure-demand SCBA or a combination pressure-demand airline and self-contained breathing apparatus that always maintains positive air pressure inside the respiratory inlet covering. Fully trained and equipped rescue must be on-site and ready to respond if needed. This is the best safety practice for confined space entry and *is required* at Company.

## RESPIRATOR PROGRAM EVALUATION

The industrial hygienist must not only ensure that his or her organization's respiratory protection program complies with the nine elements covered in the sample program presented above, but must also ensure that the eleventh element, respirator program evaluation is also accomplished. Why? Because the OSHA standard (29 CFR 1910.134) requires regular inspection and evaluation of the respirator program to determine its continued effectiveness in protecting employees. Remember that periodic air monitoring is also required, to determine if the workers are adequately protected. The overall program should be evaluated at least annually, and the written program or standard operating procedure modified if necessary.

Do you have questions about how to evaluate your respiratory protection program? Good. You should. The NIOSH guidelines in *NIOSH Guide to Industrial Respiratory Protection*, publication No. 87-116 (1987), probably provide the best answer—an evaluation checklist. In general, the respirator program should be evaluated for each job or at least annually, with program adjustments, as appropriate, made to reflect the evaluation results. Program function can be separated into administration and operation. Here is a sample from the NIOSH guide.

### Sample Respiratory Protection Evaluation Checklist

*Program Administration*

YES/NO    Is there a written policy which acknowledges employer responsibility for providing a safe and healthful workplace, and assigns program responsibility, accountability, and authority?

YES/NO    Is program responsibility vested in one individual who is knowledgeable and who can coordinate all aspects of the program at the jobsite?

YES/NO    Can feasible engineering controls or work practices eliminate the need for respirators?

YES/NO  Are there written procedures/statements covering the various aspects of the respirator program, including:

- ☐ designation of an administrator?
- ☐ respirator selection?
- ☐ purchase of OSHA/NIOSH certified equipment?
- ☐ medical aspects of respirator usage?
- ☐ issuance of equipment?
- ☐ fitting?
- ☐ training?
- ☐ maintenance, storage, and repair?
- ☐ inspection?
- ☐ use under special conditions?
- ☐ work area surveillance?

## *Program Operation*

Respiratory protective equipment selection

YES/NO  Are work area conditions and worker exposures properly surveyed?

YES/NO  Are respirators selected on the basis of hazards to which the worker is exposed?

YES/NO  Are selections made by individuals knowledgeable of proper selection procedures?

YES/NO  Are only certified respirators purchased and used; do they provide adequate protection for the specific hazard and concentration of the contaminant?

YES/NO  Has a medical evaluation of the prospective user been made to determine physical and psychological ability to wear the selected respiratory protective equipment?

YES/NO  Where practical, have respirators been issued to the users for their exclusive use, and are there records covering issuance?

Respiratory protective equipment fitting

YES/NO  Are the users given the opportunity to try on several respirators to determine whether the respirator they will subsequently be wearing is the best fitting one?

YES/NO  Is the fit tested at appropriate intervals?

YES/NO Are those users who require corrective lenses properly fitted?

YES/NO Are users prohibited from wearing contact lenses when using respirators?

YES/NO Is the facepiece-to-face seal tested in a test atmosphere?

YES/NO Are workers prohibited from wearing respirators in contaminated work areas when they have facial hair or other characteristics may cause face seal leakage?

Respirator use in the work area

YES/NO Are respirators being worn correctly (e.g., head covering over respirator straps)?

YES/NO Are workers keeping respirators on all the time while in the work area?

Maintenance of respiratory protective equipment

*Cleaning and Disinfecting*

YES/NO Are respirators cleaned and disinfected after each use when different people use the same device, or as frequently as necessary for devices issued to individual users?

YES/NO Are proper methods of cleaning and disinfecting utilized?

*Storage*

YES/NO Are respirators stored in a manner so as to protect them from dust, sunlight, heat, excessive cold or moisture, or damaging chemicals?

YES/NO Are respirators stored properly in a storage facility so as to prevent them from deforming?

YES/NO Is storage in lockers and toolboxes permitted only if the respirator is in a carrying case or carton?

*Inspection*

YES/NO Are respirators inspected before and after each use and during cleaning?

YES/NO Are qualified individuals/users instructed in inspection techniques?

YES/NO Is respiratory protective equipment designated as "emergency use" inspected at least monthly (in addition to after each use)?

YES/NO Are SCBA incorporating breathing gas containers inspected weekly for breathing gas pressure?

YES/NO Is a record kept of the inspection of "emergency use" respiratory protective equipment?

*Repair*

YES/NO   Are replacement parts used in repair those of the manufacturer of the respirator?

YES/NO   Are repairs made by manufacturers or manufacturer-trained individuals?

*Special use conditions*

YES/NO   Is a procedure developed for respiratory protective equipment usage in atmospheres immediately dangerous to life or health?

YES/NO   Is a procedure developed for equipment usage for entry into confined spaces?

Training

YES/NO   Are users trained in proper respirator use, cleaning, and inspection?

YES/NO   Are users trained in the basis for selection of respirators?

YES/NO   Are users evaluated, using competency-based evaluation, before and after training?

## SUMMARY

In previous eras, miners continuously tested the air in their underground worksites by keeping caged canaries with them. When the bird stopped singing, the miner knew the air was no longer fit to breathe, and could take action to save himself. As an indicator of poor air quality, the canary was a primitive, but necessary monitoring system.

Today, of course, we have the technology to test and monitor the air quality in our worksites, and also a measure of control over what goes into our lungs, by means of respiratory equipment. However, to use these tools effectively, we must use them safely. Careless or improper use (for whatever reason) is pointless—and dangerous. Accordingly, a properly trained program administrator must administer the respiratory protection program. The employer's responsibilities include providing respirators, training, and medical evaluations at no cost to the employee.

## REFERENCES

Mansdorf, S. Z., 1993. *Complete Manual of Industrial Safety.* Englewood Cliffs, NJ: Prentice-Hall, Inc.

National Institute for Safety and Health (NIOSH), 1987. *NIOSH Guide to Industrial Respiratory Protection*, Publication No. 87-116. Cincinnati, Ohio.

OSHA, 1995. *Code of Federal Regulations* Title 29 Parts 1900–1910 (.134). Washington, D.C.: Office of Federal Register.

OSHA, 2005. "Chemical Protective Clothing." *Technical Manual.* Section 8, chapter 1. Washington, D.C.: U.S. Department of Labor.

Ryan, Michael, 1998. "How a Little Headwork Saves a Lot of Children." *Parade* magazine. (May 24): 4–5.

## SUGGESTED READING

American National Standards Institute, *American National Standard for Respirator Protection-Respirator Use-Physical Qualifications for Personnel*, ANSI Z88.6. New York: ANSI, Inc., 1984.

Barker, R. L., and G. C. Coletta, *Performance of Protective Clothing.* Philadelphia: American Society for Testing Materials, 1986.

daRoza, R. A., and W. Weaver, *Is It Safe to Wear Contact Lenses with a Full-Facepiece Respirator?* Lawrence Livermore National Laboratory manuscript UCRL-53653, 1985.

Forsberg, K. and L. H. Keith, *Chemical Protective Clothing Performance Index Book.* New York: John Wiley & Sons, 1989.

Forsberg, K. and S. Z. Mansdorf, *Quick Selection Guide to Chemical Protective Clothing.* New York: Van Nostrand-Reinhold, 1989.

Janpuntich, D. A., "Respiratory Particulate Filtration." *J. Ind. Soc. Respira. Prot.* 2(1): 137–169, 1984.

Perkins, J. L. and J. O. Stull, eds. *Chemical Protective Clothing Performance in Chemical Emergency Response.* Philadelphia: American Society for Testing Materials, 1989.

Schwope, A. D., et al. *Guidelines for the Selection of Chemical Protective Clothing.* 3rd ed. Cincinnati, Ohio: American Conference of Governmental Industrial Hygienists, 1987.

Spellman, F. R., *Confined Space Entry.* Lancaster, Pa.: Technomic Publishing Company, 1999.

Spellman, F. R., *Surviving an OSHA Audit: A Management Guide.* Lancaster, Pa.: Technomic Publishing Company, 1998.

# 10

# Toxicology: Biological and Chemical Hazards

In my many years of teaching a basic toxicology course for industrial hygiene students at Old Dominion University, I always begin the first introductory lesson of the course with the question: What caused the Mad Hatter to go mad?

Invariably, however, I have to first explain to them who the Mad Hatter was. The most famous Mad Hatter, of course, is the one from the Mad Tea Party in Lewis Carroll's *Alice in Wonderland*, the partner of the March Hare. As the story goes, both were mad, of course.

The origin of the Hatter's madness, it is believed, is that hatters really did go mad. The chemicals used in hat-making included mercurious nitrate (an occupational hazard), used in curing felt. Prolonged exposure to the mercury vapors caused mercury poisoning. Victims, like the Mad Hatter, developed severe and uncontrollable muscular tremors and twitching limbs, called "hatter's shakes"; other symptoms included distorted vision and confused speech. Advanced cases developed hallucinations and other psychotic symptoms.

What does all this have to do with toxicology and with industrial hygienists in particular?

Good question. Since toxicology is the study of poisons—or the study of the harmful effects of chemicals—and because the Hatter was exposed to poison in his place of work, someone (an industrial hygienist, maybe?) should have been familiar with the principles of industrial toxicology to minimize workplace exposure to such toxins and to protect all workers, including the Mad Hatter.

But, of course, industrial toxicology and protecting workers from occupational exposure to workplace toxins was not a high priority in the time of Lewis Carroll and his Mad Hatter.

In this digital age, let's just hit the fast-forward key to the present. Because of present regulatory compliance mandates, requiring protection of workers from all order of workplace hazards—including workplace toxins capable of causing bodily harm—a professionally trained person, a staff industrial hygienist, for example, who understands the many factors affecting toxicity is critical to ensuring a safe workplace environment.

In a sense, we can say that one of the industrial hygienist's primary functions is to prevent workers from turning into the Mad Hatters of this digital age (or, for that matter, of any other age).

## BASIC TOXICOLOGY

Toxicology is the study of poisons. Stated differently, toxicology is the study of a substance's ability (usually a chemical substance) to cause damage to the body.

Toxicology can affect us on a daily basis. Toxicology-related events occur that impact how we regulate and manage chemicals. Many of these regulations and management decisions have resulted from specific pollution incidents. Some of these incidents have been one-time events, while others occur daily, in and out of the workplace.

All matter on our planet consists of chemicals. We are made up of a few thousand different types of chemicals, some of which are considered toxic.

As Stelljes (2000) points out,

> the vast majority of these chemicals are natural; in fact, the most potent chemicals on the planet are those occurring naturally in plants and animals.

Natural chemicals are sometimes presented by the media as being "safe" relative to manufactured chemicals. As a result, they may also be considered "safe" by much of the public.

For example, "organic" produce and livestock are becoming more popular. But does this mean pesticides are unsafe? Actually, manufactured pesticides used on crops and animals are heavily regulated and rarely contain enough chemicals to be harmful at typically encountered levels.

### What Makes a Substance Toxic?

Many factors affect how toxic a substance may become, including:

- Route of exposure (how the substance enters the body)
- Dose (the amount of toxic substance taken into the body)
- The frequency and duration of exposure
- Target organ (the site of action within the body, e.g., lung, kidney, liver, etc.)
- The exposed person's age, health, weight, etc.

## What Is a poison?

Through its chemical action, a poison is a substance that impairs, injures, or kills an organism; it inhibits the activity, course of a reaction, or process of another substance.

## Toxicology and Risk

The study of the effects of chemicals on workers' health combines toxicology with risk assessment. *Risk* is the probability that a substance will produce harm under specified conditions of use. Risk assessment evaluates the relative safety of chemicals from an exposure approach, considering that both contact with a chemical and the inherent toxicity of the chemical are needed to have an effect. We can't eliminate the toxicity of a chemical, but we can limit worker exposure to it.

## The Industrial Hygienist and Limiting Exposure

As mentioned, the industrial hygienist's primary function in the workplace is to minimize workplace exposures and protect workers. To accomplish this the industrial hygienist incorporates the tenets of safety. Safety is the probability that harm will not occur under specified conditions of use. The industrial hygienist must identify a substance's probability to cause harm to workers and specify the conditions of use and/or exposure to protect the worker.

In order to identify a substance's probability to cause harm and then to specify its safe use, the industrial hygienist must study the applicable material safety data sheet (MSDS) and/or manufacturer's information and substance labels to determine the risks. He or she must also be able to recognize the signs and symptoms of exposure. *Signs of exposure* are objective evidence of disease and *symptoms* are subjective evidence of disease perceived by the worker. In determining signs and diagnosing symptoms, of course, the expert help of an examining occupational health professional is needed.

## LD50/LC50

A common measure of acute toxicity is the lethal dose (LD50) or lethal concentration (LC50) that causes death (resulting from a single or limited exposure) in 50% of the treated animals, known as the population. LD50 is generally expressed as the dose, milligrams (mg) of chemical per kilogram (kg) of body weight. LC50 is often expressed as mg of chemical per volume (with results expressed in terms of an air concentration, e.g., ppm) the organism is exposed to. Chemicals are considered highly toxic when the LD50/LC50 is small and practically nontoxic when the figure is large.

### Dose-Response

The dose-response relationship, which defines the potency of a toxin, is typically the primary thrust of basic toxicology training. This is the case because the dose-response relationship is the most fundamental and pervasive concept in toxicology. To understand the potential hazard of a specific toxin (chemical), toxicologists must know both the type of effect it produces and the amount, or dose, required to produce that effect.

The relationship of dose to response can be illustrated as a graph called a *dose-response curve*. There are two types of does-response curves: one that describes the graded responses of an *individual* to varying doses of the chemical and one that describes the distribution of response to different doses in a *population* of individuals. The dose is represented on the x-axis. The response is represented on the y-axis.

An important aspect of dose-response relationships is the concept of *threshold*. For most types of toxic responses, there is a dose, called a threshold, below which there are no adverse effects from exposure to the chemical. This is important because it identifies the level of exposure to a toxin at which there is no effect.

### Routes of Entry

A toxin is not toxic to a living system unless there is exposure. Chemical exposure is defined as contact with a chemical by an organism. For humans, there are three primary routes of chemical exposure, in order of importance:

- Inhalation (i.e., through breathing)
- Dermal contact (i.e., by touching)
- Ingestion (i.e., by eating)

Exposures to toxic chemicals can be in the form of liquids, gases, mists, fumes, dusts, and vapors.

Generally, toxic agents are classified in terms of their target organs, use, source, and effects.

## Other Pertinent Toxicological Terms

**absorption**: Passage of a chemical across a membrane and into the body.

**acaricide**: A pesticide that targets spiders.

**acceptable risk**: Risk that is so low that no significant potential for toxicity exists, or a risk society considers is outweighed by benefits.

**acute**: A single or short-term exposure period.

**alkaloid**: Adverse group of structurally related chemicals naturally produced by plants; many of these chemicals have high toxicity.

**Ames assay**: Popular laboratory *in vitro* test for mutagenicity using bacteria.

**anesthetic**: A toxic depressant effect on the central nervous system.

**bioassay**: Toxicity study in which specific toxic effects from chemical exposure are measured in the laboratory using living organisms.

**carcinogen**: A cancer-causing substance.

**chronic**: An exposure period encompassing the majority of the life span for a laboratory animal species, or covering at least 10 percent of a human's life span.

**dermal contact**: Exposure to a chemical through the skin.

**dose**: The amount of a chemical entering the body; it is a function of the amount of and exposure to that chemical. In the 16th century, Paracelsus said, "All things are poison and nothing [is] without poison. Solely the dose determines that a thing is not a poison." Paracelsus' quotation is key to understanding what toxicity means. Knowing that a chemical is toxic does not automatically imply that adverse effects will result from exposure to any dose.

**exposure**: Contact with a chemical by a living organism.

**hazard**: Degree of likelihood of noncancer adverse effects occurring from chemical exposure.

**insecticide**: Pesticide that targets insects.

**mutagen**: A change in normal DNA structure.

**pesticide**: Chemical used to control pests.

**potency**: The relative degree of toxic effects caused by a chemical at a specific dose.

**risk**: The probability of an adverse effect resulting from an activity or from chemical exposure under specific conditions.

**sensitivity**: The intrinsic degree of an individual's susceptibility to a specific toxic effect.

**target organ**: Primary organ where a chemical causes noncancer toxic effects.

**teratogen**: Chemical causing a mutation in the DNA of a developing offspring.

**threshold dose**: Dose below which no adverse effects will occur.

Table 10.1. Scale of Relative Toxicity

| Category | Concentration | Amount for Average Adult | Example |
| --- | --- | --- | --- |
| Extremely toxic | <1 mg/kg | taste | botulinum |
| Highly toxic | 1–50 mg/kg | 7 drops–teaspoon | nicotine, Cyanide |
| Moderately toxic | 50–500 mg/kg | teaspoon–ounce | DDT |
| Slightly toxic | 500–5,000 mg/kg | ounce–pint | salt |
| Practically nontoxic | 5,000–15,000 mg/kg | pint–quart | ethanol |
| Relatively harmless | >15,000 mg/kg | >1 quart | water |

*Source*: Adapted from Stelljes, 2000.

**toxicity**: The degree to which a chemical inherently causes adverse effects. The concept of toxicity is not clear-cut, however. Cyanide, for example, is more toxic than table salt because toxic effects can result from much smaller amounts of cyanide. However, as Stelljes (2000) points out, we are not usually exposed to cyanide, but we are exposed to table salt daily.

Which chemical is more dangerous to us? See table 10.1.

**toxin**: A biologically-produced chemical that has toxicity.

**xenobiotic**: A chemical foreign to a living organism.

## Significant Chemical and Biological Toxins and Effects

There are several hundred thousand chemicals either naturally produced or manufactured that have some use for humans. We have adequate human toxicology information for less than 100 of these, and adequate animal toxicology information for less than 1,000, therefore, we do not have adequate information for almost all chemicals. (Stelljes, 2000)

Note that from the industrial hygiene point of view, toxicology is primarily concerned with exposure to and the harmful effects of chemicals in the occupational setting. In the definitions that follow we also include key biological agents along with certain chemical agents because of the nature of our present circumstances in regards to world-wide terrorism. Typically, under normal circumstances (whatever that may be), the industrial hygienist does not concern him- or herself with smallpox, anthrax, botulinum, and so on. The problem is that these are not "normal" times. One of the industrial hygienist's principal tasks is to "anticipate" hazards in the workplace. The possibility of and the potential for deliberate exposure to both chemical and biological toxins in today's workplace (e.g., anthrax in the post office) is very real. Moreover, exposure to toxins from bloodborne pathogens, foodborne disease, and the like can occur anywhere, including in the workplace.

(Concerning the foodborne disease issue, note that the landmark Delaney Clause, adopted in 1958 in the United States as part of the Food Additives Amendment, required that any food additive be found "safe" before the FDA approves it for use in food. This means that no chemical can be used as a food additive if there is a known potential for it to cause cancer.) The industrial hygienist must be aware of eating habits in the workplace because of the possibility of inadvertent mixing of workplace chemicals (contaminants) with the worker's brown-bag lunch.

The bottom line: at present, we think it is prudent for today's industrial hygienist to "anticipate" (and expect) worker exposure to workplace contaminants—no matter the source.

**anthrax**: Anthrax is an acute infectious disease caused by a spore-forming bacterium called Bacillus anthracis. It is generally acquired following contact with anthrax-infected animals or anthrax-contaminated animal products. Can also be acquired following a deliberate terrorist act.

**asbestos**: Asbestos is the name given to a group of six different fibrous minerals (amosite, chrysotile, crocidolite, and the fibrous varieties of tremolite, actinolite, and anthophyllite) that occur naturally in the environment. Asbestos minerals have separable long fibers that are strong and flexible enough to be spun and woven and are heat resistant. Because of these characteristics, asbestos has been used for a wide range of manufactured goods, mostly in building materials (roofing shingles, ceiling and floor tiles, paper products, and asbestos cement products), friction products (automobile clutch, brake, and transmission parts), heat-resistant fabrics, packaging, gaskets, and coatings. Some vermiculate or talc products may contain asbestos. Exposure to asbestos fibers can cause mesothelioma or asbestosis.

**avian flu**: Avian influenza is a highly contagious disease of birds which is currently epidemic amongst poultry in Asia. Despite the uncertainties, poultry experts agree that immediate culling of infected and exposed birds is the first line of defense for both the protection of human health and the reduction of further losses in the agricultural sector.

**bloodborne pathogens and needlestick prevention**: OSHA estimates that 5.6 million workers in the health care industry (sharps) and related occupations are at risk of occupational exposure to bloodborne pathogens, including human immunodeficiency virus (HIV), hepatitis B virus (HBV), hepatitis C virus (HCV), and others.

**botulism**: Cases of botulism are usually associated with consumption of preserved foods. However, botulinum toxins are currently among the most common compounds explored by terrorists for use as biological weapons.

**carbon monoxide**: Carbon monoxide is a vapor that can pass across the alveoli into the lungs through inhalation. Carbon monoxide causes carboxyhemoglobin formation (CO binds strongly with hemoglobin), replacing oxygen in red blood cells, leading to asphyxiation.

**cotton dusts:** Cotton dust is often present in the air during cotton handling and processing. Cotton dust may contain many substances including ground-up plant matter, fiber, bacteria, fungi, soil, pesticides, noncotton matter, and other contaminants that may have accumulated during growing, harvesting, and subsequent processing or storage periods. Occupational exposure to cotton dusts can cause byssinosis (tightness in chest, chronic bronchitis).

The OSHA Cotton Dust Standard 1910.1043 specifically lists the operations that are covered; operations not specifically listed are not covered by the standard. Covered operations include: yarn manufacturing, textile waste houses, slashing and weaving operations, waste recycling, and garneting.

**cyanide:** Cyanide is a rapidly acting, potentially deadly chemical that can exist in various forms. Cyanide can be a colorless gas, such as hydrogen cyanide (HCN) or cyanogens chloride (CNCl), or a crystal form such as sodium cyanide (NaCN) or potassium cyanide (KCN). Cyanide sometimes is described as having a "bitter almond" smell, but it does not always give off an odor, and not everyone can detect this odor. Cyanide is released from natural substances in some foods and in certain plants such as cassava. Cyanide is contained in cigarette smoke and the combustion products of synthetic materials such as plastics. Combustion products are substances given off when things burn. In manufacturing, cyanide is used to make paper, textiles, and plastics. It is present in the chemicals used to develop photographs. Cyanide salts are used in metallurgy for electroplating, metal cleaning, and removing gold from its ore. Cyanide gas is used to exterminate pests and vermin in ships and buildings. If accidentally ingested, chemicals found in acetonitrile-base products that are used to remove artificial nails can produce cyanide.

**foodborne disease:** Foodborne illnesses are caused by viruses, bacteria, parasites, toxins, metals, and prions (microscopic protein particles). Symptoms range from mild gastroenteritis to life-threatening neurologic, hepatic, and renal syndromes.

**hantavirus:** Hantaviruses are transmitted to humans from the dried droppings, urine, or saliva of mice and rats. Animal laboratory workers and persons working in infested buildings are at increased risk to this disease.

**isocyanates:** Methyl isocyanate (MIC) is used to produce carbarnate pesticides. Methyl isocyanate is extremely toxic to humans from acute (short-term) exposure. In Bhopal, India, accidental acute inhalation exposure to methyl isocyanate resulted in the deaths of several thousand people and adverse health effects in greater than 170,000 survivors. Pulmonary edema was the probable cause of death in most cases, with many deaths resulting from secondary respiratory infections.

**Legionnaires' disease***:* Legionnaires' disease is a bacterial disease commonly associated with water-based aerosols. It is often the result of poorly maintained air condition cooling towers and portable water systems.

**mercuric nitrate**: "Mad Hatter's" downfall; attacks the central nervous system (CNS).

**methyl alcohol**: Methanol; wood alcohol; Columbian spirits; Carbinol. Causes eye, skin, upper respiratory irritation; headache; drowsiness, dizziness; nausea, vomiting; dilation of the pupils, visual disturbance, blindness; excessive sweating; and dermatitis. If ingested, causes acute: abdominal pain; shortness of breath; vomiting; cold clammy extremities; blurring of vision, and hyperemia of the optic disc.

**methylene chloride**: Employees exposed to methylene chloride are at increased risk of developing cancer; adverse effects on the heart, central nervous system, and liver; and skin or eye irritation. Exposure may occur through inhalation, by absorption through the skin, or through contact with the skin. Methylene chlorine is a solvent which is used in many different types of work activities, such as paint stripping, polyurethane foam manufacturing, cleaning, and degreasing.

**molds and fungi**: Molds and fungi produce and release millions of spores small enough to be air-, water-, or insect-borne which may have negative effects on human health including allergic reactions, asthma, and other respiratory problems.

**organochlorine insecticides**: Any of the many chlorinated insecticides (e.g., DDT, dieldrin, chlordane, BHC, Lindane)—neurotoxins.

**organophosphate insecticides**: Chlorpyrifos, dimethoate, Malathion and trichlorfon are organophosphate (OP) insecticides. These insecticides interfere with nerve-impulse transmission, blocking the action of cholinesterase enzymes essential to proper nerve function. Symptoms of OP poisoning include headache, sweating, nausea and vomiting, diarrhea, loss of coordination and, in extreme cases, death.

**paradichlorobenzene**: Paradichlorobenzene is a mild respiratory irritant and hepatotoxic and is used in tobacco growing as a plant bed treatment for disease control. It is also used as a fumigant for clothes moths in fabric, and for ant control. Used on apricots, cherries, nectarines, peaches and plums for insect control. Also used as a fumigant and repellent in combination with other materials to control squirrels, moles, gophers, and rats and to repel cats and dogs.

**PCBs**: PCBs belong to a family of organic compounds known as chlorinated hydrocarbons. Most PCBs were sold for use as dielectric fluids (insulating liquids) in electrical transformers and capacitors. When released into the environment, PCBs do not easily break apart and form new chemical arrangements (i.e., they are not readily biodegradable). Instead they persist for many years, and bioaccumulate and bioconcentrate in organisms. Exposure to PCBs in humans can cause chloracne (a painful, disfiguring skin ailment), liver damage, nausea, dizziness, eye irritation, and bronchitis.

**plague**: The World Health Organization reports 1,000 to 3,000 cases of plague every year. A bioterrorist release of plague could result in a rapid spread of the pneumonic form of the disease, which could have devastating consequences.

**ricin**: Ricin is one of the most toxic and easily produced plant toxins. It has been used in the past as a bioterrorist weapon and remains a serious threat.

**severe acute respiratory syndrome** (SARS): SARS is an emerging, sometimes fatal, respiratory illness. According to the Centers for Disease Control and Prevention (CDC), the most recent human cases of SARS were reported in China in April 2004 and there is currently no known transmission anywhere in the world.

**silica**: Silicosis (fibrosis).

**smallpox**: A highly contagious disease unique to humans. It is estimated that no more than 20 percent of the population has any immunity from previous vaccination.

**thalidomide**: Thalidomide is probably one of the most well known teratogens. Teratogens are agents that cause offspring to be born with abnormalities (e.g., heart malformation, cleft palate, undeveloped or underdeveloped limbs). Teratogens cause their damage when the fertilized embryo is first forming an organ. At that time the teratogen interferes with the proper development of that organ. (By contrast, mutagens cause birth defect by altering sperm or egg cell DNA before the egg is fertilized.)

**tri-ortho-cresyl phosphate** (TOCP) (Jamaica Ginger): According to the American Botanical Council (1995), early in the year 1930, a strange new paralytic illness was affecting relatively large numbers of individuals. Victims of the disease would typically notice numbness in the legs, followed by weakness and eventual paralysis with "foot drop." In most cases, this was followed within about a week by a similar process in the arms, resulting in many cases in "wrist drop." The disease was rarely fatal, but recovery was very slow and in many cases the damage to the nervous system left the patient with permanent disabilities. It did not take long after the first appearance of the illness to link it to the consumption of fluid extract of Jamaica ginger, commonly referred to as "Jake" by many who used the product. During investigation of the incidents, chemists soon discovered the presence of a cresol compound, a substance that they had never before encountered in adulterated fluid extracts of ginger. Later, it was certain that the compound was TOCP and that this substance was present to the extent of about 2% in samples allegedly associated with paralysis.

**tularemia**: Also known as "rabbit fever" or "deer fly fever." Tularemia is a zoonotic disease and is extremely infectious. Relatively few bacteria are required to cause the disease, which is why it is an attractive weapon for use in bioterrorism.

**vinyl chloride**: Most vinyl chloride is used to make polyvinyl chloride (PVC) plastic and vinyl products. Acute (short-term) exposure to high levels of vinyl chloride in air has resulted in central nervous system (CNS) effects, such as dizziness, drowsiness, and headaches in humans. Chronic (long-term) exposure to vinyl chloride through inhalation and oral exposure in humans has resulted in liver damage. Cancer is a major concern from exposure to vinyl chloride via inhalation, as vinyl chloride exposure has been shown to increase the risk of a rare form of liver cancer in humans. USEPA has classified vinyl chloride as a Group A human carcinogen.

**viral hemorrhagic fevers** (VHFs): A group of illnesses that are caused by several distinct families of viruses. In general, the term "viral hemorrhagic fever" is used to describe a severe multisystem syndrome. Characteristically, the overall vascular system is damaged, and the body's ability to regulate itself is impaired. These symptoms are often accompanied by hemorrhage (bleeding); however, the bleeding itself is rarely life-threatening. While some types of VHFs can cause relatively mild illnesses, many of these viruses cause severe, life-threatening disease.

## Factors Affecting Toxicity

The amount of a toxin that reaches the target tissue is dependent upon four factors: absorption, distribution, metabolism, and excretion. These four in combination govern the degree of toxicity, if any, from chemical exposure.

*Absorption* is defined as passage of a chemical across a membrane into the body. There are four major factors that affect absorption and subsequent distribution, metabolism, and excretion: (1) size of the molecule, (2) lipid solubility, (3) electrical charge, and (4) cell membrane carrier molecules (Kent, 1998). Until a chemical is absorbed, toxic effects are only rarely observed and then only at points of contact with the body, for example, acid burns on the skin (Stelljes, 2000).

Once a chemical is absorbed into the body, it is *distributed* to certain organs via the blood stream (circulatory system). The rate of distribution to each organ is related to the blood flow through the organ, the ease with which the chemical crosses the local capillary wall and the cell membrane, and the affinity of components of the organ for the toxin (Lu and Kacew, 2002).

*Metabolism* is the sum of all physical and chemical changes that take place in an organism, includes the breakdown of substances, formation of new substances, and changes in the energy content of cells. Metabolism can either increase or decrease the toxicity, but typically increases the water solubility of a chemical, which leads to increased excretion (Stelljes, 2000).

*Excretion* is defined as elimination from the body, either as urine, feces, or through sweat or tears. The rate at which excretion of toxic substances occurs is important in determining the toxicity of a substance. The faster a substance is eliminated from the body, the more unlikely a biological effect will be (Kent, 1998).

*Other factors* affecting toxicity include:

- Rate of entry and route of exposure; that is, how fast is the toxic dose is delivered and by what means.
- Age, which can affect the capacity to repair damage.
- Previous exposure, which can lead to tolerance, increased sensitivity, or make no difference.
- State of health, medications, physical condition, and life style. Pre-existing disease can result in increased sensitivity.
- Host factors including genetic predisposition and the sex of the exposed individual.

## Classifications of Toxic Materials by Physical Properties

*Gas* is a form of matter that is neither solid nor liquid. In its normal state at room temperature and pressure, it can expand indefinitely to completely fill a container. A gas can be changed to its liquid or solid state under the right temperature and pressure conditions.

*Vapor* is the gaseous phase of a material which is ordinarily a solid or a liquid at room temperature and pressure. Vapors may diffuse. Evaporation is the process by which a liquid is changed into the vapor state and mixed with the surrounding air. Solvents with low boiling points will volatize readily.

*Aerosol* is liquid droplets or solid particles dispersed in air that are of fine enough size (less than 100 micrometers) to remain dispersed for a period of time. The toxic potential of an aerosol is only partially described by its concentration in $mg/m^3$. For a proper assessment of the toxic hazard, the size of the aerosol's particles is important. Particles between 5 and 10 micrometers ($\mu m$) will only deposit in the upper respiratory tract. Those between 1 and 5$\mu m$ will deposit in the lower respiratory tract. Very small particles ($<0.5\mu m$) are generally not deposited.

*Dust* is solid particles suspended in air produced by some physical process such as crushing, grinding, abrading, or blasting. Dusts may be an inhalation, fire, or dust-explosion hazard.

*Mist* is liquid droplets suspended in air produced by some physical process such as spraying, splashing, boiling, or by condensation of vapor.

*Fume* is an airborne dispersion of minute solid particles arising from the heating of a solid (such as molten metal). Gases and vapors are not fumes, although the terms are often mistakenly used interchangeably.

*Smoke* is minute airborne particles either liquid or solid (but usually carbon or soot), generated as result of incomplete combustion of an organic material.

## Target Systems and Organs Commonly Affected by Toxins

Once a toxin enters the body it is distributed by the circulatory system, where it may be absorbed by and accumulated in a variety of tissues. The composition of the tissue and the physio-chemical properties of the toxicant will determine where toxic substances will concentrate and exert their effects (Kent, 1998).

*Central Nervous System* (CNS) *effects*: The effects of toxic substances on neurons in the CNS may be placed into two categories: (1) those that affect the neuron structure, and (2) those that affect the neurotransmission between the presynaptic terminal and postsynaptic membrane. In each situation normal neuron functioning is disrupted. Damage to the structure of the neuron can be caused by a variety of toxic substances. For example, toxins in alcohol, carbon disulfide, and organomercury compounds cause damage in the CNS.

*Hematoxic (Blood) effects*: Toxic substances may have a direct effect on mature blood cells; further, they may affect the bone marrow, where the blood cells are produced.

*Immune System effects*: Immunotoxins affect the cellular component of the immune system by inhibiting the production of leukocytes in the bone marrow, or by inhibiting their proliferation in response to an antigen.

*Cardiotoxic effects*: Toxic substances affect the heart in several ways. The toxicant may have a direct effect on the cardiac tissues by affecting the cell membrane integrity or cellular metabolism, such as enzyme synthesis, or ATP production.

*Pulmonary (lung) effects*: Toxic substances that are inhaled can be divided into two categories: particulates and nonparticulates. These toxic substances have various effects on different regions of the respiratory tract, including irritation, sensitization, scar formation, cancer, pneumonia, and emphysema.

*Hepatotoxic (liver) effects*: The liver is one of the primary organs of the body involved in the detoxification of harmful substances that enter the bloodstream. When some toxic substances are retained within the liver cells they cause intracellular damage. They interfere with normal protein synthesis and enzyme functions. This interference, as well as damage to structural components of the cell results in death of the liver cells.

*Nephrotoxic (kidney) effects*: Because the kidney filters unwanted toxins out of our bloodstream and excretes them in urine, it is the target of unwanted toxins. Further, the kidneys have a high amount of blood flowing through them. This facilitates the accumulation of toxic substances.

*Reproductive system effects*: Some toxins impact the ability to reproduce. Chemicals may act on both sexes, or many only affect one sex. For example, DBCP, a fumigant, impacts the reproductive ability of only one sex. Formaldehyde is an example of a toxin that impacts the reproductive ability of females only.

## Carcinogen, Mutagens, and Teratogens

A *carcinogen* is an agent which may produce cancer (uncontrolled cell growth), either by itself or in conjunction with another substance.

A *suspect carcinogen* is an agent which is suspected of being a carcinogen based on chemical structure, animal research studies, or mutagenicity studies.

The International Agency for Research on Cancer (IARC) classifies carcinogens in the following manner:

1, carcinogenic to humans with sufficient human evidence.

2A, probably carcinogenic to humans with some human evidence.

2B, probably carcinogenic to humans with no human evidence.

3, sufficient evidence of carcinogenicity in experimental animals.

The National Institute for Occupational Safety and Health (NIOSH) classifies carcinogens as either carcinogenic or noncarcinogenic with no further categorization.

The National Toxicology Program (NTP) classifies carcinogens in the following manner:

a, carcinogenic with human evidence.

b, carcinogenic with limited human evidence but sufficient animal evidence.

The American Conference of Governmental Industrial Hygienists (ACGIH) classifies carcinogens in its TLVs (threshold limit values) as:

A1, confirmed human carcinogen. The agent is carcinogenic to humans based on the weight of evidence from epidemiologic studies.

A2, suspected human carcinogen. Human data are accepted as adequate in quality but are conflicting or insufficient to classify the agent as a confirmed human carcinogen; *or*, the agent is carcinogenic in experimental animals at doses, by routes of exposure, at sites, of histologic types, or by mechanisms considered relevant to worker exposure. The A2 is used primarily when there is limited evidence of carcinogenicity in humans and sufficient evidence of carcinogenicity in experimental animals with relevance to humans.

A3, confirmed animal carcinogen with unknown relevance to humans. The agent is carcinogenic in experimental animals at a relatively high dose, by routes of administration, at sites, or histologic types, or by mechanisms that may not be relevant to the worker. Available epidemiological studies do not confirm an increased risk of cancer in exposed humans. Available evidence does not suggest that the agent is likely to cause cancer in humans except under uncommon or unlikely routes or levels of exposure.

A4, not classifiable as a human carcinogen. These agents cause concern that they could be carcinogenic for humans but cannot be assessed conclusively because of a lack of data.

A5, not suspected as a human carcinogen. The agent is not suspected to be a human carcinogen on the basis of properly conducted epidemiologic studies in humans.

A *Teratogen* (literal translation is "monster making") is a substance which can cause physical defects in a developing embryo.

A *Mutagen* is a material that induces genetic changes (mutations) in the DNA. Mutations may or may not lead to cancer.

## EXPOSURE SENSITIVE OCCUPATIONS

In order to effect proper controls to protect workers, industrial hygienists must not only understand the basics of exposure sensitive industrial occupations and the processes involved but also associated exposure hazards. Materials used and exposure times are also important considerations.

In the following, we have highlighted a few exposure sensitive occupations to illustrate many of the hazards involved and the controls and work practices that might be enacted to protect the worker from lead exposure. The described operations are based on information provided in OSHA's Technical Manual, section V, chapter 3 (2005).

### Open Abrasive Blast Cleaning

*The Process*

The most common method of removing lead-based paints is open abrasive blast cleaning.

The abrasive medium, generally steel shot/grit, sand, or slag, is propelled through a hose by compressed air. The abrasive material abrades the surface of the structure, exposing the steel substrate underneath. The abrasive also conditions the substrate, forming a "profile" of the metal, which improves the adherence of the new paint.

The work is generally organized so that blasting proceeds for approximately one-half day, followed by compressed air cleaning of the steel and application of the prime coat of paint. Prime coat painting must follow blasting immediately to prevent surface rust from forming. Intermediate or finish coats of paint are applied later.

Structures that are typically cleaned by open abrasive blasting are bridges, tanks and towers, locks and dams, pipe racks, pressure vessels and process equipment, supporting steel and metal buildings.

### Engineering Controls

Containment/ventilation systems should be designed and operated so as to create a negative pressure within the structure, which reduces the dispersion of lead into the environment. The containment/ventilation system should be designed to optimize the flow of ventilation air past the worker(s), thereby reducing the airborne concentration of lead and increasing visibility.

Mini-enclosures, which have smaller cross-sectional areas than conventional enclosures, can be erected. Mini-enclosures have advantages over larger conventional enclosures because the same size fan and dust collector can achieve much higher velocities past the helmets of the workers. Mini-enclosure containment structures are usually light-weight, low wind-loading structures that isolate that area where blasting and surface priming is taking place on a given day.

The risk of silicosis is high among workers exposed to abrasive blasting with silica-containing media, and this hazard is difficult to control. NIOSH has therefore recommended since 1974 that silica sand be prohibited as abrasive blasting material.

Blast cleaning with recyclable abrasives such as steel grit or aluminum oxide requires specialized equipment for vacuuming or collecting the abrasive for reuse, separating the lead dust and fines from the reusable abrasive, and, in the case of steel grit, maintaining clean, dry air to avoid rusting of the abrasive.

When site conditions warrant, less dusty methods should be used in place of open abrasive blast cleaning. These include:

- Vacuum-blast cleaning
- Wet abrasive blast cleaning
- High-pressure water jetting
- High-pressure water jetting with abrasive injection
- Ultrahigh-pressure water jetting
- Sponge jetting

- Carbon-dioxide (dry-ice) blasting
- Chemical stripping
- Power-tool cleaning

### Work Practice Controls

Construction employers engaged in open abrasive blast cleaning operations should implement the following control measures:

1. Develop and implement a good respiratory protection program in accordance with OSHA requirements in 29 CFR 1926.103.
2. Provide workers with Type CE abrasive-blast respirators; these are the only respirators suitable for use in abrasive-blasting operations. Currently there are only three models of Type CE abrasive blast respirators certified by MSHA/NIOSHA:
   - A continuous-flow respirator with a loose-fitting hood that has a protection factor of 25
   - A continuous-flow respirator with a tight-fitting face-piece that has a protection factor of 50
   - A pressure-demand respirator with a tight-fitting face-piece that has a protection factor of 2000.

The first two models, the continuous-flow respirators, should be used only for abrasive blast operations where the abrasive materials do not include silica sand and the level of contaminant in the ambient air does not exceed 25 or 50 times the recommended exposure limit, respectively. The third model, which is a pressure-demand respirator, must be worn whenever silica sand is used as an abrasive material (NIOSH, 1993). In addition, ensure to the extent possible that workers are upstream from the blasting operations to reduce their exposure to lead dust entrained in the ventilation air.

## Welding, Burning, and Torch Cutting

### The Processes

Welding is typically thought of as the electric arc and gas (fuel gas/oxygen) welding process. However, welding can involve many types of processes. Some of these other processes include inductive welding, thermite welding, flash welding, percussive welding, plasma welding, and others. McElroy (1980) points out that the most common type of electric arc welding also has many variants including gas shielded welding, metal arc welding, gas-metal arc welding, gas-tungsten arc welding, and flux cored arc welding.

Welding, cutting, and brazing are widely used processes. OSHA's Subpart Q contains the standards relating to these processes in all of their various forms. The primary health and safety concerns are fire protection, employee personal protection, and ventilation. The standards contained in this subpart are as follows:

1910.251 Definitions

1910.252 General Requirements

1910.253 Oxygen-fuel gas welding and cutting

1910.254 Arc welding and cutting

1910.255 Resistance welding

1910.256 Sources of standards

1910.257 Standards organization

In taking a look back on an OSHA study (reported in *Professional Safety*, 1989) on deaths related to welding/cutting incidents, it is striking to note that of 200 such deaths over an eleven-year period, 80% were caused by failure to practice safe work procedures. Surprisingly, only 11% of deaths involved malfunctioning or failed equipment, and only 4% were related to environmental factors. The implications of this study should be obvious: equipment malfunctions or failures are not the primary causal factor of hazards presented to workers. Instead, the industrial hygienist's emphasis should be on establishing and ensuring safe work practices for welding tasks. In the section below, we discuss these safe work practices controls.

## Work Practice Controls: Fire Prevention and Protection, PPE, Ventilation

The *fire prevention and protection* element of any welding, burning, and torch cutting safety program begins with basic precautions. These basic precautions include the following:

1. *Fire hazards*: if the material or object cannot be readily moved, all movable fire hazards in the area must be moved to a safe location.
2. *Guards*: if the object to be welded or cut can't be moved, and if all the fire hazards can't be removed, then guards are to be used to confine the heat, sparks, and slag, and to protect the immovable fire hazards.
3. *Restrictions*: if the welding or cutting can't be performed without removing or guarding against fire hazards, then the welding and cutting should not be performed.
4. *Combustible material*: wherever floor openings or cracks in the flooring can't be closed, precautions must be taken so that no readily combustible materials on the floor below will

be exposed to sparks that might drop through the floor. The same precautions should be taken with cracks or holes in walls, open doorways, and open or broken windows.

5. *Fire extinguishers*: suitable fire extinguishing equipment must be maintained in a state of readiness for instant use. Such equipment may consist of pails of water, buckets of sand, hoses or portable extinguishers, depending upon the nature and quantity of the combustible material exposed.

6. *Fire watch*: firewatchers are required whenever welding or cutting is performed in locations where other than a minor fire might develop. Firewatchers are required to have fire-extinguishing equipment readily available, and must be trained in its use. They must be familiar with facilities for sounding an alarm in the event of fire. They must watch for fires in all exposed areas, try to extinguish them only when obviously within the capacity of the equipment available, or otherwise sound the alarm. A fire watch must be maintained for at least a half-hour after completion of welding or cutting operations to detect and extinguish possible smoldering fires.

7. *Authorization*: before cutting or welding is permitted, the individual responsible for authorizing cutting and welding operations must inspect the area. The responsible individual must designate precautions to be followed in granting authorization to proceed preferably in the form of a written permit (Hot Work Permit).

8. *Floors*: where combustible materials such as paper clippings, wood shavings, or textile fibers are on the floor, the floor must be swept clean for a radius of at least 35 feet (OSHA requirement). Combustible floors must be kept wet, covered with damp sand, or protected by fire-resistant shields. Where floors have been wet down, personnel operating arc welding or cutting equipment must be protected from possible shock.

9. *Prohibited areas*: welding or cutting must not be permitted in areas that are not authorized by management. Such areas include: in sprinklered buildings while such protection is impaired; in the presence of explosive atmospheres, or explosive atmospheres that may develop inside uncleaned or improperly prepared tanks or equipment which have previously contained such materials, or that may develop in areas with an accumulation of combustible dusts; and in areas near the storage of large quantities of exposed, readily ignitable materials such as bulk sulfur, baled paper, or cotton.

10. *Relocation of combustibles*: where practicable, all combustibles must be relocated at least 35 feet from the worksite. Where relocation is impracticable, combustibles must be protected with fireproofed covers, or otherwise shielded with metal or fire-resistant guards or curtains.

11. *Ducts*: ducts and conveyor systems that might carry sparks to distant combustibles must be suitably protected or shut down.

12. *Combustible walls*: where cutting or welding is done near walls, partitions, ceilings, or roofs of combustible construction, fire-resistant shields or guards must be provided to prevent ignition.
13. *Noncombustible walls*: if welding is to be done on a metal wall, partition, ceiling, or roof, precautions must be taken to prevent ignition of combustibles on the other side from conduction or radiation, preferably by relocating the combustibles. Where combustibles are not relocated, a fire watch on the opposite side from the work must be provided.
14. *Combustible cover*: welding must not be attempted on a metal partition wall, ceilings or roofs that have combustible coverings, nor on any walls or partitions, ceilings or roofs that have combustible coverings, or on walls or partitions of combustible sandwich-type panel construction.
15. *Pipes*: cutting or welding on pipes or other metal in contact with combustible walls, partitions, ceilings or roofs must not be undertaken if the work is close enough to cause ignition by conduction.
16. *Management*: management must recognize its responsibility for the safe usage of cutting and welding equipment on its property, must establish areas for cutting and welding, and must establish procedures for cutting and welding in other areas. Management must also designate an individual responsible for authorizing cutting and welding operations in areas not specifically designed for such processes. Management must also insist that cutters or welders and their supervisors are suitably trained in the safe operation of their equipment, and the safe use of the process. Management has a duty to inform contractors about flammable materials or hazardous conditions of which they may not be aware.
17. *Supervisor*:
    - is responsible for the safe handling of the cutting or welding equipment, and the safe use of the cutting or welding process.
    - must determine the combustible materials and hazardous area present or likely to be present in the work location.
    - must protect combustibles from ignition by whatever means necessary.
    - must secure authorization for the cutting or welding operations from the designated management representative.
    - must ensure that the welder or cutter secures his or her approval that conditions are safe before going ahead.
    - must determine that fire protection and extinguishing equipment are properly located at the site.

- where fire watches are required, he or she must ensure that they are available at the site.
18. *Fire prevention precautions*: cutting and welding must be restricted to areas that are or have been made fire safe. When work can't be moved practically, as in most construction work, the area must be made safe by removing combustibles or protecting combustibles from ignition sources.
19. *Welding and cutting used containers*: no welding, cutting, or other hot work is to be performed on used drums, barrels, tanks, or other containers until they have been cleaned so thoroughly as to make absolutely certain that no flammable materials are present, or any substances such as greases, tars, acids, or other materials that when subjected to heat, might produce flammable or toxic vapors. Any pipelines or connections to the drum or vessel must be disconnected or blanked.
20. *Venting and purging*: all hollow spaces, cavities, or containers must be vented to permit the escape of air or gases before preheating, cutting, or welding. Purging with inert gas (e.g., nitrogen) is recommended.
21. *Confined Spaces*: to prevent *accidental contact* in confined space operations involving hot work, when arc welding is to be suspended for any substantial period of time (such as during lunch or overnight), all electrodes are to be removed from the holders and the holders carefully located so that accidental contact can't occur. The machine must be disconnected from the power source. To eliminate the possibility of gas escaping through leaks or improperly closed valves, when gas welding or cutting, the *torch valves* must be closed and the gas supply to the torch positively shut off at some point outside the confined area whenever the torch is not to be used for a substantial period of time (such as during lunch hour or overnight). Where practicable, the torch and hose must also be removed from the confined space.

*Note:* The industrial hygienist should use the proceeding information as guidance in preparing the organizational Welding Safety Program.

### PPE and Other Protection

Personnel involved in welding or cutting operations must not only learn and abide by safe work practices, but also must be aware of possible bodily dangers during such operations. They must learn about the personal protective equipment (PPE) and other protective devices/measures designed to protect them.

*Railing and welding cable*: A welder or helper working on platforms, scaffolds, or runways must be protected against falling. This may be accomplished by the use of railings, safety harnesses, lifelines, or other equally effective safeguards. Welders must place welding cable and other equipment so that it is clear of passageways, ladders, and stairways.

*Eye protection*: Helmets or hand shields must be used during all arc welding or arc cutting operations (excluding submerged operations). Helpers or attendants must be provided with the same level of proper eye protection.

Goggles or other suitable eye protection must be used during all gas welding or oxygen cutting operations. Spectacles without side shields with suitable filter lenses are permitted for use during gas welding operations on light work, for torch brazing, or for inspection.

Operators and attendants of resistance welding or resistance brazing equipment must use transparent face shields or goggles (depending on the particular job) to protect their faces or eyes as required.

Helmets and hand shields must meet certain specifications, including being made of a material which is an insulator for heat and electricity. Helmets, shields, and goggles must not be readily flammable and must be capable of sterilization. Helmets and hand shields must be so arranged to protect the face, neck, and ears from direct radiant energy from the arc. Helmets must be provided with filter plates and cover plates designed for easy removal. All parts must be constructed of a material which will not readily corrode or discolor the skin.

Goggles must be ventilated to prevent fogging of the lenses as much as possible. All glass for lenses must be tempered and substantially free from striae, air bubbles, waves, and other flaws. Except when a lens is ground to provide proper optical correction for defective vision, the front and read surfaces of lenses and windows must be smooth and parallel. Lenses must also bear some permanent distinctive marking by which the source and shade may be readily identified.

All filter lenses and plates must meet the test for transmission of radiant energy prescribed in ANSI Z87.1–1968, the American National Standard Practice for Occupational and Educational Eye and Face Protection.

Where the work permits, the welder should be enclosed in an individual booth painted with a finish of low reflectivity (such as zinc oxide and lamp black), or must be enclosed with noncombustible screens similarly painted. Booths and screens must permit circulation of air at floor level. Workers or other persons adjacent to the welding areas must be protected from the rays by noncombustible or flameproof screens or shields, or must be required to wear appropriate eye protection.

*Protective clothing*: Employees exposed to the hazards created by welding, cutting, or brazing operations must be protected by PPE, including appropriate protective clothing required for any welding operation.

*Confined spaces*: For welding or cutting operations conducted in confined spaces (in spaces that are relatively small, or restricted spaces such as tanks, boilers, pressure vessels, or small compartments of a ship) PPE and other safety equipment must be provided.

The following must be ensured to protect personnel performing hot work in confined spaces:

- There must be proper ventilation (see below).
- Gas cylinders and welding machines must be left on the outside and secured to prevent movement.
- Where a welder must enter a confined space through a manhole or other small opening, means (lifelines) must be provided for quickly removing him or her in case of emergency.
- When arc welding is to be suspended for any substantial period of time, all electrodes must be removed from the holds, the holders carefully located so that accidental contact can't occur, and the machine disconnected from the power source.
- To eliminate the possibility of gas escaping through leaks of improperly closed valves, when performing gas welding or cutting, the torch valves must be closed and the fuel-gas and oxygen supply to the torch positively shut off at some point outside the confined area whenever the torch is not to be used for a substantial period of time.
- After welding operations are completed, the welder must mark the hot metal or provide some other means of warning others.

### Ventilation and Health Protection

All welding should be accomplished in well-ventilated areas. There must be sufficient movement of air to prevent accumulation of toxic fumes or possible oxygen deficiency. Adequate ventilation becomes extremely critical in confined spaces where dangerous fumes, smoke, and dust are likely to collect.

Where considerable hot work is to be performed, an exhaust system is necessary to keep toxic gases below the prescribed health limits. An adequate exhaust system is especially necessary when hot work is performed on zinc, brass, bronze, lead, cadmium, or beryllium-bearing metals. This also includes galvanized steel, and metal painted with lead-bearing paint. Fumes from these materials are toxic—they are very hazardous to health.

What does OSHA require for ventilation for hot work operations? Ventilation must be provided when:

- hot work is performed in a space of less than 10,000 cubic feet per welder.
- hot work is performed in a room having a ceiling height of less than 16 feet.

- hot work is performed in confined spaces where the hot workspace contains partitions, balconies, or other structural barriers to the extent that they significantly obstruct cross ventilation.

The minimum rate of ventilation must be 2,000 cubic feet per minute per welder, except where local exhaust hoods and booths are provided, or where approved airline respirators are provided.

### *Arc Welding Safety*

In 29 CFR 1910.254 (Arc Welding & Cutting), OSHA specifically lists various safety requirements that must be followed when arc welding. For example, in equipment selection, OSHA stipulates the welding equipment must be chosen for safe application to the work to be done. Welding equipment must also be installed safely as per manufacturer's guidelines and recommendations. Finally, OSHA specifies that work persons designated to operate arc-welding equipment must have been properly trained and qualified to operate such equipment. *Training* and *Qualification* procedures are important elements that must be included in any Welding Safety Program.

Along with OSHA's requirements above, the safety engineer must ensure the facility's Welding Safety Program includes written *Safe Work Practices* detailing and explaining safety requirements that must be followed whenever arc welding is performed. In the following section, we summarize OSHA and Industry requirements and recommendations for performing arc-welding operations safely.

### *Safe Work Practice: Arc Welding*

1. Ensure all welding equipment is installed according to provisions of the National Electrical Code (NEC) and regulatory bodies.
2. Ensure the welding machine is equipped with a power disconnect switch, conveniently located at or near the machine so the power can be shut off quickly.
3. Ensure that the range switch is not operated under load. The range switch, which provides the current setting should be operated only while the machine is idling and the current is open. Switching the current while the machine is under a load will cause an arc to form between the contact surfaces.
4. Repairs to welding equipment must not be made unless the power to the machine is shut off. The high voltage used for arc welding machines can inflict severe and fatal injuries.
5. Ensure welding machines are properly grounded in accordance with the NEC. Stray current may develop which can cause severe shock when ungrounded parts are

touched. Ensure the ground to your work is securely attached. Grounds are not to be attached to pipelines carrying gases or flammable liquids.

6. Ensure electrode holds do not have loose cable connections. Keep connections tight at all times. Avoid using electrode holders with defective jaws or poor insulation.
7. The polarity switch is not to be changed when the machine is under a load. Ensure you wait until the machine idles and the circuit is open. Otherwise, the contact surface of the switch may be burned and the person throwing the switch may receive a severe burn from the arcing.
8. Ensure welding cables are not overloaded, and do not operate a machine with poor connections.
9. Ensure welding is conducted in dry areas, and that hands and clothing are dry.
10. Ensure an arc is not struck whenever someone nearby is without proper eye protection.
11. Ensure pieces of metal that have just been welded or heated are allowed to cool before picking them up.
12. Always wear protective safety glasses.
13. Ensure hollow (cored) castings have been properly vented before welding.
14. Ensure press-type-welding machines are effectively guarded.
15. Ensure suitable spark shields are used around equipment in flash welding.
16. When welding is completed, turn off the machine, pull the power disconnect switch, and hang the electrode holder in its designated place.
17. Inspect cables for cuts, nicks, or abrasion.

### Safe Work Practice: Gas Welding and Cutting

Specific safety requirements for oxygen-fuel gas welding and cutting are covered under 29 CFR 1910.253, and are listed in the units involving oxyacetylene welding. These safety requirements (precautions) cover proper handling of cylinders, operation of regulators, use of oxygen and acetylene, welding hose, testing for leaks, and lighting a torch. All of these safety requirements are extremely important, and should be followed with the utmost care and regularity.

Along with the normal precautions to be observed in gas welding operations, a very important safety procedure involves the piping of gas. All piping and fittings used to convey gases from a central supply system to work stations must withstand a minimum pressure of 150 psi. Oxygen piping can be of black steel, wrought iron, copper, or brass.

Only oil-free compounds should be used on oxygen threaded connections. Piping for acetylene must be of wrought iron. (***Note:*** Acetylene gas must never come into contact with unalloyed copper, except in a torch—any contact with it could result in a violent explosion.) After assembly, all piping must be blow out with air or nitrogen to remove foreign materials.

According to Giachino and Weeks (1985) five basic rules contribute to the safe handling of oxyacetylene equipment:

1. Keep oxyacetylene equipment clean, free of oil, and in good condition.
2. Avoid oxygen and acetylene leaks.
3. Open cylinder valves slowly.
4. Purge oxygen and acetylene lines before lighting torch.
5. Keep heat, flame, and sparks away from combustibles.

### Cutting Safety

Whenever torch-cutting operations are conducted, the possibility of fire is very real, because proper precautions are often not taken. Torch cutting is particularly dangerous because sparks and slag can travel several feet and can pass through cracks out of sight of the operator. The safety engineer must ensure the persons responsible for supervising or performing cutting of any kind follow accepted safe work practices:

1. Use of a cutting torch where sparks will be a hazard is prohibited.
2. If cutting is to be over a wooden floor, the floor must be swept clean and wet down before starting the cutting.
3. A fire extinguisher must be kept in reach any time torch cutting operations are conducted.
4. Cutting operations should be performed in wide-open areas so sparks and slag will not become lodged in crevices or cracks.
5. In areas where flammable materials are stored and cannot be removed, suitable fire-resistant guards, partitions, or screens must be used.
6. Sparks and flame must be kept away from oxygen cylinders and hose.
7. Never perform cutting near ventilators.
8. Firewatchers with fire extinguishers should be used.
9. Never use oxygen to dust off clothing or work.
10. Never substitute oxygen for compressed air.

## Spray Painting with Lead-Based Paint

### The Process

In the construction field, the primary source of lead exposure in painting is red lead primers, although many finish coatings continue to contain a small percentage of lead. For most interior or exterior construction painting projects, workers employ conventional compressed-air spray equipment. Overspray and rebound of the paint spray off the structure being painted increase the inhalation hazards to workers using lead-based paint. The magnitude of the painter's particulate exposure to lead is dependent on the product used, its lead content, and the quantity of paint applied.

## Engineering Controls

The following engineering controls will reduce or eliminate worker exposures to lead during painting:

- Applying non–lead containing paints and primers
- To the extent possible, replacing lead chromate with zinc
- Hand-applying lead-based paint by brush or roller coating methods rather than spray methods
- Using local exhaust ventilation with proper filtration

## Soldering and Brazing

### The Process

Soldering and brazing are techniques that are used to join metal pieces or parts. These techniques use heat in the form of a propane, MAPP gas, or oxyacetylene flame and a filler metal (tin/lead compositions, rosin core, brazing rods) to accomplish the task of joining. This activity is usually performed by workers in the plumbing trades. The potential exposure source is the filler metal that contains lead.

Soldering and brazing operations present similar health hazards (airborne lead fumes) but to a different degree. Most soldering operations occur at temperatures that are less than 800°F. The melting point of the filler metals is usually quite low (<600°F) and the activity does not generate significant concentrations of metal fumes. Brazing operations usually occur at temperatures in excess of 800°F. The temperature of the operation is of major importance because temperature determines the vapor pressure of the metals that are heated and therefore the potential concentration of metal fumes to which the employee may be exposed.

Because most field soldering and brazing work is conducted with a torch, it is difficult to regulate operating temperatures to within recommended limits to reduce the amount of metal fumes generated. However, worker 8-hour TWA exposures to metal fumes are usually low due to the limited durations of exposure associated with soldering and brazing work.

Electricians soldering electrical connections, plumbers soldering nonpotable water lines, or roofers repairing tin flashing could all experience these short-term and intermittent lead exposures.

### Engineering Controls

In confined areas, portable local exhaust ventilation can be used to remove metal fumes and gases associated with this type of work.

## HAZARDOUS WASTE

> The most alarming of all man's assaults upon the environment is the contamination of air, earth, rivers, and sea with dangerous and even lethal materials. This pollution is for the most part irrecoverable; the chain of evil it initiates not only in the world that must support life but in living tissues is for the most part irreversible. In this now universal contamination of the environment, chemicals are the sinister, and little-recognized partners of radiation in changing the very nature of the world—the very nature of life.
>
> —Rachel Carson

In 1990, R. B. Smith reported that the U.S. Environmental Protection Agency (EPA) estimated that the United States generates 570 million tons of hazardous waste annually. This waste includes toxic, biologic, and radioactive waste. But the broader human interaction, the safety and health considerations we have with wastes most concerns the industrial hygienist. These may overlap and directly interface with the classic environmental spans of control alluded to by Rachel Carson.

In ways no environment writer had before, Carson combined the insight and sensitivity of a poet with the realism and observations of science in her classic and highly influential book *Silent Spring* (1962). For that work, Carson was ostracized, vilified, laughed at, and lambasted (particularly by chemical manufacturers who had a vested interest in the book's failure).

Examined with an unbiased eye, her message was clear: the chemicals we use commonly, in quantity—if not properly handled, treated, and disposed of—not only pose a short-term threat to human life, but also pose a long-term threat to the environment as a whole. Her plea is also clear: to end the poisoning of earth. With the clarity of vision provided by 20/20 hindsight, we see that Rachel Carson was well ahead of her time. The concerns that *Silent Spring* addressed in 1962, while based on limited data, have since been confirmed.

In this section we discuss the hazards of handling hazardous materials, especially hazardous wastes, all of which should be the focus of the industrial hygienist. We illustrate the nature of hazardous waste, the problem, and the possible consequences.

## America: A Throwaway Society

America as a whole has lost the habit of earlier generations to "use it up, wear it out, and make it do, or do without." A new American characteristic, one that might be further described as habit, trend, custom, practice, or tendency—is to discard those objects we no longer want, whether or not they still have useful life. We have become a "throwaway society."

While many of us conscientiously recycle our bottles, cans, newspapers, and plastic containers, we often simply discard other, larger items we have no more use for, simply because throwing them away is easier than finding an avenue to recycle or reuse them. When an item loses its value to us because it is broken, shabby, no longer fashionable, or no longer needed for whatever reason, discarding it should not be an insurmountable problem. But it is—especially whenever the item we throw away is a hazardous substance, one that is persistent, nonbiodegradable, and poisonous.

What is the magnitude of the problem with hazardous substance and waste disposal? Let's take a look at a few facts.

- Hazardous substances—including industrial chemicals, toxic waste, pesticides, and nuclear waste—are entering the marketplace, the workplace, and the environment in unprecedented amounts.
- The United States produces almost 300 million metric tons of hazardous waste each year—with a present population of 260,000,000+, this amounts to more than one ton for every person in the country.
- Through pollution of the air, the soil and water supplies, hazardous wastes pose both short- and long-term threats to human health and environmental quality.

## What Is a Hazardous Substance?

Hazardous wastes can be informally defined as a subset of all solid and liquid wastes that are disposed of on land rather than being shunted directly into the air or water, and which have the potential to adversely affect human health and the environment. We often believe that hazardous wastes result mainly from industrial activities, but households also play a role in the generation and improper disposal of substances that might be considered hazardous wastes. Hazardous wastes (via Bhopal and other disastrous episodes) have been given much attention, but surprisingly little is known of their nature and of the actual scope of the problem. In this section, we examine definitions of hazardous materials, substances, wastes, etc., and attempt to bring hazardous wastes into perspective both as a major environmental and as a safety and health concern.

Unfortunately, defining a hazardous substance is largely a matter of choice between the definitions offered by the various regulatory agencies and pieces of environmental legislation, each defining it somewhat differently. Many different terms are used interchangeably. Even experienced professional certified hazardous materials managers (CHMMs) have been known to interchange these terms, even though they are generated by different official sources, and have somewhat different meanings, dependent upon the nature of the problem being addressed. To understand the scope of the dilemma in defining a hazardous substance let's take a look at the terms that are in common use today, used interchangeably, and often thought to mean the same thing.

A *hazardous material* is a substance (gas, liquid, or solid) capable of causing harm to people, property, and the environment. The U.S. Department of Transportation (DOT) uses the term hazardous materials to cover nine categories identified by the *United Nations Hazard Class Number System*:

- Explosives
- Gases (compressed, liquefied, dissolved)
- Flammable Liquids
- Flammable Solids
- Oxidizers
- Poisonous Materials
- Radioactive Materials
- Corrosive Materials
- Miscellaneous Materials

The term *hazardous substance* is used by the USEPA for chemicals which, if released into the environment above a certain amount, must be reported, and depending on the threat to the environment, for which federal involvement in handling the incident can be authorized. USEPA lists hazardous substances in its 40 CFR Part 302, table 302.4.

OSHA uses the term *hazardous substance* in 29 CFR 1910.120 (which resulted from Title I of SARA and covers emergency response) differently than does the EPA. Hazardous substances (as defined by OSHA) cover every chemical regulated by both DOT and the EPA.

*Extremely hazardous substance* is a term used by the EPA for chemicals that must be reported to the appropriate authorities if released above the *threshold reporting quantity* (RQ). The list of extremely hazardous substances is identified in Title III of the *Superfund Amendments and Reauthorization Act* (SARA) of 1986 (40 CFR Part 355). Each substance has a threshold reporting quantity.

EPA uses the term *toxic chemical* for chemicals whose total emissions or releases must be reported annually by owners and operators of certain facilities that manufacture, process, or otherwise use listed toxic chemicals. The list of toxic chemicals is identified in Title III of SARA.

EPA uses the term *hazardous wastes* for chemicals regulated under the Resource, Conservation and Recovery Act (RCRA-40 CFR Part 261.33). Hazardous wastes in transportation are regulated by DOT (49 CFR Parts 170–179).

For our purposes in this text, we define a hazardous waste as any hazardous substance that has been spilled or released to the environment. For example, chlorine gas is a hazardous material. When chlorine is inadvertently released to the environment, it becomes a hazardous waste. Similarly, when asbestos is in place and undisturbed, it is a hazardous material. When it is broken, breached, or thrown away, it becomes a hazardous waste.

OSHA uses the term *hazardous chemical* to denote any chemical that poses a risk to employees if they are exposed to it in the workplace. Hazardous chemicals cover a broader group of chemicals than the other chemical lists.

## RCRA Hazardous Substance

For the purposes of this text, to form the strongest foundation for understanding the main topic of this chapter (hazardous waste handling) and because RCRA's definition for a hazardous substance can also be used to describe a hazardous waste, we use RCRA's definition.

RCRA defines a substance as hazardous if it possesses any of the following four characteristics: reactivity, ignitability, corrosiveness, or toxicity. Briefly,

*Ignitability* refers to the characteristic of being able to sustain combustion, and includes the category of flammability (ability to start fires when heated to temperatures less than 140°F or less than 60°C).

*Corrosive* substances (or wastes) may destroy containers, contaminate soils and groundwater, or react with other materials to cause toxic gas emissions. Corrosive materials provide a specific hazard to human tissue and aquatic life where the pH levels are extreme.

*Reactive* substances may be unstable or have tendency to react, explode, or generate pressure during handling. Pressure-sensitive or water-reactive materials are included in this category.

*Toxicity* is a function of the effect of hazardous materials (or wastes) that may come into contact with water or air and be leached into the groundwater or dispersed in the environment.

The toxic effects that may occur to humans, fish, or wildlife are our principal concerns here. Toxicity (until 1990) was tested using a standardized laboratory test, called the extraction procedure (EP toxicity test). The EP toxicity test was replaced in 1990 by the toxicity characteristics leaching procedure (TCLP), because the EP test failed to adequately simulate the flow of toxic contaminants to drinking water. The TCLP is designed to identify wastes likely to leach hazardous concentrations of particular toxic constituents into the surrounding soils of groundwater as a result of improper management.

TCLP extracts constituents from the tested waste in a manner designed to simulate the leaching actions that occur in landfills.

## So, What Is a Hazardous Waste?

As mentioned, a general rule of thumb states that any hazardous substance that is spilled or released to the environment is no longer classified as a hazardous substance but as a hazardous waste. The EPA uses the same definition for hazardous waste as it does for hazardous substance. The four characteristics described in the previous section (reactivity, ignitability, corrosivity, or toxicity) can also be used to identify hazardous substances as well as hazardous wastes.

Note that the EPA "lists" substances that it considers hazardous wastes. These lists take precedence over any other method used to identify and classify a substance as hazardous (i.e., if a substance is listed in one of the EPA's lists described below, it is a hazardous substance, no matter what).

## EPA Lists of Hazardous Wastes

EPA developed these lists by examining different types of wastes and chemical products to determine whether they met any of the following criteria:

- Exhibits one or more of the four characterizations of a hazardous waste.
- Meets the statutory definition of hazardous waste.
- Is acutely toxic or acutely hazardous.
- Is otherwise toxic.

All listed wastes are presumed to be hazardous, regardless of their concentrations. EPA-listed hazardous wastes are organized into three categories: nonspecific source wastes, specific source wastes, and commercial chemical products.

*Nonspecific source wastes* are generic wastes commonly produced by manufacturing and industrial processes. Examples from this list include spent halogenated solvents used in degreasing and wastewater treatment sludge from electroplating processes, as well as dioxin wastes, most of which are "acutely hazardous" wastes because of the danger they present to human health and the environment.

*Specific source wastes* are from specially identified industries such as wood preserving, petroleum refining, and organic chemical manufacturing. These wastes typically include sludges, still bottoms, wastewaters, spent catalysts, and residues, such as wastewater treatment sludge from pigment production.

*Commercial chemical products* (also called "P" or "U" list wastes because their code numbers begin with these letters) include specific commercial chemical products, or manufacturing chemical intermediates. This list includes chemicals such as chloroform and creosote, acids such as sulfuric and hydrochloric, and pesticides such as DDT and kepone (40 CFR 261.31,32,33).

Note that the EPA ruled that any waste mixture containing a listed hazardous waste is also considered a hazardous waste—and must be managed accordingly. This applies regardless of what percentage of the waste mixture is composed of listed hazardous wastes. Wastes derived from hazardous wastes (residues from the treatment, storage, and disposal of a listed hazardous waste) are considered hazardous waste as well (USEPA, 1990).

## Where Do Hazardous Wastes Come From?

Hazardous wastes are derived from several waste generators. Most of these waste generators are in the manufacturing and industrial sectors and include chemical manufacturers, the printing industry, vehicle maintenance shops, leather products manufacturers, the construction industry, metal manufacturing, and others. These industrial waste generators produce a wide variety of wastes, including strong acids and bases, spent solvents, heavy metal solutions, ignitable wastes, cyanide wastes, and many more.

## Why Are We Concerned about Hazardous Wastes?

From the industrial hygienist's perspective, any hazardous waste release that could alter the environment and/or impact the health and safety of employees in any way is a major concern. The specifics of the industrial hygienist's concern lie in acute and chronic toxicity to organisms, bioconcentration, biomagnification, genetic change potential, etiology, pathways, change in climate and/or habitat, extinction, persistence, esthetics such as visual impact, and most importantly, the impact on the health and safety of employees.

Remember, we have stated consistently that when a hazardous substance or hazardous material is spilled or released into the environment, it becomes a hazardous waste. This is important because specific regulatory legislation has been put in place regarding hazardous wastes, responding to hazardous waste leak/spill contingencies, and for proper handling, storage, transportation, and treatment of hazardous wastes—the goal being, of course, protecting the environment, and ultimately protecting the health and safety of our employees and the surrounding community.

Why are we so concerned about hazardous substances and hazardous wastes? This question is relatively easy to answer based on experience, publicity, and actual hazardous materials incidents, which have resulted in tragic consequences to the environment and to human life.

## Hazardous Waste Legislation

Humans are strange in many ways. We may know that a disaster is possible, is likely, could happen, is predictable. But do we act before someone dies? Not often enough. We often ignore the human element—we forget the victim's demise. We simply do not want to think about it, because if we think about it, we must come face to face with our own mortality. The industrial hygienist, though, must think about all potential disasters—constantly, and before such travesties occur—to prevent them from ever occurring.

Because of Bhopal and other similar (but less catastrophic) chemical spill events, the U.S. Congress (pushed by public concern) developed and passed certain environmental laws and regulations to regulate hazardous substances/wastes in the United States. This section focuses on the two regulatory acts most crucial to the current management programs for hazardous wastes. The first (mentioned several times throughout the text) is the Resource Conservation and Recovery Act (RCRA). Specifically, RCRA provides guidelines for prudent management of new and future hazardous substances/wastes. The second act (more briefly mentioned) is the Comprehensive Environmental Response, Compensation, and Liability Act (CERCLA), otherwise known as *Superfund*, which deals primarily with mistakes of the past: inactive and abandoned hazardous waste sites.

### Resource Conservation and Recovery Act

The Resource Conservation and Recovery Act (RCRA) is the United State's single most important law dealing with the management of hazardous waste. RCRA and its amendment Hazardous and Solid Wastes Act (1984) deal with the ongoing management of solid wastes throughout the country—with emphasis on hazardous waste. Keyed to the waste side of hazardous materials, rather than broader issues dealt with in other acts, RCRA is primarily concerned with land disposal of hazardous wastes. The goal is to protect groundwater supplies by creating a "cradle-to-grave" management system with three key elements: a tracking system, a permitting system, and control of disposal.

1. The *tracking system* requires that a manifest document accompany any waste that is transported from one location to another.
2. The *permitting system* helps assure safe operation of facilities that treat, store, or dispose of hazardous wastes.
3. The *disposal control system* provides controls and restrictions governing the *disposal* of hazardous wastes onto, or into, the land (Masters, 1991).

RCRA regulates five specific areas for the management of hazardous waste (with the focus on treatment, storage, and disposal):

1. Identifying what constitutes a hazardous waste and providing classification of each.
2. Publishing requirements for generators to identify themselves, which includes notification of hazardous waste activities and standards of operation for generators.
3. Adopting standards for transporters of hazardous wastes.
4. Adopting standards for treatment, storage, and disposal facilities.
5. Providing for enforcement of standards through a permitting program and legal penalties for noncompliance (Griffin, 1989).

Arguably, RCRA is our single most important law dealing with the management of hazardous waste—it certainly is the most comprehensive piece of legislation that the EPA has promulgated to date.

## CERCLA

The mission of the *Comprehensive Environmental Response, Compensation, and Liabilities Act of 1980* is to clean up hazardous waste disposal mistakes of the past, and to cope with emergencies of the present. More often referred to as the *Superfund Law,* as a result of its key provisions a large trust fund (about $1.6 billion) was created. Later, in 1986, when the law was revised, this fund was increased to almost $9 billion. The revised law is designated as the *Superfund Amendments and Reauthorization Act of 1986* (SARA). The key requirements under CERCLA are listed in the following. Briefly,

1. CERCLA authorizes the EPA to deal with both short-term (emergency situations triggered by a spill or release of hazardous substances), as well as long-term problems

involving abandoned or uncontrolled hazardous waste sites for which more permanent solutions are required.

2. CERCLA has set up a *remedial* scheme for analyzing the impact of contamination on sites under a hazard ranking system. From this hazard ranking system, a list of prioritized disposal and contaminated sites is compiled. This list becomes the *National Priorities List* (NPL) when promulgated. The NPL identifies the worst sites in the nation, based on such factors as the quantities and toxicity of wastes involved, the exposure pathways, the number of people potentially exposed, and the importance and vulnerability of the underlying groundwater.

3. CERCLA also forces those parties who are responsible for hazardous waste problems to pay the entire cost of cleanup.

4. Title III of SARA requires federal, state, and local governments and industry to work together in developing emergency response plans and reporting on hazardous chemicals. This requirement is commonly known as the Community Right-To-Know Act, which allows the public to obtain information about the presence of hazardous chemicals in their communities and releases of these chemicals into the environment.

### OSHA

Moretz (1989) points out that OSHA's hazardous waste standard specifically addresses the safety of the estimated 1.75 million workers who deal with hazardous waste: hazardous waste workers in all situations including treatment, storage, handling, and disposal; firefighters; police officers; ambulance personnel; and hazardous materials response team personnel.

Moretz summarizes the requirements of this standard:

- Each hazardous waste site employer must develop a safety and health program designed to identify, evaluate, and control safety and health hazards, and provide for emergency response.
- There must be preliminary evaluation of the site's characteristics prior to entry by a trained person to identify potential site hazards and to aid in the selection of appropriate employee protection methods.
- The employer must implement a site control program to prevent contamination of employees. At a minimum, the program must identify a site map, site work zones, site communications, safe work practices, and the location of the nearest medical assistance. Also required in particularly hazardous situations is the use of the two-person rule (buddy system) so that employees can keep watch on one another and provide quick aid if needed.

- Employees must be trained before they are allowed to engage in hazardous waste operations or emergency response that could expose them to safety and health hazards.
- The employer must provide medical surveillance at least annually and at the end of employment for all employees exposed to any particular hazardous substance at or above established exposure levels and/or those who wear approved respirators for thirty days or more on site.
- Engineering controls, work practices, and PPE, or a combination of these methods, must be implemented to reduce exposure below established exposure levels for the hazardous substances involved.
- There must be periodic air monitoring to identify and quantify levels of hazardous substances and to ensure that proper protective equipment is being used.
- The employer must set up an information program with the names of key personnel and their alternates responsible for site safety and health, and the requirements of the standard.
- The employer must implement a decontamination procedure before any employee or equipment leaves an area of potential hazardous exposure; establish operating procedures to minimize exposure through contact with exposed equipment, other employees, or used clothing; and provide showers and change rooms where needed.
- There must be an emergency response plan to handle possible on-site emergencies prior to beginning hazardous waste operations. Such plans must address personnel roles; lines of authority; training and communications; emergency recognition and prevention; safe places of refuge; site security; evacuation routes and procedures; emergency medical treatment; and emergency alerting.
- There must be an off-site emergency response plan to better coordinate emergency action by local services and to implement appropriate control actions.

## Hazardous Waste Safety Program

For the purposes of this text, hazardous waste handling includes work activities that include the collection, storage, treatment, disposal and cleanup of hazardous waste materials. We also focus on standard industrial wastes and their handling. Industrial wastes include:

- Acids
- Abrasives
- Bases
- Animal products/byproducts

- Biologic substances
- Carcinogenic substances
- Explosives
- Solvents
- Salts
- Pesticides
- Oils
- Combustible materials
- Metals
- Reactive materials
- Organic materials

From the above list, as an industrial hygienist, it is clear that you must perform a comprehensive system safety analysis. This will allow you to recognize, evaluate, and control a wide variety of hazards and associated risks, and to provide this information to all employees affected by exposure via a written Hazardous Waste Safety Program.

The comprehensive site characterization and safety analysis is required to identify specific site hazards, and to determine the appropriate safety and health control measures needed to protect employees from the identified hazards.

As mentioned, the industrial hygienist must ensure that appropriate site control measures are implemented to control employee exposure to hazardous substances before clean–up work begins. As a minimum, site control should include:

- A site map
- Site work zones
- The use of a "buddy system"
- Site communications including alerting means for emergencies
- The standard operating procedures or safe work practices
- Identification of the nearest medical assistance

All employees working on-site who have the potential for exposure to hazardous substances, health hazards, or safety hazards must receive appropriate training and obtain certification before they can be allowed to engage in hazardous waste operations that could expose them to hazardous substances and safety or health hazards.

Employees engaged in hazardous waste operations must be included in a company medical surveillance program.

Engineering controls, work practices, and PPE for employee protection must be implemented to protect employees from exposure to hazardous substances and safety and health hazards.

Monitoring must be performed to assure proper selection of engineering controls, work practices (such as confined space entry) and PPE so that employees are not exposed to levels which exceed permissible exposure limits (or published exposure levels if there are no permissible exposure limits) for hazardous substances.

Decontamination procedures for all phases of decontamination must be developed and implemented.

An emergency response plan must be developed and implemented by all employers who engage in hazardous waste operations. The plan must be written and available for inspection by employees and appropriate regulatory agencies.

## REFERENCES

American Botanical Council, 1995. "The Jamaica Ginger Paralysis Episode of the 1930s." *Journal of the American Botanical Council* 36: 34, 28.

Carson, R., 1962. *Silent Spring*. Boston: Houghton Mifflin.

49 CFR-170–179, Hazardous Materials Regulations. U.S. Department of Transportation.

Giachino, J. and W. Weeks, 1985. *Welding Skills*. Homewood, Ill.: American Technical Publications

Griffin, R. D., 1989. *Principles of Hazardous Materials Management*. Chelsea, Mich.: Lewis Publishers.

Kent, C., 1998. *Basics of Toxicology*. New York: John Wiley & Sons.

Lu, F. C. and S. Kacew, 2002. *Lu's Basic Toxicology*, 4th ed. Boca Raton, Fla.: Taylor & Francis.

Masters, G. M., 1991. *Introduction to Environmental Engineering and Science*, New York: Prentice Hall.

McElroy, F. E. ed., 1980. *NSC Accident Prevention Manual for Industrial Operations: Engineering and Technology*, 8th ed. Merrifield, Va.: International Fire Chiefs Association.

Moretz, S., 1989. "Industry Prepares for OSHA's Final Hazardous Waste Role," *Occupational Hazards*, (November): 39–42.

NIOSH, 1993. *Health Hazard Evaluation Report*. DHHS NIOSH Publication HETA 90-075. Boston Edison Company.

OSHA, 2005. *Controlling Lead Exposures in the Construction Industry: Engineering and Work Practice Controls*. OSHA Technical Manual, Section V: chapter 3. Washington, D.C.: U.S. Department of Labor.

OSHA, 1973. *Arc Welding & Cutting*. 29 CFR 1910.254. Washington, D.C.: U.S. Department of Labor.

OSHA, 1989. "OSHA News: OSHA Studies Workplace Deaths Involving Welding," *Professional Safety* (Feb. 8)

RCRA, Public Law 98-616, Hazardous and Solid Wastes Act, amendments PL 94–580 (42 USC 6901), 1984.

SARA (CERCLA), Public Law 99–499, Superfund Amendments and Reauthorization Act of 1986. (Amended 142 USC 9601), 1990.

Smith, R. B., 1990. "Manufacturing Companies of All Sizes Benefit from Waste-Reduction Policy." *Occupational Health & Safety*. (Jan).

Stelljes, M. E., 2000. *Toxicology for Non-Toxicologists*. Lanham, Md.: Government Institutes.

USEPA, 1990. Identification and listing of hazardous waste. 40 CFR 261.24, .31, .32, .33.

USEPA, 1990. *RCRA Orientation Manual*, Washington, D.C.: U.S. Environmental Protection Agency.

## SUGGESTED READING

Blackman, W. C. *Basic Hazardous Waste Management*. Boca Raton, Fla.: Lewis Publishers, 1993.

Coleman, R. J. and K. H. Williams. *Hazardous Materials Dictionary*. Lancaster, Pa.: Technomic Publishing Company, 1988.

Goetsch, D. L. *Occupational Safety and Health*, 2nd ed. Englewood Cliffs, N.J.: Prentice Hall, 1996.

Kavianian, H. R. and C. A. Wentz. *Occupational and Environmental Safety Engineering and Management*. New York: Van Nostrand Reinhold, 1990.

Kharbanda, O. P. and E. A. Stallworthy. *Waste Management*. UK: Grower, 1990.

Knowles, P-C., ed. *Fundamentals of Environmental Science and Technology*. Rockville, Md.: Government Institutes, Inc., 1992.

Lindgren, G. F. *Managing Industrial Hazardous Waste: A Practical Handbook*. Chelsea, Mich.: Lewis Publishers, 1989.

Portney, P. R., ed. *Public Policies for Environmental Protection*. Washington, D.C.: Resources for the Future, 1993.

Wentz, C. A. *Hazardous Waste Management*. New York: McGraw-Hill, 1989.

# 11

# Ergonomics

All sedentary workers . . . suffer from the itch, are a bad color, and in poor condition . . . for when the body is not kept moving the blood becomes tainted, its waste matter lodges in the skin, and the condition of the whole body deteriorates.

—Bernardino Ramazzini, 1700

## WHAT IS ERGONOMICS?

I have found that few people understand the meaning of the word—and even fewer understand what ergonomics is all about. Until recently, the term, *ergonomics* was used primarily in Europe and elsewhere in the world to describe *human factors engineering* (the synonymous term most commonly used in the United States). Common practice in the United States has now also adopted the term ergonomics—thus in this chapter we use the term ergonomics as the appropriate and accepted substitute for human factors engineering, human engineering, and engineering psychology.

So, again, what is ergonomics? The Greek *ergon* means work and *nomos* means law. Thus, ergonomics means the laws of work?

Let's further define ergonomics by pointing out that it relates to the interface between people and a variety of elements: equipment, environments, facilities, vehicles, printed materials, and so forth (Brauer, 1994). Grimaldi and Simonds define ergonomics as "the measurement of work" (1989, 512). Ergonomics could be defined as how human physical considerations affect work.

So, we find that ergonomics means the laws of work, the interface between people and a variety of elements, and/or the measurement of work. However, the definition is still incomplete. The best definition we have been able to find to date is derived, though slightly modified for our purposes, from Chapanis (1985): "Ergonomics discovers and applies information about human behavior, abilities, limitations, and other characteristics to the design of tools, machines, systems, tasks, jobs, and environments for productive, safe, comfortable, and effective human use." Simply put, ergonomics considers the total physiological and psychological demand of a worker's job, rather than just productivity, safety, and health.

Okay—Chapanis's definition for ergonomics seems logical (and it is the definition we use to describe ergonomics in this text), but what is the goal of ergonomics? Stated simply, the goal of ergonomics is to protect the worker, to minimize worker error, and to maximize worker efficiency—while providing a bit of comfort to the worker while he or she performs job tasks.

## Social Significance of Ergonomics

Dul and Weerdmeester (2003) make the following points in regards to the social significance of ergonomics:

- Ergonomics can contribute to the solution of a large number of social problems related to safety, health, comfort and efficiency.
- Daily occurrences such as accidents at work, in traffic, and at home, as well as disasters involving cranes, airplanes, and nuclear power stations can often be attributed to human error.
- From the analysis of these failures it appears that the cause is often a poor and inadequate relationship between operators and their task.
- The probability of accidents can be reduced by taking better account of human capabilities and limitations when designing work and everyday-life environments.
- Many work and everyday-life situations are hazardous to health.
- In western countries diseases of the musculoskeletal system (mainly lower back pain) and psychological illnesses (for example, due to stress) constitute the most important cause of absence due to illness, and of occupational disability.
- These conditions can be partly ascribed to poor design of equipment, technical systems, and tasks.
- Here, too, ergonomics can help reduce the problems by improving the working conditions.

- Therefore, in a number of countries, occupational health services are obliged to employ ergonomists.
- Ergonomics can contribute to the prevention of inconveniences and also, to some considerable degree, help improve performance. In the design of complex technical systems such as process installations, (nuclear) power stations and aircraft, ergonomics has become one of the most important design factors in reducing operator error.
- Some ergonomic knowledge has been compiled into official standards whose objective is to stimulate the application of ergonomics.
- A range of ergonomic subjects is covered internationally by International Standardization Organization (ISO) standards, and in the United States by the American National Standards Institute (ANSI).

## An Effective Ergonomics Program

In the opening statement from Health, Education, and Human Services (HEHS, 1997), we see that to have an effective ergonomics program, certain elements must be included:

- Hazard identification
- Program evaluation
- Training
- Medical management
- Management commitment and employee participation
- Hazard prevention and control

It is important to point out that before the industrial hygienist takes any kind of remedial action to correct a workplace ergonomic problem an important ergonomic principle to consider is that equipment, technical systems, and tasks have to be designed in such a way that they are suited to every user.

The variability within populations is such that most designs are suited to only 95% of the population. This means that the design is less than optimum for 5% of the users, who then require special, individual ergonomic measures. Examples of groups of users, who from an ergonomic perspective require additional attention, are short or tall persons, people who are overweight or disabled, the old and the young, and pregnant women.

In this chapter, we discuss each of the elements of the ergonomics proposal that are "doable" (common sense practice—remember, common sense is not so common, but is critical to success in anything we attempt to do as professional practitioners).

## WHAT ARE CUMULATIVE TRAUMA DISORDERS?

> Muscle aches and pains are common to many sedentary jobs . . . when the body is still, circulation is slowed and as a result fewer nutrients are delivered to the muscles, and fewer wastes are removed from the muscles, blood vessels and spinal discs.
> —VDT Guidelines, N.J. Department Health and Senior Services, 1989

Cumulative trauma disorders (CTDs) are injuries of the musculoskeletal and nervous systems that may be caused by repetitive tasks, forceful exertions, vibrations, mechanical compression (pressing against hard surfaces), or sustained or awkward positions. CTDs are also called repetitive motion disorders (RMDs), overuse syndromes, regional musculoskeletal disorders, repetitive motion injuries, or repetitive strain injuries.

These painful and sometimes crippling disorders develop gradually over periods of weeks, months, or years. They include the following disorders which may be seen in office workers.

- *Carpal tunnel syndrome*: a compression of the median nerve in the wrist that may be caused by swelling and irritation of tendons and tendon sheaths.
- *Tendonitis*: an inflammation (swelling) or irritation of a tendon. It develops when the tendon is repeatedly tensed from overuse or unaccustomed use of the hand, wrist, arm, or shoulder.
- *Tenosynovitis*: an inflammation (swelling) or irritation of a tendon sheath associated with extreme flexion and extension of the wrist.
- *Low back disorders*: these include pulled or strained muscles, ligaments, tendons, or ruptured disks. They may be caused by cumulative effects of faulty body mechanics, poor posture, and/or improper lifting techniques.
- *Synovitis*: an inflammation (swelling) or irritation of a synovial lining (joint lining).
- *DeQuervain's disease*: a type of Synovitis that involves the base of the thumb.
- *Bursitis*: an inflammation (swelling) or irritation of the connective tissue surrounding a joint, usually of the shoulder.
- *Epicondylitis*: elbow pain associated with extreme rotation of the forearm and bending of the wrist. The condition is also called tennis elbow or golfer's elbow.
- *Thoracic outlet syndrome*: a compression of nerves and blood vessels between the first rib, clavicle (collar bone), and accompanying muscles as they leave the thorax (chest) and enter the shoulder.
- *Cervical radiculopathy*: a compression of the nerve roots in the neck.
- *Ulnar nerve entrapment*: a compression of the ulnar nerve in the wrist.

- *Trigger finger:* another tendon disorder, which is attributed to the creation of a groove in the flexing tendon of the finger. If the tendon becomes locked in the sheath, attempts to move the finger will cause snapping and jerking movements. The palm side of the fingers is the usual site for trigger finger. This disorder is often associated with using tools that have handles with hard or sharp edges. Meatpackers, poultry workers, electronic assemblers, and carpenters are at risk of developing trigger finger.
- *Raynaud's syndrome:* or white finger, occurs when the blood vessels of the hand are damaged as a result of repeated exposure to vibration for long periods of time. The skin and muscles are unable to get the necessary oxygen from the blood and eventually die. Common symptoms include intermittent numbness and tingling in the fingers; skin that turns pale, ashen, and cold; and eventual loss of sensation and control in the fingers and hands. This condition is also intensified when the hands are exposed to extremely cold temperatures. This illness is associated with the use of vibrated tools over time—e.g., pneumatic hammers, electric chain saws, and gasoline powered tools. After long-term exposure—perhaps 10 to 15 years working 6 to 7 hours a day with vibrating tools—the blood vessels in the fingers may become permanently damaged. There is no medical remedy for white finger. If the fingers are fairly healthy, the condition may improve if exposure to vibration stops or is reduced.

In addition to the potential affects of Raynaud's Syndrome and other segmental body part ailments, vibration may affect the entire body, producing overall fatigue and potential permanent damage. Vibration in conjunction with prolonged sitting may also result in degenerative changes in the spine. For example, drivers of tractors, trucks, buses, construction machines, and other heavy equipment may suffer from low back pain, and permanent abdominal, spinal and bone damage.

Cumulative trauma disorders can also result from other than work activities that involve repetitive motions or sustained awkward positions such as sports or hobbies. Work and nonwork activities may together contribute to cumulative trauma disorders. These disorders can also be aggravated by medical conditions such as diabetes, rheumatoid arthritis, gout, multiple myeloma, thyroid disorders, amyloid disease, and pregnancy.

## Symptoms of Cumulative Trauma Disorders

Symptoms of CTDs may involve the back, shoulders, elbows, wrists, or fingers. **Note:** If symptoms last for at least a week or if they occur on many occasions, a doctor should be consulted.

- Numbness
- Decreased joint motion
- Swelling
- Burning
- Pain
- Aching
- Redness
- Weakness
- Tingling
- Clumsiness
- Cracking or popping of joints

## What Does the Scientific Literature Tell Us about CTDs? (OROSHA, 2005)

### Low Back Disorders

The epidemiologic literature supports a relationship between the development of low back disorders and each of the following workplace risk factors: (1) lifting and forceful movements, (2) bending and twisting in awkward postures, and (3) whole-body vibration.

A National Institute for Occupational Safety and Health (NIOSH) study in a grocery warehouse in Ohio illustrates the problem of low back disorders. Warehouse workers, performing long hours of repetitive, heavy manual lifting, had a rate of workers' compensation claims for back injuries of 16 per 100 workers, compared to the national average rate of workers' compensation claims for back injuries of one to two cases per 100 full-time workers. In this warehouse, workers sometimes lifted a total of more than 3000 pounds in less than one hour. Clearly, this is a high risk workplace. As a matter of fact, workers' compensation data from the National American Wholesale Grocers' Association and the International Foodservice Distributors Association for the years 1990 to 1992, found that back strains/sprains accounted for 30% of all injuries for warehouse workers. Data from the same report indicated that more than a third of all workers experience an annual injury in warehouse operations, accounting for a cost of $0.61 per worker-hour. Many other workplaces like this one experience high rates of work-related musculoskeletal disorders.

### Disorders of the Neck and Shoulders

For disorders of the neck and neck/shoulder region, the literature identifies two primary workplace factors: (1) sustained postures causing static contractions of the neck and shoulder muscles (for example, working overhead in automobile assembly or in construction), and (2) combinations of highly repetitive and forceful work involving the arm and hand, which also affect the musculature of the shoulder and neck region.

Analysis of job components can show how workers develop musculoskeletal disorders. One study describes the job of a carbon setter, an important job in aluminum processing, which provides us with an example of repetitive and forceful work. Aluminum ore is melted at very high temperatures in large electric pots, about the size of a conference table. These pots develop a hard crust that must be broken in order to add materials to the melting aluminum. To break the crust, carbon setters frequently throw a large bar weighing twenty-five pounds into the pot. This report documented increased shoulder disorders among the carbon setters in the study.

The basic shoulder motion when throwing the bar down into the pot is the same motion involved in baseball pitching, (and we've all heard about million-dollar rotator cuff injuries in the major leagues). In a typical game, a major league baseball pitcher throws between 90 and 120 pitches. A carbon setter throws the equivalent of between one and two doubleheaders every shift. The bar does not travel 90 mph like a baseball, but a baseball doesn't weigh 25 pounds.

### Disorders of the Hand, Wrist, and Elbow

There are several conditions to consider within the hand and wrist region. Combined work factors of forceful and repetitive use of the hands and wrists are associated with carpal tunnel syndrome. Vibration from hand tools like chainsaws (those that do not have vibration controlling mechanisms) also contributes to carpal tunnel syndrome.

Vibrating tool use has also been strongly linked to hand and arm vibration syndrome, a separate condition of the hand and wrist that affects the nerves and blood vessels.

Workers in industries such as meatpacking, garment work, and fish and poultry processing spend their workdays performing forceful exertions and repetitive movements of the hand and wrist. The combination of these has been found to have a strong association with tendonitis of the wrist and has also been associated with disorders of the elbow, such as epicondylitis.

### Length and Intensity of Exposure

The epidemiological literature indicates that the greater the level of exposure to a single risk factor or combination of factors, the greater the risk of having a work-related musculoskeletal disorder. The literature also indicates that an important factor is the time between each episode of exposure. With adequate time to recover or adapt, and particularly when lower forces are involved, there may be less harm to the body from repeated exposures. You will recall the carbon setter/major league baseball pitcher analogy; it is important to remember that a major league pitcher plays only every third or fourth game, while the carbon setter throws the equivalent of one or two double-headers five days a week, fifty weeks a year. The intensity as well as the extended length of the exposure to forceful, repetitive work plays a substantial role in the risk of work-related musculoskeletal disorder in many traditional occupational settings.

### *Psychological Factors*

In workplaces with high rates of work-related musculoskeletal disorders there is little scientific evidence that the principal reason for the excess number of injuries or illnesses is the workers' psychological reaction to their workplace. However, there is evidence, particularly in office settings, suggesting that both physical and psychosocial [work organization] factors may be important contributors to musculoskeletal disorders. Work organization refers to the way work processes are structured and managed, and it deals with subjects such as scheduling of work, job design, interpersonal aspects of work, career concerns, management style, and organizational characteristics. We know more about how physical factors contribute to musculoskeletal disorders than we do about work organization factors. NIOSH is supporting research projects in both areas.

## LOWER BACK INJURY PREVENTION PROGRAMS

Putnam (1988) relates the following statistics concerning workplace back injuries:

- Lower back injuries account for 20 to 25 percent of all workers' compensation claims.
- Thirty-three to 40 percent of all workers' compensation costs are related to lower back injuries.
- Each year there are approximately 46,000 back injuries in the workplace.
- Back injuries cause 100 million lost workdays each year.
- Approximately 80 percent of the population will experience lower back pain at some point in their lives. (48–49)

Because of lower back injury statistics, and because back injuries that result from improper lifting are among the most common injuries in industrial settings (accounting for approximately $12 billion in workers' compensation costs annually), the industrial hygienist's main focus is almost always diverted from environmental concerns to dealing with the implementation and management of a health and safety program that places back injury prevention programs at the forefront.

Improper lifting, reaching, sitting, and bending are typical causes of back injuries in the workplace. Personal lifestyles, ergonomic factors, and poor posture also contribute to back problems.

## Causes of Back Injuries

In the not too distant past, when a worker injured his or her back while working on the job, often the worker maintained silence about the injury. Why? Because the worker feared losing his or her job. Only when such an injury caused extraordinary pain and suffering or was debilitating did the worker complain or remain off the job.

There is much controversy over what proportion of back injuries should be attributed to work-related causes (or even if the worker has actually sustained such an injury), and how much should be ascribed to normal degeneration, off-the-job causes, and, unfortunately, fraud. Despite this controversy, however, the safety engineer should recognize that job-related back injuries can and do occur.

Whose fault is it when a worker injures his or her back while attempting to lift some object or material at work?

One thing is certain—in the past, too much emphasis, and most of the blame, has been put on workers for the resulting injuries (CoVan, 1995). We believe that the emphasis should instead be placed on the workplace. Egonomic conditions or the task at hand place the worker at risk; incentive programs in which the worker is pressured to accomplish a certain amount of work to qualify for pay bonuses encourage workers to assume additional personal risk.

Contrary to popular view, back injuries and complaints are widespread among all people and all occupations. They are not limited to construction or industrial activities. Our experience has shown us that back injuries are common among hospital workers, for example, who injure their backs while lifting patients. Office workers are another group with prevalent back complaints—national surveys have shown that more than 50% of all office workers have back complaints at some time. Another study noted that as many as 4 out of 5 Americans will suffer at least 1 episode of lower back pain between the ages of 20 and 60. In 1981 Caillet estimated that 70 million Americans suffered back injury, and that this number will increase by 7 million people annually. Annual statistics compiled by the National Safety Council (NSC) and the Bureau of Labor Statistics (BLS) since 1981 support Caillet's estimates (BLS, 1997–2000; NSC, 2002).

Why so many back injuries? What are the causes? Why have back injuries related to on-the-job-activities shown only a marginal decline since 1972, despite improved medical care, increased automation in industry, and more extensive use of preemployment physical examinations? Many professionals are currently performing various types of research in an effort to find the answers to these questions.

## Approaches to Controlling On-the-Job Back Injuries

Materials handling techniques, job design, and individual physical conditions and characteristics all contribute to back injuries. Much of the recent research in this area has focused on material handling capabilities and on setting recommended workload limits. In the following, we discuss the approach taken by NIOSH and others who have spent several years conducting studies on back injury prevention. Later we discuss our thoughts on the topic, based on real-world observations over the years.

NIOSH (1981) and others since have taken three distinct approaches for assessing manual material handling capabilities: the biomechanical approach, the physiological approach, and the psychological approach. (***Note***: Biomechanics is the subdiscipline of ergonomics concerned with the mechanical properties of human tissue, particularly the resistance of tissue to mechanical stresses.)

*Biomechanical approaches* of manual material handling capabilities have brought under scrutiny many of the long-cherished maxims about proper lifting techniques. The driving force behind the scrutiny is, of course, the lack of improvement in back injury rates, despite the long-term emphasis that has been placed on the straight-back lifting method.

Biomechanical approaches view the body as a system of links and connecting joints corresponding to segments of the body, such as upper arm (link), elbow (joint), and forearm (link). Physical principles are used to determine the mechanical stresses on the body and the muscle forces needed to counteract the stresses (Saunders and McCormick, 1993).

According to Brauer (1994),

> biomechanical analysis of lifting gives us insight into some of the problems. When a person lifts and carries an object, the back muscles must counteract the load. The spine is the fulcrum and the back muscles are a fixed, short distance from the spine. The load in front of the body is much farther from the spine, at minimum nearly the thickness of the trunk. The moment (i.e., the force acting over a distance: Moment = (Force) × (Distance)) is greater when a load is held far from the body compared to holding it close to the body, whether standing, sitting or stooping. The moment created by the load must be counteracted by the back muscles. (193)

From Brauer's analysis, we can list a number of significant points related to lifting which are important to us.

1. Stooping to raise a load creates even greater moments; to keep the moment small, the load must be held close to the body.
2. The size of the load can contribute to the moment.
3. The length of vertical distance (lift) can increase the potential for injury.
4. The weight of the object being lifted is also important.
5. Frequency of lift is also important (Note: The biochemical approach has been limited to analyzing infrequent lifting tasks).
6. The human body is not well suited to asymmetrical loads or rotation; twisting or lifting with one hand during a lift adds to the likelihood of injury.

Computerized biomechanical analyses indicate that the familiar straight-back, bent-knees lift method may cause increased loads at vulnerable L5/S1 lumbosacral disc, and demand muscle strength beyond the capabilities of many lifters (CoVan, 1995). For example, Anderson and Catteral (1987) point out that a computerized biochemical analysis of lifting a 100-pound object from single pallet height above floor level created a 1500 lb L5/S1 compressive force, which relates to an eightfold risk level of low back pain. We must also factor in that all people are different (i.e., they vary in size, weight, strength, physical condition in general, physical condition of muscles, condition of joints, and other factors), thus predicting where and under what conditions an individual will experience pain, a strain of a muscle or other form of injury is difficult.

When workers are required to perform lifting on a frequent or a continuous basis (e.g., throughout an entire 8-hr shift), the *physiological approach* is best suited. Energy consumption and stresses acting on the cardiovascular system is the focus of the physiological approach.

In the *psychophysical approach* both biomechanical and physiological approaches are integrated in the thought process involved with the subjective evaluation of perceived stress. What does this mean? Good question. What we are saying is that in the psychophysical approach, those required to do the lifting must assess the task first—then adjust the weight of a load or the frequency of handling a load to the maximum amount (commonly known as the *maximum acceptable weight of load*, or MAWL) they can sustain without strain or discomfort, and without becoming unusually tired, weakened, overheated, or out of breath. Sounds like common sense, doesn't it? Just about everything involving personal and worker safety is, at least to a point.

## Back Injury Prevention and the Industrial Hygienist

Where and how does the industrial hygienist fit into the attempt to reduce the incidence of on-the-job back injuries? First of all, the safety engineer needs to keep in mind the sound advice given by Ayoub and Mital (1989): "The most useful rule regarding safe lifting is that there is no single, correct way to lift." You might ask, if this is the case (and it is), then what is the safety engineer to do? After gaining an appreciation for Ayoub and Mital's advice, the industrial hygienist's next move should be to review NIOSH's *Work Practices Guide for Manual Lifting* (1981). Although somewhat dated, this manual is still the definitive text on the subject of safe lifting practices, and recommends controls for minimizing lifting injuries.

NIOSH suggests both administrative controls and engineering controls. Let's take a look at each one of these recommended controls.

## Administrative Controls

For back injury prevention, administrative controls involve the selection and training of workers. *Selection* involves preemployment screening—through physical assessment, strength testing, and testing for aerobic work capacity—to identify people who already have back problems when they apply. Let's take a closer look at the use of this administrative control, through examination of an example currently used by a large sanitation district in southeastern Virginia. This organization prescreens all potential new-hires. This screening process includes substance abuse testing, a regular physical examination, and fitness to wear a respirator on the job determined by a medical doctor. The purpose of the preemployment physical examination is multifaceted. First, potential employees are screened for substance abuse—those failing this examination are not hired for employment. Second, the general physical examination is designed to determine the potential employee's suitability to work in wastewater treatment operations. This includes the ability to lift a minimum of 100 pounds. Third, the medical facility not only examines the potential employee for physical fitness, but also attempts to determine his or her past medical history—an important factor, meant to ensure the hiring of an employee who not only is capable of performing work required by the job classification they are filling, but also to protect the organization against future medical claims for medical conditions that preexisted employment at the firm. This is one of the main purposes of the sanitation district's general medical examination—to protect itself from litigation resulting from an injury that occurred in the past at some other place of employment.

Is this really a problem? Yes it is. For example, hearing acuity tests are generally part of the physical examination process. The employer (who has equipment, machines, or processes that generate noise) must ensure that the new employee has not damaged his or her hearing acuity somewhere else prior to their employment at your firm, and that the degree of hearing loss is documented prior to employment. If not, the new employee may file a claim against you in the future, claiming their hearing loss resulted from working in your workplace. This can be a real headache generator—and costly, too.

In addition to hearing loss, the employee's pulmonary function should be checked, especially if the new employee will be required to wear a respirator in the normal performance of his or her duties. Workers are sometimes exposed to asbestos, silica, or other agents before they apply for work with your organization. You need to ensure that any loss of pulmonary function from previous exposure(s) is well documented, and backed up by competent medical testing—before you hire.

Obviously, these requirements also apply to those potential employees with a history of back injury. If you make the mistake, for example, of hiring an employee with chronic back problems (as a result of previous injury anywhere else), you will expose your company to the possibility of increased medical costs and lost time.

The preemployment physical examination used by the sanitation district includes physical assessment and strength testing. A written report of findings is submitted to the Human Resource Office and the industrial hygienist for their review and records. If the results of these tests indicate that the employee is incapable of performing lifting operations required for his or her job classification, then he or she should not be hired for the job.

We cannot overstress the need for preemployment physical examination screenings. Medical costs are on the rise. Lost time incidents contribute to the cost of doing business. Hiring someone who is not medically fit to perform his or her job function is a serious mistake, one that can be very costly. According to Dr. Alex Kaliokin (1988) it is important for another reason—40% of back injuries occur in the first year of employment.

*Training* involves recognition of the dangers of manual lifting, how to avoid unnecessary stress, and assessment of what a person can handle safely. Our experience has demonstrated that companies that provide back safety training report a significant decrease in back injuries. The best back injury prevention safety training includes training designed to help employees understand how to lift, bend, reach, stand, walk, and sit safely (see table 11.1).

## Engineering Controls

Engineering controls used to minimize lifting injuries include container design, handle and handhold designs, and floor-worker interface. Container design, obviously, would be employed in companies where lifting is a standard work activity—where standard items are

**Table 11.1. Recommended Lifting Procedures**

1. Size up the load before attempting to lift it.

***Note***: A common mistake workers make is to bend down, grab hold of some object, then attempt to jerk it up to the carrying position. We recommend that in the case of objects where the actual weight is not known, the worker should be taught to try to push it, move it by hand (without exerting a great deal of pressure), or take their foot and try to move the object, to try to gauge how heavy it is. If the object won't budge, or is obviously too heavy, the worker should employ other means to lift it.

2. Keep the load as close to the body as possible.
3. Get a firm footing. Make sure the floor is not slippery.
4. Avoid rapid jerking of the load.
5. Spread your feet for a stable stance.
6. Get a firm grip. Use handles or gripping or other lifting tools that will help.
7. Make sure the load is free, not locked down or stuck.
8. Avoid twisting or bending with the load during lifting.
9. Keep your back straight. Keeping your chin tucked in will help keep your back straight.
10. Avoid lifting above shoulder height.
11. Control the pace of lifting.
12. Lift with your legs.
13. Tighten your stomach muscles.
14. Limit the time you are holding the weight (the weight (>23kg) should not be held until it hurts to hold it any longer).
15. Be careful of your fingers and hands when carrying a load, so you don't strike them against something.
16. Anticipate the need for unexpected movements.
17. Set the load down gently. Use your legs. Keep your back straight.
18. Avoid pinching your fingers.

*Source*: Spellman and Whiting, 2005.

manually lifted each workday on a continuous basis. From the manufacturing point of view, designing a container for your manufactured product that is user- and back-friendly is also a goal. Designing containers that will protect products from damage during shipping and handling operations is important, and so is designing containers that protect handlers and customers from being injured by the products.

Along with container design, environmental working conditions such as lighting, color, and labeling should be considered.

Material handling system alternatives should also be looked at. The question you should ask yourself is, "Is it safer to lift certain objects using materials handling equipment and job aids such as hooks, bars, rollers, forklifts, and overhead cranes than to use manual brute force?" Seems like a logical question—and it is. Unfortunately, this question is often asked only after an injury occurs. Remember that the key to good industrial hygiene practice is prevention and prediction—not reaction.

# ERGONOMICS

## When to Employ Administrative and Engineering Controls

NIOSH's *Work Practices Guide for Manual* Lifting (1981) provides assistance for managers and designers of lifting tasks to determine when administrative and engineering controls should be used to minimize the potential for injury. The guide describes various risk factors associated with lifting, the procedures for evaluating lifting tasks, and methods for reducing lifting hazards. To use the NIOSH guide, you must first measure the following six task variables:

1. *Object weight* (L): Measured in pounds or kilograms.
2. *Horizontal distance* (H): The location of the object's center of gravity measured in the sagittal plane (which bisects the body, dividing it into symmetric left and right sections) from a point midway between the ankles. This measurement should be made from the origin and the destination of the lift, in inches or centimeters.
3. *Vertical location* (V): The location of the hands at the origin of the lift, measured vertically from the floor or working surface in inches or centimeters.
4. *Lifting distance* (D): The vertical displacement of the object (origin to destination) measured in inches or centimeters.
5. *Lifting frequency* (F): The number of lifts per minute, averaged over the time that manual lifting is performed.
6. *Duration of lifting*: Classified as *occasional* if lifting activities can be performed for less than one hour, or *continuous* if lifting activities are performed for more than one hour.

NIOSH has developed two equations to assess safe lifting practice. Using the above information (specifically variables 2 through 5), two limits, an *acceptable lift* (sometimes called the *action level*, or AL) and a *maximum permissible lift* (MPL), can be determined. The equations and their use apply only to:

1. smooth lifting
2. two-handed, symmetric lifting in the sagittal plane (no twisting)
3. moderate width tasks (less than 30 in.)
4. unrestricted lifting posture
5. good handles, grips, shoes, and flooring
6. favorable ambient environments

The magnitude of the acceptable lift (AL) (in lbs) is determined algebraically using the formula:

$$AL = 90 \text{ LB} \times HF \times VF \times DF \times FF \tag{11.1}$$

where:

>HF is a discounting factor based on horizontal location
>
>VF is a discounting factor based on the vertical location of the object at the origin of the lift
>
>DF is the discounting factor based on the lift distance
>
>FF is a discounting factor based on the lift frequency

All of the discounting factors in equation 11.1 range between 0 and 1 and can be estimated using the graphs in the NIOSH guide. Because the discounting factors are multiplicative, the maximum value of the AL is 90 lbs—that is, when all factors are equal to 1. This situation occurs when a lift is ergonomically ideal (close to the body, comfortable initial height, short travel distance, and low frequency). As conditions deviate from the ideal, the corresponding values of the discounting factors decrease, thus reducing the magnitude of AL. The computed value of the AL is particularly sensitive to the horizontal distance. (Increasing the horizontal distance from 6 to 12 inches reduces the horizontal discount factor from 1.0 to 0.5.) Highly frequent lifting also substantially reduces the value of AL.

Once the AL has been determined, it is easy to compute the maximum permissible lift (MPL) using the following formula:

$$MPL = 3 \times Al \qquad (11.2)$$

For additional details on using and interpreting equations 11.1 and 11.2, refer to the NIOSH *Work Practices Guide for Manual Lifting*.

Keyserling (1988) points out that after the AL and MPL have been determined, the job can be classified into one of three risk categories:

1. *Acceptable*: If the weight of the lifted object is less than the AL, the job is considered acceptable (i.e., most of the workers in the workforce could perform the job with only a minimal risk of injury).
2. *Administrative controls required*: If the weight of the objects falls between the AL and the MPL, the job is assigned to this category, implying that some individuals in the workforce would have difficulty in performing the job. Because of their limited strength and increased risk of injury, action should be taken to protect these individuals.
3. *Hazardous*: If the lifted object weight more than the MPL, the job is considered hazardous. The only acceptable approach to resolving this situation is redesigning the job to eliminate or reduce lifting stresses.

## Low Back Pain and Standing

Carson (1994) points out that "prolonged standing or walking is common in industry and can be very painful. Low back pain, . . . and other health problems have been associated with prolonged standing." For minimizing standing hazards, Carson recommends:

*Anti-fatigue mats*, to provide cushioning between the feet and hard working surfaces such as concrete floors. This cushioning effect can reduce muscle fatigue and lower back pain.

*Shoe inserts*, when anti-fatigue mats are not feasible because employees must move from area to area and, correspondingly, from surface to surface.

*Foot rails* that allow employees to elevate one foot at a time four or five inches. The elevated foot rounds out the lower back, thereby reliving some of the pressure on the spinal column.

*Workplace design* that allows workers to move about while they work.

*Sit/Stand chairs* that are higher-than-normal chairs to allow employees who typically stand while working to take quick mini-breaks, and return to work without the hazards associated with getting out of lower chairs.

*Proper footwear*, with a comfortable fit, that grips the work surface and allows free movement of the toes.

## Other Considerations

When an employee sustains back injuries while performing manual labor, the worker injured usually (1) performed the task in an unsafe manner, or (2) was in poor physical condition when he or she attempted to lift the object that caused the injury.

We have spent much of this discussion covering the first characteristic. The second characteristic has been discussed in terms of hiring workers for jobs regularly requiring lifting or other strenuous work. But what about workers at jobs that do not require much physical exertion? For example, office workers, whose job classification requires them to sit at a computer terminal and enter data for the greater part of an 8-hour shift. We have noticed that workers who have sedentary jobs are not always in the best physical condition, especially when it comes to performing manual lifting they do not do on a daily or routine basis. This type of person seems most susceptible to on-the-job back injuries from manual lifting.

Industrial hygienists have their work cut out for them in their attempt to reduce on-the-job back injuries. The statistics prove this to be the case. What else should the industrial hygienist do to help reduce such injuries? One practice that helps is to display poster illustrations in strategic areas (loading dock, storeroom, etc.) where lifting usually occurs or could occur. Displaying posters that illustrate proper lifting, reaching, sitting, and bending techniques can help.

Conducting regularly scheduled safety inspections or audits is another tool the safety engineer can employ in the workplace to identify potential problem areas, so that corrective action can be taken immediately.

*Note:* An important ergonomic principle to remember (whether dealing with lower back disorders or any other environmental/workplace problem) is that equipment, technical systems, and tasks have to be designed in such a way that they are suited to every user. Remember, as stated earlier, that most designs can only be suited to 95% of the population. The industrial hygienist will have to anticipate the needs of the 5%, who from an ergonomic perspective require special attention.

## ELEMENTS OF AN ERGONOMICS PROGRAM

### Hazard Identification

Considering the effects of ergonomic problems in today's workplace (e.g., reduction in work quality and productivity, increase in worker fatigue, poor work performance, increase in workers' compensation claims, etc.), the first and most obvious step in devising an organizational ergonomics program is to conduct a worksite hazard analysis to identify all hazards.

The best way to conduct a hazard identification procedure is to use the checklist approach. For example, standard checklists, derived from information and recommendations of the National Safety Council, Food and Drug Administration, and the National Institute of Occupational Safety and Health, have had years of actual use, and have proven their worth. Checklists are especially effective when used with the written guidelines provided in the sample computer Video Display Terminal (VDT) Evaluation and Recommendation procedure, described later along with a sample VDT checklist.

### Hazard Prevention and Control

Once the worksite has been thoroughly surveyed and the hazards identified, hazard prevention and control measures need to be put in place. As with all hazards, the safety engineer either attempts to eliminate the hazard or to engineer it out. When this is not possible, other prevention and control measures must be adopted. For example, when evaluating a particular workstation, the safety engineer may discover that an employee who operates a VDT complains about persistent neck pain. When the industrial hygienist evaluates that particular VDT station, he or she determines that the VDT monitor is not adjusted to the proper height. If this is the case, then the remedial action (the prevention and control phase) is easily put into place by adjusting the monitor correctly—to fit the individual.

## Management Commitment and Employee Participation

We have stated throughout this text that without proper management commitment and employee participation, the effort to include safety in any workplace is an empty effort. You must have top management support and include employee participation in the organization's safety program. The same holds true for the incorporation of an ergonomics program into the organization.

## Medical Management

When an employee who operates a VDT complains that he or she begins to notice a loss of feeling in the hands, then a tingling in the arms, and complains that at night he or she can't sleep because of a burning sensation in the wrists, this may be symptomatic of *Cumulative Trauma Disorder* (CTD) or *repetitive strain injury* (RSI). The question is, What is the employer to do about it—what is the employer "required" to do about it? The employer and the industrial hygienist are going to be confronted with many such medical questions.

So, what is to be done about it? At the present, this question is difficult to answer. Many insurance companies do not recognize CTD or RSI as a compensable injury. Even where insurance companies do recognize CTD or RSI as compensable, as in Virginia, CTD or RSI injuries are still not recognized as compensable injuries under Workers' Compensation laws.

This will change whenever OSHA's Ergonomics Standard is promulgated (the Federal Standard is on hold mainly because the original was unmanageable and, in the author's opinion, basically a buffet of dysfunction); most workplaces will have to recognize CTD and/or RSI and other similar repetitive motion-type injuries as legitimate compensable on-the-job injuries.

Whenever the new Standard is mandated, organizations will be required to set up a medical surveillance program that includes monitoring affected employees while on the job. That is, if an employee complains of wrist pain or other problems related to their workplace functions (e.g., VDT operators), not only will the organization need to attempt to mitigate the causal factors, they must also take whatever steps medical authorities recommend to alleviate the employee's pain—and to monitor the employee's progress toward alleviating the pain.

## Program Evaluation

Any on-line safety program must periodically undergo evaluation. This should be accomplished not only to verify the effectiveness of the program and remedial actions taken, but also the currency and efficacy of the program. The effectiveness of any safety program can be measured by the results. For example, suppose an employee who operates a VDT 40 hours per week complains about wrist pain. An ergonomic evaluation indicates that the employee's arms and wrists are not properly supported. The remediation was to adjust the employee's position by whatever means determined. Continued employee complaints would indicate, obviously, that the remedial actions taken were not correct. A different approach might need to be taken. The results in this particular case illustrate the importance of evaluation.

The program should also be evaluated to ensure that it is current with applicable regulatory requirements. Regulations are often dynamic (changing constantly). The only way to ensure compliance with the latest requirements is to evaluate the program regularly.

## Training

To aid in reducing ergonomically related hazards in the workplace, employee participation is critical. Employee participation is normally increased whenever employees are properly trained on both program requirements and on those elements that make up the program. As with almost all safety and health provisions, training is an essential, required ingredient. Employees need to be aware of the organization's efforts, not only to reduce, eliminate, evaluate, and control ergonomic hazards, but also to be aware of the types of workplace situations and practices that lead to ergonomic problems. Your training program should enable the employee to answer the question, "What do I do if I experience eyestrain, wrist pain, back pain, neck pain, and other discomforts from the performance of my day-to-day work activities?" Your training program should enable each employee to know how to go about reporting, and to whom to report, suspected ergonomic problems in the workplace.

## SAMPLE VDT ERGONOMICS PROGRAM

The following describes a VDT ergonomics program that has been in use since 1990 (Spellman and Whiting, 2005). This program has proven its worth. While true that your overall ergonomics program will probably need to include other work activities, we have chosen this VDT program to provide you a representative sample of the type of ergonomics program that may be beneficial for use in your workplace. (*Note:* The program that follows is used in conjunction with the Visual Display Terminal Evaluation and Recommendation Checklist.)

I. VDT Work Stations
  A. Keyboard
    1. Height
       The correct height of a keyboard allows the operator's upper and lower arms to be at a 90-degree angle. If this angle is more than 15% either way, one study estimates that as much as a 50% loss of productivity may occur. The incorrect height can cause pain in the shoulders, back, arms, and wrists.
    2. Angulation
       Incorrect angulation or elevation of the keyboard can cause pain in the wrists and fingers. The keyboard should be elevated in the rear approximately 2.5 inches, which places the operator's wrist at a slight (natural) angle. Elevating the keyboard should allow the operator's wrists to rest on the work surface. Should the work surface become uncomfortable, use of a wrist rest is recommended. If the keyboard is not adjustable, placing a sturdy object (piece of wood, book, etc.) under the rear should be sufficient.
    3. Position
       The keyboard should be positioned directly in front of the operator. Putting the keyboard to either side of the operator causes twisting of the upper body, which causes fatigue and pain of the neck, shoulders, and back.
    4. Wrist Rest
       The use of a wrist rest is imperative. The purpose of a wrist rest is to support the hands, wrists, arms, shoulders, and upper back. Without a wrist rest, employees may complain of pain and tension to parts of the body as mentioned. To increase productivity while enhancing the posture of employees, a wrist rest should be used. Employees who are not accustomed to wrist rests will generally object to using them. Remind the employees that this change is for their well being and that familiarization with the wrist rest will only take time. In essence, it is simply a change of attitude.
  B. Visual Display Terminal (VDT)
    1. Height
       The correct height of the VDT screen is very important. The top of the screen should be at eye level. If the screen is lower than the recommended height, the

operator is forced to look downward, which causes fatigue, stress, and aches of the neck, back, etc. Raising the height of the VDT terminal can be accomplished by placing unused books, files, etc., under the screen. Raising the screen allows the operator to sit with correct body posture.

2. Angulation

   Recent studies have shown that eliminating any angulation of the screen is most effective for worker comfort. If the screen is tilted or angled backwards, it picks up reflective glare from overhead lights, windows, etc.

3. Position

   The VDT should be directly in front of the operator. The document and keyboard should also be directly in front of the operator. This positioning eliminates considerably undue turning and twisting, which reduces fatigue, stress, strains, and pain of the eyes, neck, shoulders, back, and arms.

4. Distance

   The correct viewing distance should be 18 to 24 inches from the worker's face. Any operator who is closer or farther than the recommended distance could experience headaches, fatigue, and stress, and place undue stain on the eyes.

5. Brightness

   The cathode ray tube (CRT) should have a brightness control knob. If reflective glare is on the CRT screen, the brightness control can help relieve, reduce, or possibly eliminate the glare.

6. Image

   Because computers are relatively new, worn or out-dated parts have not yet become a problem. In time, however, the phosphorous lining which covers the inside of the CRT screen wears out and causes nonsharp images; to eliminate undue stress and strains to the eyes, screens should be replaced every few years.

7. Color

   Color terminals are preferred over black and white terminals because they are easier on the eyes. When replacing a computer, keep this in mind. Color terminals help relieve or reduce stress and strain on the eyes, and general fatigue and stress on the body.

8. Screen

   Keeping the screen clean is important. Dirty screens force employees to strain their eyes to read images. This eyestrain also causes aches of the head, neck, and back. A clean screen will produce less strain on the employee and improve productivity.

C. Chair

The most important component of computer and desk set-up equipment is the chair. In many instances, several employees operate the same terminal. In light of this, the pneumatic-type chair is recommended because of its adjustability.

1. Height

When you don't feel excessive pressure on your legs from the edge of the seat, the chair is at a comfortable working height. Pressure from the seat's front can cause loss of circulation in the legs, resulting in discomfort and loss of productivity. When seated, the purpose of your thighs (femur muscles) is to support your upper body weight. The average adult has approximately one hundred pounds of upper body (torso) weight. When an employee sits too high, the thigh muscles do not support this weight, and the lower back muscles are forced to support the upper body, resulting in lower back complications. In one study, the majority of employees who sat too high in their chairs experienced lower back pain. Approximately 25% to 30% of these employees had serious back problems. Insist that employees sit one to three inches lower (top seat buttocks measurement) than the hollow of the knee measurement. Two measurements are taken to determine the correct height of the chair.

   a. Sit with the soles of your shoes flat on the floor. Keep your shins perpendicular to the floor and relax your thigh muscles.

   b. Measure the distance from the hollow of the knee to the floor. This measurement generally ranges 13 to 19 inches for women and 14.5 to 22 inches for men.

   c. Next measure from the point at the back of the knee to the back of the buttocks. This measurement usually ranges from 15 to 23 inches for men and about 14 to 22 for women.

   d. The chair chosen should give you a measured result within 2 inches of your personal measured result.

2. Adjustability

As stated earlier, the adjustability of the chair is very important, especially if two or more employees use the same chair. If the chair is not easily adjustable, employees will instinctively use it as found. If the chair is not the correct height for the operator, aches and pain of the feet, legs, and lower back will probably occur, resulting in loss of productivity.

3. Backrest

The backrest should fit comfortably at the lower back (lumbar) to provide good back support. Again, the backrest must be easily adjustable for the previously stated reasons. A backrest that is improperly positioned can cause stress, strain, and fatigue to the shoulders and back.

4. Arm Rest

Armrests were designed to support the arms, shoulders, and upper back. At first, most employees do not prefer arm rests. Some armrest chairs do not permit the employee to get close to their desks, because the arm rests block how close the chair can pull in to the desk. However, with careful preparation and foresight, in most cases, the correct arm rest chairs can be obtained.

D. Footrest

Should any changes be made to the VDT workstation, adjustments may be required to compensate for the change. For example, if your desktop is too high, you will probably have to raise your chair beyond the recommended height to be in the right position. Unfortunately, your legs may dangle. If this occurs, you will need a footrest to minimize pressure from the seat's front on your legs. If employees have to operate the foot-peddle to a dictaphone, we recommend a footrest with rubber matting on the top. The usage of the footrest should reduce or eliminate aches and pains of the legs and feet and increase productivity.

E. Document

The document should not be to the side lying flat on the desk. This position causes operators to consistently turn their head and refocus their eyes, which places undue stress on the eyes, neck, shoulders, and upper back. Headaches are a common result. To eliminate these symptoms, place the document to the side of the screen at the same height as the CRT. The purchase of a document holder is highly recommended.

The document holder should be placed at the same distance from the operator's face as that of the screen. When this position is achieved, the refocusing and readjusting of the operator's eye will be eliminated. Employee complaints of eyestrain should be reduced or eliminated.

F. Desk

1. Height

The height should be the same as for the keyboard, which allows a 90-degree angle of the upper and lower arms. In the event the desktop is too low, a new desk is not needed. Simply raise the desk and place blocks of wood under the legs. This procedure should be sufficient. Should the desk height be too high, raise the operator's chair and use a footrest.

2. Area

The work surface area should be adequate to house the VDT terminal, the document, and the keyboard. Should the desk area be inadequate, we recommend replacing the desk or making other arrangements.

3. Legroom

Most people are comfortable if they have about 25- to 28-inches of space from the underside of the work surface to the floor. Obviously, taller employees may need more legroom. Operators should have enough room to move their legs from side to side without accidental contact. Storage of any materials under the desk should not be permitted.

II. Work Environment

A. Noise

In the office environment, many causes of noise exist (telephones, loud conversations, printers, radios, noisy equipment, etc.). What may be music to one person may be noise to another. Basically, noise is any unwanted sound. Therefore, in an open office environment, employees should be considerate of each other in controlling the noise they generate voluntarily.

Other sources of noise are inherent to the environment, but steps can be taken to reduce these noises as well. The signal volume on telephones can be adjusted. Covers to printers are usually beneficial in controlling noise. Generally, placing a thick rubber mat under a noisy machine will help eliminate the noise. Another method of controlling noise is by isolating or absorbing the noise with portable partitions. This method is very effective, and creates a quieter office for the employees. Another method of partitioning noise is relocating file cabinets, etc.

B. Traffic

Traffic in the work area or someone performing another operation near your VDT screen is often distracting, decreasing visual attention or concentration and causing fatigue and error. Several methods of controlling traffic are commonly used; (1) Re-route the traffic, (2) rearrange the furniture, (3) add partitions, (4) in private offices, close the door whenever possible.

C. Wall Paint

Wall paint is an excellent method of controlling reflective glare. White paint reflects light. This reflection could appear on the CRT screen. When this happens, notify the industrial hygienist. When it's time to repaint, the safety engineer will request something like a flat tan or light blue paint. These paints have been found to absorb light rays, and should not create a reflective glare.

D. Glare

Glare is undoubtedly a nuisance and troublesome factor to the operator. The majority of CRT screens will have some type of nuisance glare. When glare appears on the CRT, the operator should adjust their eyes and reposition him- or herself to reduce or eliminate the glare. Constantly refocusing the eyes and repositioning the body to avoid glare causes stress, fatigue, and aches of the head, eyes, neck, shoulders, back, and arms.

Overhead lighting is a major contributor of glare. In many cases, if the overhead lighting is the cause of glare, attaching a piece of cardboard or similar object on top of the VDT terminal with a few inches of overhang may eliminate it. Every effort should be made to eliminate glare, to make the work environment more pleasant, healthier, and more productive.

1. Direct

Direct glare is produced when an intense light shines directly into the operator's eyes. It usually occurs from placing the VDT terminal in front of a window, or aligning the terminal with overhead light fixtures which are directly in front of the operator's view. When light enters the eyes, it reduces the amount of contrast the operator can perceive, making reading the screen difficult.

2. Indirect

Indirect glare results when the reflection of a window or an overhead light causes bright spots on the VDT screen. Reflections of the walls or other objects can also affect the screen. Too much light usually causes these reflections. Glare decreases the contrast between the characters and their background, so the operator is forced to strain to read the screen. This strain causes visual fatigue, and because the operator is likely to change positions to try to see better, posture is also affected.

E. Lighting

Lighting can enhance or hinder an employee's work environment. There are many concerns pertaining to this subject, such as (1) type of lighting, (2) location of lighting, and (3) location of VDT. Due to these concerns, one type of lighting or control of lighting may be preferred over another. Should the office be rearranged, the lighting could change dramatically. If you feel that the lighting is not adequate or is more than adequate, consider rearranging the furniture. Computer environments are different from the standard office environment—the lighting and arrangements of the VDT terminal must be controlled to eliminate reflective glare. Because a VDT incorporates its own light source, the ambient lighting can be slightly dimmer than the usual office lighting. Generally speaking, most VDT work areas should have a lower lighting environment. This reduction in lighting usually reduces or eliminates reflective glare and enhances the sharpness of the images on the CRT screen. Unfortunately, computer operators have become accustomed to bright office lighting, so reducing the lighting will probably create complaints. Reduced lighting is for the employees' benefit. In time they will become accustomed to the difference, and welcome the change. In a computer environment, 2- to 5-foot-candles is preferred. In an office environment, 70 to 100-foot candles are preferred.

In summary, the correct order of controlling the lighting in the computer work area is: (1) rearrange or relocate the computer workstation, (2) reduce the amount of light, or (3) change the type of light control. By following these steps and controlling the ambient lighting, stress and strains of the eyes should be reduced and productivity increased.

1. Fluorescent

    Fluorescent lighting, generally speaking, is not the preferred lighting in the computer environment, because it produces an over-abundance of light. However, this does not necessarily mean that fluorescent lighting is unsatisfactory or unsuitable for all computer operations. If fluorescent lighting is not suitable for your computer work area, controlling the abundant light may be the solution. This, however, should only be done after rearranging the computer workstation.

2. Full Spectrum

    The full-spectrum natural light tube (C-50) is another type of lighting control. This lighting is a little softer and more pleasing to the eyes, and may possibly reduce reflective glare.

3. Task

    In the event that the lighting in the work area has been reduced, a task light may be needed. The task light attaches to the desk and is maneuverable. Caution should be taken to ensure that the operator does not put the light where it causes reflective glare on the VDT screen.

4. Parabolic Diffusers

    Parabolic diffusers (grates) are an excellent method of controlling reflective and disability glare. The parabolic diffuser attaches to light fixtures and reflects or diffuses the light downward, reducing glare. However, because of the expense, this should be the last method of controlling lighting.

5. Windows

Just about every employee wants a window near their workstation—however; windows are one of the greatest causes of glare. When operating the VDT terminal, window blinds should be closed. This should eliminate the glare. When the operator takes a break, the blinds can be reopened. Positioning the VDT terminal so that it is perpendicular to the window may eliminate nuisance glare.

F. Temperature

The preferred office temperature, generally speaking, should be approximately 72 to 75 degrees. Unfortunately, many factors concerning comfort levels must be considered. When an office environment includes both male and female employees, sometimes an agreeable office temperature is difficult to maintain. For example, on average, men's anatomy and blood vessels are larger than women's. Therefore, men generally prefer cooler temperatures than women. The solution? Within the temperature range of 72 to 75 degrees, the majority of the employees should be pleased. Studies have shown that when employees are subjected to an uncomfortable environment, an increase of complaints and errors are inevitable, as well as loss of concentration and productivity. If you maintain the office temperature within the 72 to 75 degree temperature range, the majority of the employees should be happier and more productive—but remember, you are simply never going to be able to please all of the employees when it comes to environmental temperature.

G. Work Load

Determining the average workload is important. A heavy workload could cause fatigue, stress, and possible cumulative trauma disorders of the wrist and fingers. A moderate or light workload (generally speaking) should not cause the same signs and symptoms as a heavy workload. However, there are some exceptions. Typically, employees who operate terminals for more than four hours are more susceptible to cumulative trauma disorders. Therefore, if possible, employees should be encouraged to pace their workstation duties as opposed to working at their terminal stations in long spurts.

H. Breaks

Breaks are essential to employees to reduce or eliminate cumulative trauma disorders, fatigue, and stress. However, taking a break doesn't necessarily mean stopping work to have a snack or smoke a cigarette. Employees can alternate their work duties (i.e., circulate mail, do filing, etc.). Questions on how often an employee should take a break are often raised. If the workload is heavy, we recommend a fifteen-minute break every hour. If the workload is moderate, a fifteen-minute break every two hours is recommended.

## Computer Workstation Ergonomic Checklist (OSHA, 2005)

This checklist can help you create a safe and comfortable computer workstation. A "NO" response indicates that a problem may exist.

| | |
|---|---|
| **WORKING POSTURES**—The workstation is designed or arranged for doing computer tasks so it allows your | |
| YES/NO | **Head** and **neck** to be upright, or in-line with the torso (not bent down/back). |
| YES/NO | **Head, neck**, and **trunk** to face forward (not twisted). |
| YES/NO | **Trunk** to be perpendicular to floor (may lean back into backrest but not forward). |
| YES/NO | **Shoulders** and **upper arms** to be in-line with the torso, generally about perpendicular to the floor and relaxed (not elevated or stretched forward). |
| YES/NO | **Upper arms** and **elbows** to be close to the body (not extended outward). |
| YES/NO | **Forearms, wrists,** and **hands** to be straight and in-line (forearm at about 90 degrees to the upper arm). |
| YES/NO | **Wrists** and **hands** to be straight (not bent up/down or sideways toward the little finger). |
| YES/NO | **Thighs** to be parallel to the floor and the **lower legs** to be perpendicular to floor (knees may be slightly elevated above thighs). |
| YES/NO | **Feet** rest flat on the floor or are supported by a stable footrest. |
| **SEATING**—Consider these points when evaluating the chair: | |
| YES/NO | **Backrest** provides support for your lower back (lumbar area). |
| YES/NO | **Seat width** and **depth** accommodate the specific user (seat pan not too big/small). |
| YES/NO | **Seat front** does not press against the back of your knees and lower legs (seat pan not too long). |
| YES/NO | **Seat** has cushioning and is rounded with a "waterfall" front (no sharp edge). |
| YES/NO | **Armrests**, if used, support both forearms while you perform computer tasks and they do not interfere with movement. |

**KEYBOARD/INPUT DEVICE**—Consider these points when evaluating the keyboard or pointing device. The keyboard/input device is designed or arranged for doing computer tasks so the

| | |
|---|---|
| YES/NO | **Keyboard/input device platform(s)** is stable and large enough to hold a keyboard and an input device. |
| YES/NO | **Input device** (mouse or trackball) is located right next to your keyboard so it can be operated without reaching. |
| YES/NO | **Input device** is easy to activate and the shape/size fits your hand (not too big/small). |
| YES/NO | **Wrists** and **hands** do not rest on sharp or hard edges. |

**MONITOR**—Consider these points when evaluating the monitor. The monitor is designed or arranged for computer tasks so the

| | |
|---|---|
| YES/NO | **Top** of screen is at or below eye level so you can read it without bending your head or neck down/back. |
| YES/NO | **User with bifocals/trifocals** can read the screen without bending the head or neck backward. |
| YES/NO | **Monitor distance** allows you to read the screen without leaning your head, neck or trunk forward/backward. |
| YES/NO | **Monitor position** is directly in front of you so you don't have to twist your head or neck. |
| YES/NO | **Glare** (for example, from windows, light) is not reflected on your screen which can cause you to assume an awkward posture to clearly see information on your screen. |

**WORK AREA**—Consider these points when evaluating the desk and workstation. The work area is designed or arranged for doing computer tasks so the

| | |
|---|---|
| YES/NO | **Thighs** have sufficient clearance space between the top of the thighs and your computer table/keyboard platform (things are not trapped). |
| YES/NO | **Legs** and **feet** have sufficient clearance space under the work surface so you are able to get close enough to the keyboard/input device. |

**ACCESSORIES**—Check to see if the

| | |
|---|---|
| YES/NO | **Document holder**, if provided, is stable and large enough to hold documents. |
| YES/NO | **Document holder**, if provided, is placed at about the same height and distance as the monitor screen so there is little head movement, or need to refocus, when you look from the document to the screen. |

| | |
|---|---|
| YES/NO | **Wrist/palm rest**, if provided, is padded and free of sharp or square edges that push on your wrists. |
| YES/NO | **Wrist/palm rest**, if provided, allows you to keep your forearms, wrists, and hands straight and in-line when using the keyboard/input device. |
| YES/NO | **Telephone** can be used with your head upright (not bent) and your shoulders relaxed (not elevated) if you do computer tasks at the same time. |
| **GENERAL** | |
| YES/NO | Workstation and equipment have sufficient adjustability so you are in a safe working posture and can make occasional changes in posture while performing computer tasks. |
| YES/NO | Computer workstation, components and accessories are maintained in serviceable condition and function properly. |
| YES/NO | Computer tasks are organized in a way that allows you to vary tasks with other work activities, or to take micro-breaks or recovery pauses while at the computer workstation. |

In the future, OSHA may mandate an Ergonomics Standard that will require specific requirements for compliance by certain industries. These regulations will undoubtedly affect enormous numbers of workplaces and workers. Implementing the ergonomics program will create special problems for industrial hygienists, because of the difficulties in dealing with the sheer number of workers involved, as well as the problems inherent in changing deeply ingrained work habits.

The sample visual display terminal program we presented here provides you with a guideline to use for evaluation and recommendations designed to enhance productivity, and more importantly to ensure a safe and healthy work environment for employees. From this sample program, the industrial hygienist is able to identify any areas of question or concern relating to any possible hazards of utilizing a VDT. However, ergonomics are the shared responsibility of management and employees. Unless the recommended procedures are used, an ergonomics program, like any other, is useless. The program is designed to benefit both management and employees. Thus, it is logical for both management and employees to work together to ensure an ergonomically safe and healthy work environment.

## REFERENCES

Anderson, C. K and M. J. Catteral, 1987. "A Simple Redesign Strategy for Storage of Heavy Objects." *Professional Safety*. (Nov.): 35–38.

Ayoub, M. and A. Mital, 1989. *Manual Materials Handling.* London: Taylor & Francis.

Brauer, R. L., 1994. *Safety and Health for Engineers.* New York: Van Nostrand Reinhold.

Bureau of Labor & Statistics (BLS), 1997–2000. Back Injury Statistics 1997–2000. National Institute for Occupational Safety & Health.

Caillet, R., 1981. *Low Back Pain Syndrome.* Philadelphia: F. A. Davis.

Carson, R., 1994. "Stand By Your Job," *Occupational Health & Safety* (April): 38.

Chapanis, A., 1985. "Some Reflections on Progress." *Proceedings of the Human Factors Society 29th Annual Meeting.* Santa Monica, Calif.: Human Factors Society.

CoVan, J., 1995. *Safety Engineering.* New York: John Wiley & Sons.

Dul, J., and B. Weerdmeester, 2003. *Ergonomics for Beginners.* Boca Raton, Fla.: Taylor & Francis.

Grimaldi, J. V. and R. H. Simonds, 1989. *Safety Management,* 5th ed., Homewood, Ill.: Irwin.

HEHS-97-163. Worker Protection: Private Sector Ergonomics Programs Yield Positive Results. Health, Education, and Human Services: 1997.

Kaliokin, A., 1988. "Six Steps Can Help Prevent Back Injuries and Reduce Compensation Costs," *Safety & Health* 138, no. 4 (October): 50.

Keyserling, W. M., 1988. "Occupational Ergonomics: Designing the Job to Match the Worker," B. S. Levy and D. H. Wegman, eds. *Occupational Health: Recognizing and Preventing Work-Related Disease,* 2nd ed., Boston: Little, Brown and Company.

National Safety Council, 2002. *Ergonomics.* At www.nsc.org/issues/ergotop.html (accessed November 2005).

NIOSH, 1981. *Work Practices Guide for Manual Lifting.* Cincinnati, Ohio: National Institute for Occupational Safety and Health.

OROSHA, 2005. *Cumulative Trauma Disorders (CTDs).* www.Orosha.org/educate/training/pages/201disorders.html (accessed July 22, 2005).

Putnam, A., 1988. "How to Reduce the Cost of Back Injuries," *Safety & Health* 138, no. 4 (October): 48–49.

Saunders, M. S. and E. J. McCormick, 1993. *Human Factors in Engineering and Design,* 7th ed., New York: McGraw-Hill.

Spellman, F. R., and N.E. Whiting, 2005. *Safety Engineering: Principles and Practices,* 2nd ed. Lanham, Md.: Government Institutes.

## SUGGESTED READING

American Industrial Hygiene Association. *Ergonomics Guides,* including Kroemer, K.H.E. *Ergonomics of VDT Workplaces.* Akron, Ohio, 1984.

Goetsch, D. L., *Occupational Safety and Health: In the Age of High Technology for Technologists, Engineers, and Managers.* Englewood Cliffs, N.J.: Prentice Hall, 1996.

Morris, B. K., ed. "Ergonomic Problems at Paper Mill Prompt Fine," *Occupational Health & Safety Letter* 20, no. 16 (August 8, 1990): 127.

Owen, B. D. "Exercise Can Help Prevent Low Back Injuries." *Occupational Health & Safety* (June 1986): 33–37.

Sorock, G. "A Review of Back Injury Prevention and Rehabilitation Research: Suggestions for New Programs." For the National Safety Council Back Injury Committee. (May 2, 1981).

Snook, S. H., "Approaches to the Control of Back Pain in Industry: Job Design, Job Placement, and Education/Training." *Professional Safety* (Aug. 23–31, 1988).

# 12

# Engineering Design and Controls

An engineer is charged with the responsibility for designing a new seatbelt that is comfortable, functional, inexpensive, and easy for factory workers to install. He designs a belt that meets all these requirements and it is installed in 10,000 new cars. As the cars are bought and accidents begin to occur, it becomes apparent that the new seatbelt fails in crashes involving speeds over 36 miles per hour. The engineer that designed the belt took all factors into consideration except one: *safety*.

—D. L. Goetsch

The question is "What is the best way employers can ensure the safety and health of their employees?" Based on our experience, the most commonly provided answer is "Provide them as much safety protective equipment as possible." When we ask specifically "What type of safety protective equipment are you referring to?" We generally hear the same answer over and over again. "You know, safety protective equipment such as eye and face protection, head protection, hand protection, respiratory protection, fall protection, and electrical protection."

By now you should know that the above respondents are referring to personal protective pquipment (PPE). You should also know by now that notwithstanding the efficacy of PPE in protecting employees on the job, PPE is protection of the last resort. That is, engineering and/or administrative controls and/or safe work practices are always the preferred and recommended methods of protecting workers on the job. PPE is to be used only when the others are not possible or feasible.

Again, PPE should be used only when it is impossible or impractical to eliminate a hazard or control it at its source through engineering design. Remember, wearing PPE does not eliminate the hazardous condition. PPE is used to establish a barrier between the exposed employee and the hazard to reduce the probability and severity of an injury.

The question remains "What is the best way employers can ensure the safety and health of their workers?"

Following the guidelines provided in OSHA's standards is the best way. But let's take a broad overview of the problem(s) involved.

Since most companies operate with one primary goal in mind (to make a profit), it stands to reason that these companies will attempt to operate in the most cost-effective manner possible. But what would that be? What is the most cost-effective manner possible? Some would answer, quite simply, cut costs, cut costs, cut costs—since obviously, costs are the steady state of concern of the business world. Typically, the cost of ensuring the safety of a workforce is seen as an add-on cost—a cost with no return—a cost that does not contribute to the bottom line. Though this view is shortsighted and somewhat warped, it is a view commonly held in industry today. But why? Why do managers feel that safety is a burden, analogous to taking money and dumping it down an endless drain? This question becomes even more complex, complicated, and compounded when you factor in other issues. For example, law (by OSHA) regulates the safety and health of workers—should the company manager think strictly in terms of OSHA or other regulations? Or is it more important to take a broader view?

Because workplace environments have become more technologically complex, we feel the need for protecting workers from safety and health hazards becomes more pressing, more complex, and more necessary—which requires the manager to take the broader view and to make choices and decisions that require a broad experience background and greater level of knowledge.

Fortunately, most company executives eventually come to share this view (i.e., that being aware of and concerned with employee safety and health is important), either of their own volition or via the results of regulatory pressure. When this occurs (if it does), the focus shifts from "we must comply, we must ensure the safety and health of our workers" to the original question—"What is the best way to ensure the safety and health of our employees?" This question naturally leads to other questions. "Shall we adopt elaborate engineering controls or trust the effectiveness of personal protective equipment?" "Should we undergo complete process or hardware redesign, or simple modification of existing systems?" In any case, costs are usually the main factor—the bottom line. The question shifts back to "What are the costs and benefits?" Others would take this question a step further and include "What are the limitations and risks of each possible approach?"

This short chapter attempts to answer some of these questions for the manager (and the industrial hygienist). To provide the broadest possible grasp of the issue, we concentrate on plant design and layout for safety, using engineering controls instead of PPE to ensure the safety and health of workers.

# ENGINEERING DESIGN AND CONTROLS

Simply put, we feel that effective industrial hygiene practice begins long before the worker appears in the workplace—a critical fact that astute planners, managers, design engineers, and industrial hygienists must remember. Why? Because it is less costly (remember, reducing costs is the bottom line) and more efficient to correct safety and health hazards (engineer them out—that is, eliminate them before they exist) before they become part of the workplace.

At this point, we can hear the doubting Thomases out there in the real world saying to themselves: "This is an idealistic, impossible approach . . . in the real world it just doesn't happen that way." We ask, "Why not?" The method for achieving it is relatively simple: proper attention to safety and health in the design phase is the answer. The industrial hygienist's primary function in the workplace is to reduce and/or eliminate hazards. This can be accomplished through proper design and layout of the plant or facility (Spellman and Whiting, 2005).

We first discuss codes and standards, then physical plant layout, illumination, high hazard areas, personal services and sanitation facilities, and finish with the concept of system engineering—all of which are important in the planning and design phase, but are especially important in ensuring effective accident prevention.

## CODES AND STANDARDS

Probably the first known written admonition regarding the need for accident prevention is contained in Hammurabi's Code, about 1750 B.C. It states: "If a builder constructs a house for a person and does not make it firm and the house collapses and causes the death of the owner, the builder shall be put to death." Today some of us would say that the justice rendered in such a case is rather severe (others may feel the punishment is just about right). Though true that the penalties have become less severe, the need for care has increased exponentially with the growth of technology since Hammurabi's time.

Countless pages have been written relating laws, standards, and codes regarding safety, health, and the environment since Hammurabi's Code. The fact of the matter is codes and standards have become essential tools in any plan of operations—and in the design of any workplace—which are used in accident control. Standards and codes have as their primary intent to prescribe minimum requirements and controls to safeguard life, property, or public welfare from hazards.

It is important to point out that building codes and standards are not intended to limit the appropriate use of materials, appliances, equipment, or methods of design or construction not specifically prescribed by the code/standard, provided the building official (OSHA compliance office) determines that the proposed alternate materials, appliances, equipment, or methods of design or construction are at least equivalent to that prescribed in the code or standard. In other words, you might be able to use alternate construction methods or materials, provided you can prove—to the satisfaction of the building official and/or regulator—that your way is as good as or better than what the code/standard prescribes.

Keep in mind that building codes are constantly changing and they can vary by state, county, city, town, and/or borough. In order to learn which codes are being used and how they will affect your operation and your construction project, contact your local building inspection department, office of planning and zoning, and/or department of permits. You might want to start by calling the most local government body that has jurisdiction over the property where you will be building. They should be able to provide you with specific information about which building codes are currently used as guidelines in your area.

To understand codes and standards used in accident control, you'll need to learn a few pertinent definitions (provided by Hammer, 1989).

**Criterion**: Any rule or set of rules that might be used for control, guidance, or judgment.

**Standard**: A set of criteria, requirements, or principles.

**Code**: A collection of laws, standards, or criteria relating to a particular subject, such as the National Electric Code (NEC), the Uniform Fire Code (UFC), or Building Officials & Code Administrators (BOCA) National Fire Prevention Code.

**Regulation**: A set of orders issued to control the conduct of persons within the jurisdiction of the regulatory authority.

**Specification**: A detailed description of requirements, usually technical.

**Practice**: A series of recommended methods, rules, or designs, generally on a single subject.

**Design handbooks, guides, or manuals**: Contain nonmandatory practices, general concepts, and examples to assist a designer or operator.

Let us point out that local or state laws also have many ordinances governing specific requirements that cover such items or systems as fire sprinklers, fire alarms, exhaust and ventilation systems, emergency lighting, and means of egress. City, county, state, and federal agencies may have specific standards for sanitation, building construction, and pollution control and prevention requirements. Criteria contained in such standards and other work rules and in building and operating permits can be extremely beneficial in accident prevention. Written standards aid in making designers' (and industrial hygienists') jobs easier, by providing useful technical information and promoting consistency to provide a basic level of safety in similar operations, material, and equipment.

# ENGINEERING DESIGN AND CONTROLS

A large number of standards and voluntary safety codes (consensus standards) have been incorporated into law. The best known example of this practice is the American National Standards Institute (ANSI)—many of the original OSHA standards originated from ANSI standards. ANSI has a wide range of standards for such items as ladders, stairs, sanitation, building load design, floor and wall openings, marking hazards, accident prevention signs, and many others. The designer and industrial hygienist must keep in mind that standards provided by ANSI, the NEC, National Fire Protection Association (NFPA), American Society of Mechanical Engineers, and others are only recommendations—a starting point for safe workplace design—but the design engineer and/or safety engineer who does not pay attention to various codes, standards, and local requirements is setting him- or herself up for admonitions that may not be quite as severe as the ones recommended by Hammurabi, but a headache generator at the very least.

## PLANT LAYOUT

During the design phase for a plant or facility, and especially for general working areas, several elements must be taken into consideration. With safety and efficient use of materials in various processes and methods as the primary goal, the location, size, shape, and layout of worksite buildings should be determined. Designers and industrial hygienists have learned from experience (generally from past mistakes) that when the worksite functions to produce a finished product, designing the worksite so that raw materials enter at one end of the worksite and the finished product is shipped at the other is most efficient, and sometimes safer. What we have basically described here is *process flow*—an important consideration that should not be overlooked. For example, if a certain process calls for robotic welding to be conducted during a product's assembly phase, a process flow diagram should indicate this—to ensure that hazardous materials such as flammable cleaning solvents, gasoline, and/or explosives are not staged or stored in such an area. Obviously, this is critical to ensuring safe operations. Process flow diagrams also aid in the proper positioning of equipment, electrical apparatus, heating, ventilation, and air conditioning (HVAC), storage spaces, and other appurtenances or add-ons.

Note that plant layout is often a compromise between a number of factors such as:

- The need to keep distances for transfer of materials between plant/storage units to a minimum to reduce costs and risks.
- The geographical limitations of the site.
- Interaction with existing or planned facilities on-site such as existing roadways, drainage, and utilities routings.

- The need for plant operability and maintainability.
- The need to locate hazardous materials facilities as far as possible from site boundaries and people living in the local neighborhood.
- The need to prevent confinement where release of flammable substances may occur.
- The need to provide access for emergency services.
- The need to provide emergency escape routes for on-site personnel.
- The need to provide acceptable working conditions for operators.

The most important factors of plant layout as far as safety aspects are concerned are those designed to:

- Prevent, limit and/or mitigate escalation of adjacent events (domino effect).
- Ensure safety within on-site occupied buildings.
- Control access of unauthorized personnel (9/11 has heightened our awareness of the need in this area).
- Facilitate access for emergency services.

## Illumination

Care must be taken with lighting design—not only to ensure that enough lighting is provided for workers to perform their work tasks safely and efficiently, but also to ensure that the lighting does not interfere with work or cause visual fatigue. To ensure that the proper quantity or amount of illumination (usually measured in foot-candles) is installed in the worksite, you must determine exactly what kind of work is to be performed in the space. The amount of illumination will vary with the job function. Experience has shown that a lack of proper illumination (poor illumination) for various industrial areas (including office areas) is listed as a common cause of accidents. ANSI in its *Practice for Office Lighting* (ANSI/IER RP7, 1-1982/1983) lists the minimum levels of illumination for various industrial areas and tasks. These are listed in table 12.1.

One aspect of lighting that is often overlooked in the design phase is emergency lighting. No one doubts that an emergency of just about any size is apt to involve the loss of electrical power, which, of course, would mean that shutdown of equipment and processes, evacuation of workers, and rescue must be performed in darkness—unless emergency lighting is provided. The design engineer should at least incorporate into the workplace design standby sources of light that come on automatically when the power fails, if only to allow for safe evacuation. Whatever type of emergency lighting system is chosen, remember that

### Table 12.1. Minimum Levels for Industrial Lighting

| Area | Foot-Candles |
|---|---|
| Assembly—rough, easy seeing | 0 |
| Assembly—medium | 100 |
| Building construction—general | 10 |
| Corridors | 20 |
| Drafting rooms—detailed | 200 |
| Electrical equipment, testing | 100 |
| Elevators | 20 |
| Garages—repair areas | 100 |
| Garages—traffic areas | 20 |
| Inspection, ordinary | 50 |
| Inspection, highly difficult | 200 |
| Loading platforms | 20 |
| Machine shops—medium work | 100 |
| Materials—loading, trucking | 20 |
| Offices—general areas | 100 |
| Paint dipping, spraying | 50 |
| Service spaces—wash rooms | 30 |
| Sheet metal—presses, shears | 50 |
| Storage rooms—inactive | 5 |
| Storage rooms—active, medium | 20 |
| Welding—general | 50 |
| Woodworking—rough sawing | 30 |

it must be designed to operate from an independent connection at the point where the main service line enters the workplace.

## High Hazard Potential Work Areas

Work areas involved with process operations typically include some areas or operations that have an inherent high hazard potential. Obviously, such areas may require special precaution measures and planning, such as the need for sprinkler systems, containment dikes, alarms, electrical interlocks, and other precautionary measures. These areas include:

1. Spray-painting areas
2. Explosives manufacturing, use, or storage
3. Manufacturing, use, or storage of flammable materials
4. Areas with process equipment of high-energy movement through a power source, such as steam, electrical, hydraulic, pneumatic, and mechanical
5. Radiation areas
6. Confined spaces
7. Chemical mixing areas

## Personal and Sanitation Facilities

Not only must the design engineer factor into any workplace design several sanitation and personal hygiene requirements (i.e., provisions for potable water for drinking and washing; sewage, solid waste, and garbage disposal; sanitary food services; and drinking fountains, washrooms, locker rooms, toilets, and showers), he or she must also design the facility for easy and correct housekeeping activities. Housekeeping and sanitation are closely related. Control of health hazards requires sanitation, and control is usually effected through good housekeeping practices. While it is true that disease transmission and ingestion of toxic or hazardous materials are controlled through a variety of sanitation practices, it is also true that if the workplace is not properly designed with correct sanitary and storm sewers, availability of safe drinking water and sanitary dispensing equipment, then sound sanitary practices are made much more difficult to include within the workplace.

## Summary

The cost-effectiveness of safety is open to constant debate. What *is* the most cost-effective way? "It depends on the individual situation" is often the answer. But in most cases, for the best long-term results, engineered controls put into place at the earliest possible stages is the most cost-effective, both in terms of dollars and in terms of worker health and safety (Spellman & Whiting, 2005).

## REFERENCES

Goetsch, D. L., 1996. *Occupational Safety and Health*. Englewood Cliffs, N.J.: Prentice-Hall.

Hammer, W., 1989. *Occupational Safety Management and Engineering*, 4th ed. Englewood Cliffs, N.J.: Prentice-Hall.

Spellman, F. R. and N. E. Whiting, 2005. *Safety Engineering: Principles and Practices*. Lanham, Md.: Government Institutes.

## SUGGESTED READING

The following publications are available at www.ansi.org or from American National Standards Institute, 1430 Broadway, New York, NY 10036.

> *Buildings and Facilities—Providing Accessibility and Usability for Physically Handicapped People*, A117.1.

# ENGINEERING DESIGN AND CONTROLS

*Life Safety Code,* A9.1 (NFPA 101)

*Manual on Uniform Traffic Control Devices for Streets & Highways,* ANSI D6.1

*Minimum Requirements for Sanitation in Places of Employment,* Z4.1

*Safety Requirements for Construction,* A10 Series

# Index

abatement period, 26
abrasive blasting, 26
absorbed dose, 161, 169
absorption, 26, 85, 269, 275
acaricide, 269
acceptable lift, 319
acceptable risk, 269
accident, 26
accident analysis, 27
accident prevention, 27
accommodation, 27
accuracy, 27, 73
acid, 27
acoustic trauma, 127
acoustics, 27
action level, 27, 71, 127, 138
activated charcoal, 27
acute, 28, 79, 269
acute exposure, 18
acute toxicity, 28
administrative controls, 21–23, 28, 319
adsorption, 28
aerodynamic diameter, 79
aerosol, 28, 251, 276
aerosol photometry, 87
Agricola, 4

air, 28
airborne particulates, 71, 77
air cleaner, 28, 212
air contaminants, 17–18
air contamination, 28
air density, 198
air-line respirator, 28
air monitoring, 28, 76
air monitoring and sampling, 72–77
air pollution, 28
air-purifying respirators, 28, 246
air sample volume, 72
air sampling, 28
air-supplying respirators, 29, 246
alkaloid, 269
allergens, 118
allergic reaction, 29
alpha particles, 29, 161
alpha radiation, 164
alpha radiation detectors, 164–165
alveoli, 29
ambient, 29
American Conference of Governmental Industrial Hygienists (ACGIH), 16
American National Conference Institute, 16
Ames assay, 269

349

amorphous, 29
analytical method for gases and vapors, 74–76
anesthetic, 269
aniometer, 198
anthrax, 271
arc welding safety, 288–289
area monitoring, 84
area sampling, 74
aromatic, 29
asbestos, 2–3, 251
asbestos exposure, 106–108; exposure monitoring, 109; methods of compliance, 109; permissible exposure limits for, 108; regulated area, 109–111
asbestosis, 29, 271
asphyxiant, 29
asphyxiation, 29
atmosphere, 29
atmosphere supplying respirator, 29
atmospheric pressure, 30
atomic weight, 30
attenuate, 30, 127
attenuation, 127
audible range, 30, 127
audiogram, 30, 127
audiologist, 127
audiometric testing, 30, 136–138, 141–142
authorized person, 30
auto-ignition temperature, 30
auxiliary body cooling, 190
avian flu, 271

background noise, 30, 127
back injuries, 313
baghouse, 30
banana oil, 251
base, 30
baseline audiograms, 127, 137
bel, 31
benchmarking, 31
beta particle, 31, 161
beta radiation, 165

Bhopal, 49, 293
bias, 73
bioaerosols, 31
bioassay, 161, 269
biohazard, 31
biological exposure limits, 71
biological half-life, 161
biological stressors, 11, 13
biomechanical approaches, 314
blasting abrasives, 251
bloodborne pathogens, 271
body temperature measurements, 183
body water loss, 240
boiling point, 31
botulism, 271
Boyle's Law, 31, 83
brake horsepower, 198
brazing, 291
breathing resistance, 251
breathing zone, 31
bremsstrahlung, 161
building-related illness, 97
bursitis, 308

calibration, 74, 85, 89
calorie, 178
canopy hood, 198
capture hood, 220
capture velocity, 31, 198
carbon monoxide, 271
carcinogen, 31, 269, 278
cardiotoxic effect, 277
carpal tunnel syndrome, 31
catalyst, 31
causal factor, 32
ceiling, 71
ceiling limit, 32
central nervous system effects, 277
Charles's Law, 32, 83
chemical change, 32
chemical hazards, 235, 251

# INDEX

chemical stressors, 11–12
chemiluminescence, 87
chilblains, 195
Chimney-Sweepers Act of 1788, 4
chromatography, 32, 75
chronic, 32, 79, 269
chronic exposure, 18
code, 341–342
coefficient of entry, 219
coefficient of friction, 32
cold hazards, 193–195
colorimetry, 32, 87
combustible gas indicator, 33
combustion, 33
commercial chemical products, 297
common airflow pathways, 105
competent person, 33, 109
*Comprehensive Environmental Response, Compensation, and Liabilities Act of 1980 (CERCLA)*, 299
computer workstation, 333
concentration, 33
conduction, 178
conductive hearing loss, 33
confined spaces, 33, 90, 285
contact dermatitis, 33
containment, 33
contaminant permeation, 237
contingency plan, 33
continuous noise, 127
continuous sampling, 85
convection, 33, 178
corrosive substances, 295
cotton dust, 272
coulometer, 33
coulometry, 87
criterion, 342
criterion sound level, 127
cumulative trauma disorder (CTD), 33, 308–309, 323
curie, 161

cutaneous, 34
cyanide, 272
cyclone, 34

Daily Noise Dose, 156
Dalton's Law of Partial Pressures, 34
decibel (dB), 34, 127, 132
decontamination, 34, 236–237
deer fly fever. *See* tularemia
degradation, 236
density, 34
DeQuervain's disease, 308
dermal contact, 269
dermatitis, 34
design guide, 342
design handbooks, 342
design load, 34
design manuals, 342
diffusive samples, 85
dike, 34
dilution ventilation, 205–206
direct-reading colorimetric devices, 88
direct reading instruments, 34, 86
discomfort, 95
dose, 34, 269
dose equivalent, 161, 169
dose-response, 268
dose-response relationship, 16, 34
dosimeter, 35, 128, 161
double hearing protection, 127
driving forces, 105
dry bulb temperature, 35
duct diameter, 210
duct velocity, 35
dust, 17, 35, 78–79, 251, 276
dyspnea, 35

effective half-life, 163
Ellenborg, Ulrich, 4
enclosing hood, 220
energy, 35

engineering controls, 21, 35, 319, 339–346
engineering design, 339–346
environmental controls, 20–24
epicondylitis, 308
equal-energy rule, 128
ergonomics, 36, 305–335
ergonomic stressors, 11, 13
etiology, 36
evaluation of cold stress, 194
evaluation testing, 91
evaporation, 36
evaporative cooling, 178
environmental heat, 179
environmental heat measurements, 183
evase, 198
exchange rate, 128
excretion, 275
exhaust ventilation, 36, 202
exposure, 36, 269
exposure ceiling, 36
exposure limits, 69
extraction procedure, 296
extremely hazardous substance, 294
extremely low frequency radiation, 170
eye and face protection, 229
eye protection, 286

face shield, 229
face velocity, 36
factors affecting indoor climate, 97
fall arresting system, 36
fan curve, 198
fan laws, 198
fans, 221
fan total pressure, 221
fiber, 17, 36
filter, 81
fire watch, 283
fit-testing, 251
flame ionization, 87
flame ionization detector, 36
flammable liquid, 36

flammable range, 37
flammable solid, 37
flash point, 37
flow rate, 198
fluid, 82
foodborne disease, 272
foot-candle, 37
foot protection, 230
forced expiratory volume, 251
forced vital capacity, 251
formaldehyde exposure, 113–114
frequency, 128
friction loss, 198
frostbite, 194
frost nip, 194
full body protection, 230
fumes, 17, 37, 276, 291
fungi, 273

gamma radiation, 37, 162, 166
gas, 17, 37, 82, 251, 276
gas absorbents, 85
gas chromatograph, 75
gas chromatography, 37
gas cutting, 289
gas welding, 289
gauge pressure, 198
general ventilation, 37, 205–206
glare, 330
globe thermometer, 37
grab sampling, 37, 85
ground-fault circuit interrupter (GFCI), 37

half-life, 38
Hamilton, Alice, 4, 14–15
hand protection, 229
hanta virus, 272
hazard, 16, 38, 269
hazard analysis, 38
hazard assessment, 38, 227
Hazard Communication Standard (HazCom), 38, 49–69

# INDEX

hazard control, 38
hazard determination, 52
hazardous chemical, 295
hazardous material, 38, 294
hazardous noise, 128
hazardous substance, 38, 293
hazardous waste, 38, 292–303
head, 199
head protection, 228–229
hearing conservation, 38
hearing conservation program evaluation, 148–153
hearing conservation programs, 125
hearing conservation written program, 138–145
hearing loss, 128
hearing protection, 142–143
hearing threshold level, 129
heat, 178
heat collapse, 181
heat cramps, 18, 38
heat exhaustion, 38, 180
heat fatigue, 181
heat rash, 39, 181
heat stress calculations, 185–186
heat stress checklist, 191–193
heat stress index, 185
heat stroke, 39, 180
hematoxic effects, 277
hepatoxic effects, 277
hertz (Hz), 129
hood, 199
hood entry loss, 39, 199
hood-face velocity, 211
hood static pressure, 199, 211
hot work, 39
human factors engineering, 39, 305
hypothermia, 194

ice vests, 190
Ideal Gas Law, 83–84
IDLH (Immediately Dangerous to Life and Health), 39, 251
ignitability, 295

illumination, 39, 344
immersion foot, 194
immune system effects, 277
impaction, 39
impervious, 39
impingement, 39
impulse noise, 39, 129
incident, 40
indoor air contaminant, 207
indoor air quality (IAQ), 40, 93–105
indoor contaminant transport, 101–105
industrial hygiene, 3–5, 40
industrial toxicology, 14–15
infrared analyzer, 87
infrared radiation, 170
ingestion, 18, 40
inhalable particulate mass, 71
inhalation, 18, 40
injection, 18
injury, 40
insecticide, 269
insoluble, 40
inspirable particles, 79
interlocks, 40
ionization, 163
ionizing radiation, 12, 40, 159
irritant smoke, 251
irritation, 40
isocyanates, 272

Jamaica Ginger, 274
job hazard analysis, 40

laminar flow, 199
laser, 41
latent period, 41
LD50/LC50, 267–268
lead-based paint, 290
lead exposure, 114–116; health effects of, 115
Legionnaires' disease, 172
lifting procedures, 318
limits of detection, 72–73

limits of quantification, 73
liquids, 17
local exhaust ventilation, 41, 199, 204–205
loudness, 129
lower back injury prevention, 312
lower explosive limit (LEL), 41

makeup air, 41
manometer, 199, 203
material safety data sheet (MSDS), 12, 41, 50–52; sample written program for, 54
maximum permissible lift, 319
medical surveillance, 126
mercuric nitrate, 273
mesothelioma, 41
metabolic heat, 42, 178
metabolism, 275
metastasis, 42
methyl alcohol, 273
methylene chloride, 273
microwave, 170
mist, 17, 42, 252, 276
mold, 42, 273
mold control, 117–122
mold prevention, 119
mold remediation, 119–120; checklist for, 120–121
monitoring noise levels, 126
multiple chemical sensitivity, 97
mutagen, 42, 269, 279
mycotoxins, 42, 118

National Institute for Occupational Safety and Health (NIOSH), 10
nephrotoxic effects, 277
noise, 12, 129, 125–157
noise controls, 126
noise dose, 129
noise dosimeter, 130, 133
noise-induced hearing loss, 130
noise safe work practices, 144
noise units, 153

nonionizing radiation, 12, 42, 159, 169
nonspecific source wastes, 296
nuisance dust, 42, 71

occupational environmental limits, 69–71
Occupational Safety and Health Act, 2, 8–10, 43
octave-band noise analyzers, 135
open abrasive blast cleaning, 279–281
optical density, 172
organochlorine insecticide, 273
organophosphate insecticide, 273
ototoxic, 130
ototraumatic, 130
oxyacetylene equipment, 290
oxygen deficient atmosphere, 43

parabolic diffusers, 331
paradichlorabenzene, 273
particle size, 79
particulate, 252
particulate collection, 80–81
particulate matter, 43
Pascal's Law, 83
PCBs, 273
penetration, 236
performance effects, 96
performance standards, 43
permeation, 236
permissible exposure limit (PEL), 43, 252
permissible noise exposure, 139
permit-required confined space, 91
personal air sampling, 84
personnel protective equipment (PPE), 23–24, 43, 223–263; levels of, 232–234
Personal Protective Equipment Standard, 225
pesticide, 269
photon, 162
physical stressors, 11–12
physiological approach, 315
pitot tube, 43, 199
plague, 273
plant layout, 343

plenum, 199
polarography, 88
polymerization, 43
population, 268
potency, 269
Pott, Percival, 4
practice, 342
precision, 43, 73
presbycusis, 130
pressure, 44
process, 50
process flow, 343
protective clothing, 286
psychophysical approach, 315
pulmonary effects, 277

rabbit fever. *See* tularemia
rad, 162
radiant heat, 13, 44, 179
radiation, 159–176
radiation area, 173
radiation dose, 169
radiation exposure controls, 175
radioactive decay, 167
radioactivity, 88
radiofrequency, 170
Ramazzini, Bernardo, 4
Raynaud's syndrome, 44, 158, 309
RCRA hazardous substance, 295
reactive chemicals, 207
reactive substances, 295
reactivity, 44
receiving hood, 220
reflective clothing, 190
regulation, 342
rem, 162
repetitive strain injury, 323
replacement air, 199
reportable quantity (RQ), 44
representative sample, 84
respirable particles, 79
respirable particulate mass, 71

respirator, 252
respiratory hazard, 252
respiratory protection, 246–263
Resource Conservation and Recovery Act (RCRA), 298
return air, 199
ricin, 274
risk, 44, 267, 269
roentgen, 162
rotameters, 45
routes of entry, 18–20, 268

saccharin, 252
Safety and Health Program Management Guidelines, 9–10
safety glasses, 229
sanitation facilities, 346
segmental vibration, 157
sensitivity, 269
sensitization, 45
sensorineural hearing loss, 131
Severe Acute Respiratory Syndrome (SARS), 274
short term exposure limit (STEL), 45, 70
sick building syndrome (SBS), 96, 200
significant threshold shift (STS), 131
silica, 274
silica exposure, 111–113; guidelines for, 113
skin absorption, 18
skin notation, 71
slot velocity, 200
smallpox, 274
smoke, 252
sociacusis, 131
soldering, 291
solvent, 46
sound intensity, 131
sound level meter (SLM), 131, 134
sound power, 131, 153
sound pressure level (SPL), 131, 154
specific gravity, 46
specific source wastes, 297
specification, 342
speed of sound, 154

spirometric evaluation, 252
spray painting, 290
stack, 200
stale air, 96
standard, 341–342
standard threshold shift (STS), 131
static pressure, 200, 203
Stoke's Law, 80
stratified atmospheres, 91
stress, 10
stuffy air, 96
sub-chronic, 79
suction pressure, 200
Superfund Amendments and Reauthorization Act of 1986 (SARA), 299
supply system, 202
synovitis, 308

target organ, 269
temporary threshold shift (TTS), 46
tendonitis, 308
tenosynovitis, 308
teratogen, 269, 279
terminal velocity, 79
terrorism, 1–2; 9/11 attack, 1–2; anthrax attack, 1–2
thalidomide, 274
thermal comfort, 178–179
thermal stress, 177–195
thermal stress control, 186
thoracic outlet syndrome, 308
thoracic particulate mass, 71
threshold, 46
threshold dose, 269
threshold level, 16
threshold limit value, 46
threshold shift, 131
tight building, 93
tight building syndrome, 200
time weighted average (TWA), 70
tobacco smoke, 215
total pressure, 200, 203
toxic chemical, 295

toxicity, 16, 46, 270, 295
toxicity characteristics leaching procedure, 296
toxicology, 15–16, 265–279
toxin, 46, 270
trench foot, 194
trigger finger, 309
tularemia, 274
turbulent flow, 200

ulnar nerve entrapment, 308
ultraviolet radiation, 171
upper explosive limit (UEL), 47

vapor, 47, 82, 252, 276
velocity, 131
velocity pressure, 200, 203
velometer, 47
ventilation, 105
ventilation codes, 208
verification testing, 91
vibration, 157–158
vinyl chloride, 274
viral hemorrhagic fevers (VHFs), 275
visible light radiation, 171
volatile organic compounds, 118
volatility, 47
volume flow rate, 200

wavelength, 132, 154
weighted measurements, 132
welding, 281–290
wet bulb globe temperature (WBGT) index, 183
wetted clothing, 190
whole-body vibration, 157
work practice controls, 21
worker lead protection program, 116
workers' compensation, 4, 47
workplace stressors, 10–14

xenobiotic, 270
x-rays, 162, 166

# About the Author

**Frank R. Spellman** is Assistant Professor of Environmental Health at Old Dominion University, Norfolk, Virginia. He has extensive experience (more than 35 years) in environmental science and engineering, in both the military and the civilian communities. Professor Spellman holds a BS/BA in Business Management/Public Administration, MBA, MS and PhD in Environmental Engineering. A professional member of the American Society of Safety Engineers, the Water Environment Federation, and the Institute of Hazardous Materials Managers, Spellman is a Board Certified Safety Professional (CSP) and Board Certified Hazardous Materials Manager (CHMM). He has authored and coauthored 47 texts on safety, occupational health, water and wastewater operations, environmental science, and concentrated animal feeding operations.